AF774264

Springer Tracts in Advanced Robotics 81

Editors

Prof. Bruno Siciliano
Dipartimento di Informatica
e Sistemistica
Università di Napoli Federico II
Via Claudio 21, 80125 Napoli
Italy
E-mail: siciliano@unina.it

Prof. Oussama Khatib
Artificial Intelligence Laboratory
Department of Computer Science
Stanford University
Stanford, CA 94305-9010
USA
E-mail: khatib@cs.stanford.edu

For further volumes:
http://www.springer.com/series/5208

Editorial Advisory Board

Oliver Brock, TU Berlin, Germany
Herman Bruyninckx, KU Leuven, Belgium
Raja Chatila, LAAS, France
Henrik Christensen, Georgia Tech, USA
Peter Corke, Queensland Univ. Technology, Australia
Paolo Dario, Scuola S. Anna Pisa, Italy
Rüdiger Dillmann, Univ. Karlsruhe, Germany
Ken Goldberg, UC Berkeley, USA
John Hollerbach, Univ. Utah, USA
Makoto Kaneko, Osaka Univ., Japan
Lydia Kavraki, Rice Univ., USA
Vijay Kumar, Univ. Pennsylvania, USA
Sukhan Lee, Sungkyunkwan Univ., Korea
Frank Park, Seoul National Univ., Korea
Tim Salcudean, Univ. British Columbia, Canada
Roland Siegwart, ETH Zurich, Switzerland
Gaurav Sukhatme, Univ. Southern California, USA
Sebastian Thrun, Stanford Univ., USA
Yangsheng Xu, Chinese Univ. Hong Kong, PRC
Shin'ichi Yuta, Tsukuba Univ., Japan

STAR (Springer Tracts in Advanced Robotics) has been promoted under the auspices of EURON (European Robotics Research Network)

Toshio Fukuda, Yasuhisa Hasegawa,
Kosuke Sekiyama, and Tadayoshi Aoyama

Multi-Locomotion Robotic Systems

New Concepts of Bio-inspired Robotics

Springer

Authors
Prof. Toshio Fukuda
Department of Micro-Nano Systems
Engineering
Nagoya University
Nagoya
Japan

Assoc. Prof. Kosuke Sekiyama
Department of Micro-Nano Systems
Engineering
Nagoya University
Nagoya
Japan

Assoc. Prof. Yasuhisa Hasegawa
Department of Information Interaction
Technologies
University of Tsukuba
Tsukuba
Japan

Assist. Prof. Tadayoshi Aoyama
Department of System Cybernetics
Hiroshima University
Higashi Hiroshima
Japan

ISSN 1610-7438
e-ISSN 1610-742X
ISBN 978-3-642-30134-6
e-ISBN 978-3-642-30135-3
DOI 10.1007/978-3-642-30135-3
Springer Heidelberg New York Dordrecht London

Library of Congress Control Number: 2012937220

© Springer-Verlag Berlin Heidelberg 2012
This work is subject to copyright. All rights are reserved by the Publisher, whether the whole or part of the material is concerned, specifically the rights of translation, reprinting, reuse of illustrations, recitation, broadcasting, reproduction on microfilms or in any other physical way, and transmission or information storage and retrieval, electronic adaptation, computer software, or by similar or dissimilar methodology now known or hereafter developed. Exempted from this legal reservation are brief excerpts in connection with reviews or scholarly analysis or material supplied specifically for the purpose of being entered and executed on a computer system, for exclusive use by the purchaser of the work. Duplication of this publication or parts thereof is permitted only under the provisions of the Copyright Law of the Publisher's location, in its current version, and permission for use must always be obtained from Springer. Permissions for use may be obtained through RightsLink at the Copyright Clearance Center. Violations are liable to prosecution under the respective Copyright Law.
The use of general descriptive names, registered names, trademarks, service marks, etc. in this publication does not imply, even in the absence of a specific statement, that such names are exempt from the relevant protective laws and regulations and therefore free for general use.
While the advice and information in this book are believed to be true and accurate at the date of publication, neither the authors nor the editors nor the publisher can accept any legal responsibility for any errors or omissions that may be made. The publisher makes no warranty, express or implied, with respect to the material contained herein.

Printed on acid-free paper

Springer is part of Springer Science+Business Media (www.springer.com)

Foreword

Robotics is undergoing a major transformation in scope and dimension. From a largely dominant industrial focus, robotics is rapidly expanding into human environments and vigorously engaged in its new challenges. Interacting with, assisting, serving, and exploring with humans, the emerging robots will increasingly touch people and their lives.

Beyond its impact on physical robots, the body of knowledge robotics has produced is revealing a much wider range of applications reaching across diverse research areas and scientific disciplines, such as: biomechanics, haptics, neurosciences, virtual simulation, animation, surgery, and sensor networks among others. In return, the challenges of the new emerging areas are proving an abundant source of stimulation and insights for the field of robotics. It is indeed at the intersection of disciplines that the most striking advances happen.

The *Springer Tracts in Advanced Robotics (STAR)* is devoted to bringing to the research community the latest advances in the robotics field on the basis of their significance and quality. Through a wide and timely dissemination of critical research developments in robotics, our objective with this series is to promote more exchanges and collaborations among the researchers in the community and contribute to further advancements in this rapidly growing field.

The monograph by Toshio Fukuda, Yasuhisa Hasegawa, Kosuke Sekiyama and Tadayoshi Aoyama focuses on a novel concept of bio-inspired robotics, namely multi-locomotion. This includes not only the classical legged locomotion which has inspired a wide host of research on humanoid robots, but also hopping, climbing, brachiation and snake locomotion. A number of challenges concerned with the various types of robotic systems are dealt with in the tract, including design, motion planning, sensing, dynamics, stabilization and control.

Rich of experimental results and discussion about the performance and the potential of multi-locomotion in robotics, this volume constitutes a fine addition to STAR!

March 2012
Naples, Italy

Bruno Siciliano
STAR Editor

Preface

Nowadays, multiple attention have been paid on a robot working in the human living environment, such as in the field of medical, welfare, entertainment and so on. In order to accomplish these kinds of robot and put them into practical use, there are quite many unsolved problems. Various types of researches are being conducted actively in a variety of fields such as artificial intelligence, cognitive engineering, sensor- technology, interfaces and motion control. In the future, it is expected to realize super high functional human-like robot by integrating technologies in various fields including these types of researches. This book deals with locomotion mobility among the issues mentioned above. Here, focusing on animals in the nature, it is seen that diversity of locomotion is important. They are capable to perform several kinds of locomotion by stand-alone and to accommodate the alteration of environment by choosing the adequate locomotion from multiple locomotion modes. Inspired by this, we introduce a novel concept of bio-inspired robotics, Multi-Locomotion Robot system. The Multi-Locomotion Robot has a high ability to ambulate by achieving several kinds of locomotion in stand-alone.

This book is organized as follows. Chapter 1 introduces robot locomotion systems such as legged locomotion, hopping, climbing, brachiation robot, and snake locomotion; and then the concept of the Multi-Locomotion Robot is also introduced. In Chapter 2, basics of robotic motion control are explained. Especially, the Passive Dynamic Autonomous Control (PDAC) are explained. Chapter 3 describes the link structure and control system including sensors and actuators of Gorilla Robot–designed as Multi-Locomotion Robot–that is employed in the experiment. In Chapter 4, multiple brachiating controllers (learning-based method and analytical method) for the Gorilla Robot are described. In Chapter 5, static quadruped walking called crawl gait is realized so that Multi-locomotion Robot can move stably. In addition, the structure of the Gorilla Robot is analyzed as quadruped hardware. Chapter 6 describes three climbing ladder motions of the Gorilla Robot. In Chapter 7, a load-allocation algorithm is proposed to balance the loads of the joint motors during transition from a ladder to another. By applying the load-allocation algorithm, the transition motion from ladder climbing to brachiation is achieved. In Chapter 8, we propose a method to choose a suitable locomotion mode by estimating the falling down risk. The falling down risk is estimated from internal conditions of the robot using Bayesian Network. A stable locomotion along some unknown test courses with transition between biped and quadruped walks is experimentally realized. In Chapter 9, the PDAC is applied to multiple motions. First applications are the 3-D biped walking of 2-D dynamics such as lateral and sagittal motions. Second one is the heel-off biped walking that makes it possible to avoid the problem for the impact shock at a foot landing. Third one is the 3-D biped walking based on 3-D dynamics without dividing into 2-D plane. Fourth one is the 3-D biped walking on uneven terrain. Fifth one is the quadruped walking, and the final application is the brachiation. Finally, we summarize and discuss perspective of these studies in Chapter 10.

Nagoya, February 2012

Toshio Fukuda
Yasuhisa Hasegawa
Kosuke Sekiyama
Tadayoshi Aoyama

Contents

1 Introduction 1
- 1.1 Robot Locomotion 1
- 1.2 Related Works of Robot Locomotion 2
  - 1.2.1 Quadruped Locomotion 2
  - 1.2.2 Hexapod Locomotion 5
  - 1.2.3 Hopping 6
  - 1.2.4 Brachiation 7
  - 1.2.5 Snake Locomotion 8
  - 1.2.6 Biped Locomotion 9
- 1.3 Bio-inspired System 28
  - 1.3.1 Foundation of Neural Network 28
  - 1.3.2 Recurrent Neural Network 34
  - 1.3.3 Feed-forward Neural Network 38
  - 1.3.4 Cerebellar Model Arithmetic Computer (CMAC) 45
  - 1.3.5 Fuzzy Neural Network 46
  - 1.3.6 Genetic Algorithms 50
  - 1.3.7 Central Pattern Generator 52
- 1.4 Multi-Locomotion Robot 53
  - 1.4.1 Bio-inspired Robot 53
  - 1.4.2 Diversity of Locomotion in Animals 54
  - 1.4.3 Multi-Locomotion Robot 55
- 1.5 Organization of This Book 58

2 Basics 61
- 2.1 Trajectory Generation Method of Robots 61
  - 2.1.1 Generation of a Desired Trajectory 61
  - 2.1.2 Basic Orbital Function 62
  - 2.1.3 Design of Basic Orbital Function Using n-Dimensional Polynomial 62
- 2.2 Limit Cycle 63
- 2.3 Passive Dynamic Autonomous Control (PDAC) 65
  - 2.3.1 Dynamics of PDAC 65

2.3.2 Control System ... 70
2.3.3 Advantage of PDAC ... 71

3 Hardware of Multi-Locomotion Robot ... 75
3.1 Brachiation Robot (Conventional Bio-inspired Robot) ... 75
3.2 Gorilla Robot (Multi-Locomotion Robot) ... 76
3.2.1 Gorilla Robot I ... 77
3.2.2 Gorilla Robot II ... 79
3.2.3 Gorilla Robot III ... 80
3.3 Summary ... 80

4 Brachiation ... 83
4.1 What Is Brachiation? ... 83
4.2 Learning Algorithm for a Gorilla Robot Brachiation ... 84
4.2.1 Motion Learning ... 84
4.2.2 Experiment ... 88
4.2.3 Summary of This Section ... 94
4.3 Continuous Brachiation Using the Gorilla Robot ... 95
4.3.1 Smooth, Continuous Brachiation ... 95
4.3.2 Controller Design ... 97
4.3.3 Experiment ... 101
4.3.4 Summary of This Section ... 106
4.4 Continuous Brachiation on the Irregular Ladder ... 106
4.4.1 Motion Design of the Brachiation ... 106
4.4.2 Locomotion Action ... 108
4.4.3 Swing Action ... 111
4.4.4 Experiment ... 114
4.4.5 Summary of This Section ... 116
4.5 Summary ... 116

5 Quadruped Walking ... 117
5.1 Realization of a Crawl Gait ... 117
5.1.1 Motion Design of a Crawl Gait ... 117
5.1.2 Joint Trajectory of the Leg ... 120
5.1.3 Estimation of Walking Energy ... 122
5.1.4 Experiment ... 124
5.2 Joint Torque Evaluation of the Gorilla Robot on Slopes as Quadruped Hardware ... 127
5.2.1 Structure of Gorilla Robot III ... 127
5.2.2 Basic Gait Pattern ... 127
5.2.3 Evaluation of Joint Torque in Quadruped Walk on a Slope ... 129
5.2.4 Simulation Analysis ... 132
5.2.5 Experiment ... 135
5.3 Summary ... 137

6 Ladder Climbing Motion 139
6.1 Model of Ladder Climbing 139
6.1.1 Basic Motion Model 139
6.1.2 Ladder Climbing Gait 140
6.1.3 Body Yawing Momentum 141
6.1.4 Error Recognition and Escape Motion 144
6.2 Experiment 145
6.2.1 Transverse Gait 146
6.2.2 Pace Gait with Constant Velocity 148
6.2.3 Trot Gait with Acceleration 148
6.3 Summary 150

7 Transition Motion from Ladder Climbing to Brachiation 153
7.1 Motion Design 153
7.1.1 Environment Statement 153
7.1.2 Motion Planning 154
7.1.3 Transition Motion 155
7.2 Contact Forces Formulation 156
7.2.1 Assumptions and Equilibrium Equations 156
7.2.2 Supporting Forces Decomposition 157
7.2.3 Brief Summary and Problem Statement 158
7.3 Load-Allocation Control 158
7.3.1 Concept of Load-Allocation Control 158
7.3.2 Objective Function and Constraints 159
7.3.3 Generation of Optimized Supporting Forces 161
7.3.4 Load-Allocation Algorithm 162
7.4 Eexperiment Results and Discussion 165
7.4.1 Validating the Assumptions and Load-Allocation Ability 166
7.4.2 Discussion of Failures with Position Control 168
7.4.3 Experiment Results with Load-Allocation Control 169
7.5 Summary 171

8 Locomotion Transition Based on Walking Stabilization Norm Using Bayesian Network 173
8.1 Introduction 173
8.2 Sensor System and Locomotion Mode 173
8.3 Locomotion Stabilization 175
8.4 Stabilization Based on External Information 176
8.4.1 Recognition of Ground 176
8.5 Stabilization Based on Internal Conditions 177
8.5.1 Estimation of Probability 177
8.5.2 Consideration of Stability Margin 179
8.5.3 Shift of Locomotion Mode 179
8.6 Experiments 180

8.6.1 Experimental Conditions ... 180
8.6.2 Experimental Result ... 180
8.7 Summary ... 185

9 Application of the Passive Dynamic Autonomous Control (PDAC) ... 187
9.1 Lateral Motion Control with PDAC ... 187
9.1.1 Lateral Motion and Dynamics ... 187
9.1.2 Control of Lateral Period ... 192
9.1.3 Stabilization ... 193
9.1.4 Experimental Results ... 195
9.1.5 Summary of This Section ... 198
9.2 Sagittal Motion Control with PDAC ... 200
9.2.1 Sagittal Motion and Dynamics ... 201
9.2.2 Stabilization ... 204
9.2.3 Sagittal Motion Period ... 209
9.2.4 Quick Convergency Method ... 210
9.2.5 Simulation ... 210
9.2.6 Experiment ... 212
9.2.7 Upper Layer Controller ... 213
9.2.8 Summary of This Section ... 214
9.3 Heel-off Walking Control with PDAC ... 215
9.3.1 Sagittal Motion Design ... 216
9.3.2 Converged Dynamics ... 219
9.3.3 Stabilization ... 220
9.3.4 Foot-Contact ... 221
9.3.5 Simulation ... 221
9.3.6 Summary of This Section ... 227
9.4 3-D Biped Walking Based on 3-D Dynamics with PDAC ... 228
9.4.1 Walking Model ... 228
9.4.2 Foot-Contact Model ... 235
9.4.3 Stabilization ... 239
9.4.4 Experiment ... 252
9.4.5 Simulation ... 252
9.4.6 Summary of This Section ... 256
9.5 3-D Biped Walking Adapted to Rough Terrain Environment ... 256
9.5.1 Foot-Contact Model ... 256
9.5.2 Stability Analysis ... 258
9.5.3 Experiment ... 261
9.5.4 Summary of This Section ... 266
9.6 Quadruped Walking with PDAC ... 266
9.6.1 Lateral Motion Control ... 266
9.6.2 Design of Pendulum Length ... 267
9.6.3 Sagittal Motion Control ... 270
9.6.4 Estimation of Walking Energy ... 272

9.6.5 Experiment 276
9.6.6 Summary of Quadruped Walking Control 277
9.7 Brachiation with PDAC 278
9.7.1 Brachiation Controller with PDAC 278
9.7.2 Analysis of the Robot Dynamics 279
9.7.3 Experiment 285
9.7.4 Summary of Brachiation Control 290
9.8 Summary 290

10 Conclusion 291
10.1 Summary 291
10.2 Future Perspective 294

References 297

Chapter 1
Introduction

1.1 Robot Locomotion

It is often pointed that the aging society resulted from declining birthrate become an inevitable problem. Its effect in production site is deterioration of workforce. If the ratio of elderly people to all population is high, it is natural to necessitate manufacturing technology that makes it possible to maintain productivity by a small number of workers. From this background, it is easily conceivable that robots play a role as substitute for human and conduct some tasks in collaboration to human. Besides, it is also conceivable that robots take an active part not only in production site such as factories but in human daily life and society such as home, medical scene, disaster relief, nursing care, and entertainment. In fact, pet robot, which is highly valued, is put on sale and highly functional animal-like robot, which is developed actively, is realized in recent years. Advancing robotics technology, it is expected to develop life-supporting robots to match with human life and cohabit with human e.g. nursing-care robots to attend elderly or sick people and home-helper robots to fulfill cleaning, washing, and cooking.

Considering that robots conduct cooperative work with human or robots live and ambulate in human residential environment to support or substitute for a care giver as mentioned above, it is plausible to reflect that humanoid robots with the same mechanical structure as human are the most suitable. However, in order to accomplish these kinds of robot and put them into practical use, there are quite many unsolved problems. Technologies for safety, mobility and intelligence, are developing and have not reached the practical level. Nowadays, to solve these problems, various types of research are being conducted intensively in a variety of fields such as artificial intelligence, cognitive engineering, sensoring technology, interfaces and motion control. In the future, it is expected to develop capable human-like robot by integrating technologies in various fields including these types of research.

This book focuses on locomotion mobility among the issues to develop a human-partner robot. Studies of this robot locomotion pursue several kinds of locomotion inspired in the nature such as multi-legged locomotion, brachiation, bipedal walking,

T. Fukuda et al.: Multi-Locomotion Robotic Systems, STAR 81, pp. 1–59.
springerlink.com © Springer-Verlag Berlin Heidelberg 2012

and flying. Fig. 1.1 shows the classification of robot locomotion controls. We describe related works of robot locomotion below.

1.2 Related Works of Robot Locomotion

1.2.1 Quadruped Locomotion

Quadruped walking of a robot is actively pursued as one of most stable locomotion on the ground, and many quadruped robots were built. The control algorithm of quadruped walking can be categorized mainly into static walking, ZMP-based method, inverted-pendulum-based method, and CPG-based method.

1.2.1.1 Static Walking

In static walking, a robot performs quadruped locomotion, keeping the static stability. Although walking velocity of static walk is low, the stability and adaptability to an environment is high.

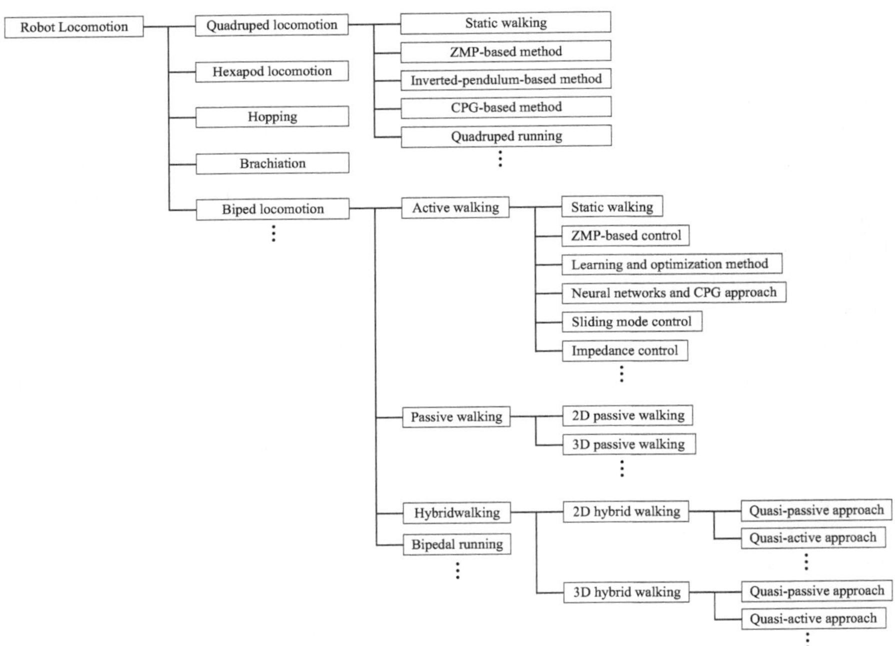

Fig. 1.1 Classification of robot locomotion control

Tomiyama *et al.* presented all-directional quadruped locomotion which combines crawl with rotational gait[277]. In their method, a robot decides the walking pattern based on a desired COG trajectory, and adjusts the landing position of swing-leg at foot-contact. By designing the transition between crawl and rotational gait pattern, they achieveed the proposed quadruped locomotion, and showed the limitation of its method by experiments.

Chen *et al.* designed the adaptive gait generation algorithm finding a sequence of suitable foothold on rough terrain so that static stability is maintained[39]. In their work, adaptive gait is planned by active compensation of stability margin according to terrain condition. They demonstrated that the static stability was maintained during walking motion by numerical simulation.

Hirose *et al.* investigated the transition from a certain static stable posture to other one, which they term it The Standing Posture Transformation Gait, and found the necessary times of leg motion to shift a standing posture into other[113]. The validity is confirmed by numerical simulations and experiment with quadruped robot, TITAN Ⅲ.

1.2.1.2 ZMP-Based Method

ZMP (Zero Moment Point) is the indicator of dynamic stability. ZMP-based method is the algorithm to control a robot so that ZMP is constantly inside the support polygon formed by more than three contact-points. Hirose *et al.* proposed the novel concept of quadruped gait named Dynamic and Static Fusion Control, which achieves static and dynamic stability, by introducing a generalized trot gait that unifies crawl and trot gait, and a sideway sway compensation trajectory that sways the body COG to lateral direction[114]. Moreover, Kurazume *et al.* developed Hirose's work and designed the quadruped gait controller using 3-D ZMP trajectory, which achieve more energy efficiency[160]. By numerical simulation, they showed that the proposed method could attain the stable dynamic walking on the flat ground, and that it was difficult to achieve the stable locomotion on the rough terrain in proposed 3-D ZMP trajectory method.

Tsukagoshi *et al.* proposed the Intermittent Crawl Gait as the quadruped gait which can maximize the static stability margin, and designed the standard foot trajectory achieving the maximum stability and walking velocity[282]. They also presented about the controller to change direction while keeping stability, and tested the validity of the proposed quadruped gait and control by experiment. Yoneda *et al.* presented Intermittent Trot Gait to achieve the control of dynamic omni-directional walk[317]. Intermittent Trot Gait iterates four-leg-supporting phase and two-leg-supporting phase in order. Besides, they designed the landing position controller based on the operator command and standard leg formation.

Lin *et al.* composed the dynamic model faithful to the realistic quadrupedal walking which consists of four cylindrical pantograph legs and twenty-nine links[164]. Investigating the dynamic stability and energy efficiency of a wave gait, they concluded that leg inertia in the transfer phase possesses significant influence on the dynamic stability, and obtained the relationship between the energy efficiency of the wave gait and some control parameters: step-length, duty factors, and walking velocity.

1.2.1.3 Inverted-Pendulum-Based Method

In inverted-pendulum-based method, a robot is modeled by assuming that the contact between a robot and the ground is assumed to be point, and controlled based on this model. Kimura *et al.* analyzed the dynamics of quadruped locomotion[149]. In trot gait, the plane formed by diagonal supporting legs was modeled as an inverted pendulum, and in a pace gait, the plane formed by legs on the same side was modeled as a double inverted pendulum. They designed the motion trajectory based on the gait, walking period, and step-length derived from the following three indicators: stability, maximum walking velocity, and energy efficiency. Finally, the validity of the proposed method was confirmed by experiment. Sano *et al.* built the controller of pace gait by use of a inverted pendulum model[249]. 3-D motion was divided into the lateral and sagittal motions, and each motion was controlled individually. In the lateral plane, by extending supporting-leg, the energy loss at foot-contact is compensated for. They modeled its dynamics as a discrete-time system, and examined PI control by using a maximum absolute value of the closed loop eigenvalues. By means of the proposed algorithm, they realized quadruped pace gait with a robot.

1.2.1.4 CPG-Based Method

CPG (Central Pattern Generator) is a neurons' network generating the rhythmic pattern without sensory feedback and signals from high-level neuronal centers. CPG is going to be described in Section 1.3.7. CPG-based method is the control architecture to realize locomotion by acquiring rhythmic motion pattern by means of CPG. Fukuoka *et al.* presented coupled-dynamics-based motion generating controller which connects a neural system and mechanical system[71]. In their method, a robot locomotion pattern is generated by CPG connecting sensory input. CPG varies the period of the active phase as responses. They applied the proposed controller to the quadruped robot, Tekken, and realized the stable dynamic walking. Tsuchiya *et al.* built the CPG-based control architecture of quadruped locomotion by connecting CPG networks with pre-designed foot trajectories[281]. They demonstrated that the variation of duty ratio that is designed as the parameter of CPG brought the energy efficient gait autonomously, which corresponds to the animal quadruped locomotion. The validity of the controller was verified by experiment.

Albiez *et al.* proposed the new control algorithm of a quadruped robot, inspired by the activation patterns in the brain and the spinal cord of animals[2]. They designed the behavior network consisting of the leg network which controls the movement of one leg and the overall network integrates the sub-networks of one leg. They demonstrated that both the stable gait on an unstructured ground and the flexible transition between standing and walking by combining aspects of classical robot control with the proposed behavior network.

1.2.1.5 Quadrupedal Running

In the field of quadruped locomotion, some work regarding running was reported. Poulakakis *et al.* built an untethered four-legged running robot only with one actuator per compliant leg, and designed a quite simple control algorithm that relies on proprioceptive feedback from motor encoders and leg length potentiometers[227]. Their running controller needs no state feedback, and relies on the passive dynamics of the mechanical system. As a result of analysis, the proposed running was shown to be highly energy efficient with a specific resistance of only 1.4.

1.2.2 Hexapod Locomotion

In the field of bio-inspired robots, along with quadruped locomotion and biped walking, the research of hexapod locomotion inspired by some insects is conducted. Basically, the hexapod robot can perform more stable locomotion than quadruped or biped. Thus, many researchers pursued the highly adaptable running locomotion by imitating the actual insects in nature.

Cham *et al.* built the running hexapod robot based on design principles obtained from biological studies of function in running animals[32], and presented the hexapod running locomotion adaptable to the irregular ground adjusting stride period[33]. In their method, the stride period is adapted based on measurements of ground contact timing acquired from sensors on the robot's feet. They examined the relationship between ground contact timing and the stride frequency, and achieved the hexapod locomotion adaptable to the variation of ground slope.

Kern *et al.* applied a fuzzy logic to a navigation controller, and achieved an autonomous locomotion of a hexapod mobile robot in unknown environment[146]. The controller they designed adjusts the speed and steering according to three input photo-sensors.

Quinn *et al.* developed the cockroach-like robot in terms of mechanical structure and actuators, and designed the controller, referring to the principles of biological movement[231]. This robot could adapt to a wide range of environments, similarly to the real insects in nature. Moreover, they manufactured the hexapod robot which is non-biomimetic but takes advantage of the essence of cockroach, and demonstrated autonomous locomotion on an irregular terrain. In addition, Quinn's group modeled how the cockroach senses and regulates loads in walking and climbing by use of finite element analysis, and consequently suggested that sensors located in the leg can detect specific force vectors, and that the system possesses the positive load feedback using this information to regulate walking movement[137].

Seipel *et al.* developed SLIP model (Spring Loaded Inverted Pendulum model proposed by Koditschek *et al.*[79]) into 3-D SLIP, and applied it to a passive bipedal kangaroo-hopper, an actuated insect model, and passive and actuated versions of a hexapod robot model[255]. They showed the hexapod locomotion achieved by thier proposed method was reasonably robust by numerical simulation.

1.2.3 Hopping

A hopping robot was initiated by the famous Raibert's work[232]. The Hopping robot possesses spring of spring-like actuator to hop. Raibert built the simple hopping robot, using the leg composed of a double-acting air cylinder. By designing three parts of controller, hopping motion control, forward speed control, and body-upright-attitude control, he demonstrated completely dynamic and stable hopping locomotion. After the Raibert's work, many researchers investigated the hopping robot in terms of dynamics and control.

Koditschek *et al.* deeply investigated the dynamics of a simplified models of Raibert's hopping robot[151]. In their work, the hopping height control was represented as a stability problem in a nonlinear discrete dynamical control system. They employed two different models that resulted in two quite different classes of nonlinear oscillators, and as a result of analysis using return map, demonstrated that their qualitative properties match those of Raibert's physical data.

Schwind *et al.* developed the hopping robot controller proposed by Raibert[254]. The simple decoupled feed-back designed by Raibert, whereas better regulation was achieved by using coupled feedback which takes into account the dynamic structure of the robot. They compared the proposed control with Raibert's controller, and demonstrated that the proposed controller gives better regulation and a larger region of attraction.

Francois *et al.* presented a new control of the planar one-legged hopping robot that determines the torque and leg-force inputs to attain local asymptotic stability, based on a one-dimensional manifold in a Poincare return map[60]. They performed the stability analysis on a simplified model of the system, and demonstrated the validity of proposed approach. In addition, the potentials and limitations of legged locomotion were described.

Cherouvim *et al.* investigated the energy efficiency of one-legged hopping robot, using the SLIP model, and obtained optimal gait, considering both mechanical and electrical losses during motion[40]. In addition, they imposed actuator-induced limitations such as the joint torque and angular velocity on the proposed gait, and re-derived the optimal hopping motion.

Vermeulen *et al.* developed the controller of a planar one-legged hopper so that it can perform locomotion on an irregular terrain[125]. In their work, the robot was modeled as a planar multibody system composed of an articulated leg, a body, and massless foot. By use of this model, they proposed the algorithm controlling step-length and stepping height by expressing conservation of angular momentum as a Caplygin form.

1.2.4 Brachiation

The first work of brachiation robot was pursued by Fukuda *et al.*[64, 67]. In robot brachiation, it is impossible to apply any torque around a bar, i.e. rotation around the bar is passive, thus the control system is underactuated. Saito *et al.* proposed the method to generate joint trajectories by heuristics and presented the control algorithm of swing-excitation and locomotion by means of Cerebellar Model Arithmetic Controller (CMAC)[243, 245], and they realized robot brachiation with 2-link robot shown in Fig. 1.2[242]. In addition, aiming at the realization of dynamic and dexterous motion like animals, they built 12 degree of freedom robot which is designed based on skeletal preparations of gibbon, Brachiator III, shown in Fig. 1.3, and developed the control method adapted to a higher degree of freedom robot[245]. Moreover, the robot is difficult to construct the controller based on physical model since it has high degree of freedom and it is driven by wire, hence Hasegawa *et al.* designed the hierarchical motion control structure based on the pre-designed basic motion and proposed the motion adjusting method to vary the weight of basic motion, and consequently realized smooth dynamic brachiation with Brachiator III[98]. In the control architectures proposed by Saito and Hasegawa *et al.*, the dynamical parameters and robot model are not necessary, however they requires try-and-error experimental interation to employ learning.

In contrast, some researchers presented analytical control method. Nakanishi *et al.* proposed Target Dynamics Method which encodes a task as the output of a lower dimensional target dynamical system[194]. They encoded brachiation as the dynamics of a harmonic oscillator, and control swing-up motion, uniform ladder brachiation, and irregular ladder brachiation of Brachiator II[193]. Furthermore, the

Fig. 1.2 Brachiator II.

Fig. 1.3 Brachiator III.

controller to achieve leap brachiation and demonstrated the validity of proposed method by numerical simulation[190].

Kajima *et al.* proposed energy-based control adjusting the quantity of swing-up to supply energy enough to achieve brachiation based on 3-link model and demonstrated continuous brachiation with Gorilla Robot shown in Fig. 3.9[131]. Moreover, they applied the proposed energy-based controller to irregular ladder problem and verified its validity by experiment[132].

1.2.5 Snake Locomotion

Hirose is a pioneer in biomechanical study and started the study observing actual snakes and designing snake-like robots in 1971. There were two reasons for the interest in this snake research: one is the scientific interest in the mechanism of the snake's motion and the other is the engineering interest in the future applications of snake-like robots [112].

Mori and Hirose developed the snake type robot ACM-R3 and showed it has possibility of the various move methods by experiments [184, 185]. The robot is composed of 20 joints, and it is 1.8 [m] in length and 12 [kg] in weight. The robot is composed of two types of bending joints, having different bending axes connected alternately at 90 [deg]. This connection makes it possible to make 3-D motion. Yamada et al. developed ACM-R5 that is an amphibious snake-like robot characterized by its hermetic dust- and waterproof body structure [309]. The Slim Slime robot was developed by Ohno and Hirose. The robot was made up of linearly connecting multiple modules that pneumatically bend and elongate. The module is 128 [mm] in diameter, from 114 to 178 [mm] in length, 1.7 [kg] in weight, and has a maximum bending angle of 30 [deg] [212]. Sugita et al. developed ACM-S1 and proposed a bending and expanding joint unit which has three degrees of freedom (3DOF) in inchworm/angleworm-like motion. It has inchworm motion over uneven ground and of angleworm motion over flat, smooth ground [269]. Yamada and Hirose developed ACM-R4 which has three functions: active wheels, dust- and waterproofing, and overload prevention. Experiments using ACM-R4 showed that a snake-like robot with active wheels operates practically in narrow spaces and on stepped terrain. It imazines that snake-like robots provide practical serviceability [310].

Ross and Choset addressed the gait design problem of snake locomotion by introducing two algorithms: Annealed chain fitting and Keyframe wave extraction [101]. Annealed chain fitting efficiently maps a continuous backbone curve describing the three-dimensional shape of the robot to a set of joint angles for a snake robot. Keyframe wave extraction takes joint angles fit to a sequence of backbone curves and identifies parameterized periodic functions that produce those sequences. Together, they allow a gait designer to conceive a motion in terms three-dimensional shapes and translate them into easily manipulated wave functions, and so unify two previously disparate gait design approaches. The algorithms are validated by using them to produce rolling and sidewinding gaits for crawling and climbing.

Transet et al. described a novel nonsmooth (hybrid) mathematical model for wheel-less snake robots, which allows the snake robot to push against external obstacles apart from a flat ground [279]. The framework of the nonsmooth dynamics and convex analysis allows us to systematically and accurately incorporate both unilateral contact forces (from the obstacles) and isotropic friction forces based on Coulomb's law using set-valued force laws. The mathematical model is verified through experiments using an actual snake robot Aiko.

1.2.6 Biped Locomotion

From the viewpoint of the number of actuation and unactuation, research of biped walking is classified in three categories—active walking, passive walking, and their hybrid walking. In active walking, all joints are fully actuated and there is no passive factor. Passive walking robots possess no actuator and completely passive. Hybrid walking, which is in between active walking and passive walking, is the control method to employ both active and passive joints.

Active Walking

Active walking can be categorized into static walking and dynamic walking. In static walking, the point on the ground where COG is projected is maintained constantly inside the foot of support-leg, hence the static stability is guaranteed during walking. Meanwhile, dynamic walking satisfies dynamic stability, i.e. there is the time that the COG-projecting point is outside the support region and that static stability is broken during walking.

1.2.6.1 Static Walking

Static walking was realized first by Kato *et al.*[141] with WABOT 1, and some research was reported. Shih designed the static gait pattern generation by use of polynomial trajectory of COG[257], and applied the proposed method to locomotion on stairs[256]. Zheng proposed the control scheme of static walking which allowed a biped robot to walk from level ground to slope[319]. In addition, Zheng *et al.* designed biped locomotion method maintaining static balance by only hip joint actuation, in order to reduce the ankle torque[314].

As far static walking, walking motion was slow and the walking velocity is quite low since it is necessary to retain static stability. Besides, since the COG-projection point on the ground needs to be inside the foot of support-leg, it was impossible to make step-length long.

1.2.6.2 ZMP-Based Control

In order to solve the limitation of static walking, the novel concept of dynamic stability instead of static stability, which is ZMP (Zero Moment Point), was proposed by Vukobratovic[289]. ZMP, which was introduced as the stability criterion, is the point on the ground at which the net moment of the inertial forces and the gravity forces has no component along the horizontal axes. ZMP can be construed as the point on the ground where the center of dynamic stability is projected. Thus, if ZMP is inside the support region, which is support polygon formed by more than contact points of a robot body with the floor surface, a robot sole does not detaching from the ground, hence a robot can walk without falling down. A great many control methods taking advantage of this characteristic have been proposed so far.

Vukobratovic *et al.* have been continuing the studies of biped walking control based on ZMP vigorously since their first development of ZMP. They proposed the hybrid control combining the model-based and fuzzy logic-based control techniques in order to solve the problem of requirement of accurate physical parameters in model-based method[290]. In their proposed hybrid controller, fuzzy logic-based tuners adjusted gains of all joints to obtain better tracking performance of nominal trajectory. They confirmed that under the same condition, the hybrid controller reduced the numerical complexity without degrading the tracking accuracy in numerical simulation. Furthermore, Vukobratovic *et al.* investigated the algorithm to enable a complex humanoid robotic system with elastic or no elastic elements to walk on the moving ground based on ZMP trajectory, and developed software package[57]. In their developed system, what the user should do is only to select the joint types.

In Japan, Takanishi's group pursued biped walking control vigorously early on[271], and focused on ZMP-based control[270, 311]. Li *et al.* proposed the control method to achieve the dynamic stable walking by use of the pre-designed ZMP trajectory and iterative learning[163]. In this research, at first, the motion pattern of lower limb is given and stable ZMP trajectory is designed so that ZMP is constantly inside the support region. Then, the motion pattern of an upper limb, which is designed to compensate for the ZMP error, and the motion pattern of the lower limb are coupled. By means of iterative learning to make the actual ZMP get closer to the desired ZMP, they achieved the dynamic walking. Furthermore, Yamaguchi *et al.* pursued ZMP-based walking control on the uneven ground with unknown shapes[311, 312]. They developed a foot sole mechanism to obtain the shape and gradient of ground surface during walking, and designed the control system adapting to the ground surface by employing the information from the developed foot-sensory system. As a result of experiment, at low velocity, the adaptive walking was achieved on unknown uneven ground surface.

In contrast with Yamaguchi and Li *et al.* employing a predetermined ZMP trajectory and designing the upper body motion to enhance the stability margin, Park proposed a fuzzy-logic based ZMP trajectory generation method, which produced the reference ZMP trajectory based on the swing-leg motion trajectory so that the trunk was maintained in the vicinity of upright position[219]. He demonstrated that the walking stability was increased, and that the small range of the trunk motion could be accepted, by the simulation of 7-DOF planar robot.

Nishiwaki and Kagami et al. realized online walking by designing basic trajectories with constraints of cycle time, height of swing leg, and height of upper body and realized biped walking with humanoid robot H6 [239, 209, 208]. Nishiwaki and et al. presented the algorithm to adjust walking pattern online so that the actual ZMP follows the desired ZMP in order to avoid the walking instability caused by the arbitrary motion of upper body [208]. In this method, ZMP trajectory is adjusted in every cycle by solving the differential equation regarding the error between actual ZMP and desired ZMP. By using this algorithm, the stable walking which is not affected by upper body motion was realized with humanoid robot H7.

Yoshino *et al.* proposed Walking Pattern Regulator, which stabilize walking by means of the optimal regulator algorithm, to solve the problem that a robot cannot walk stably in high walking velocity due to the impact shock at foot-contact or irregularity of the ground even if the walking motion is designed based on stable ZMP trajectory[318]. This method enhances walking stability by extending the optimal regulator theory to the whole system including a servo system. By means of this algorithm, the controller maintains the walking pattern if disturbance acts on a servo system. In the experiment, they succeeded in stabilization for the disturbance that is the ground irregularity of range in which vision detection is impossible.

Goswami reconsidered ZMP and redefined it as COP (Center of Pressure)[83, 226], although this definition stirred up controversy with Vukobratovic[288]. He proposed FRI Point as a point on the foot/ground-contact surface where the overall ground reaction force would have to act to keep the foot stationary, and compared FRI with COP by use of simple example of an inverted pendulum. Sardain redefined COP and ZMP, and demonstrated that COP and ZMP were coincided[250]. Based on this result, they proposed a virtual COP-ZMP able to be employed on uneven terrain.

Huang *et al.* planed the trajectory of hip and foot-tip of swing-leg, and designed the walking pattern generator based on both the designed trajectory and the stability margin derived by ZMP. In their method, it is possible to generate the walking pattern adaptable to the ground condition, taking into consideration robot specification[121]. Moreover, they proposed the feedback control, which they termed sensory reflex, to enhance the walking stability in terms of ZMP according to sensor information, and applied it to an actual 26-degree-of-freedom humanoid robot[120]. As a result of experiment, they achieved the stable biped locomotion on the unknown and uneven ground.

Kajita *et al.* proposed 3-D-LIPM (Linear Inverted Pendulum Mode) that decompose the 3-D dynamics in two planes by placing the COG trajectory under the constraint plane of constant height, and applied the approach of preview control to ZMP-based biped locomotion in order to control a foot-contact position toward a desired one[134, 135, 133]. Modeling a biped robot as an inverted pendulum model, and simplifying its dynamics by use of LIPM (Linear Inverted Pendulum Model: described later) including ZMP position, they designed the preview control of ZMP to generate the on-line corresponding COG trajectory according to ZMP trajectory.

A group of the National Institute of Advanced Industrial Science and Technology (AIST) developed humanoid robots in Japanese HRP project [139, 138, 140] and utilize mainly ZMP-based control as a biped walking control method. Nishiwaki and Kagami proposed an online walking control system that frequently generates and updates dynamically stable motion patterns with a cycle time of 20 ms. They also proposed a free leg trajectory design method that allows the landing position to be changed during a step and a future ZMP adjustment method that modifies the landing position in order to maintain the dynamic stability of the generated pattern. The proposed algorithm was implemented on HRP-2 and verified its performance [205]. Nishiwaki and Kagami then proposed an online generation method of a torso height trajectory as a part of walking pattern generation in order to satisfy the kinematic constraint of the leg [207]. Nishiwaki and Kagami also proposed strategies by which to extend adaptivity to terrain of unknown roughness by changing the future reference ZMP trajectory. The proposed methods are implemented as part of an online walking control system with short cycle pattern generation and realized 3-D biped walking on unknown rough terrain using the HRP-2 [206].

As other research, many controller were proposed as follows: the control method integrating measurement data of human walking and linguistic rules such as to maintain ZMP inside support region[321], KWALK which includes ZMP damping control[315], stabilizing control by use of shock absorbing foot[312], and Optimization method of ZMP trajectory[257].

Some big companies also developed humanoid robots and utilize ZMP-based control: HONDA's ASIMO [246], Toyota's Partner Robot [278], Sony's QRIO [124]. The development of humanoid robot by HONDA started mainly in the field of biped walking control from 1986. Biped waking robot P1 and the world's first autonomous humanoid robot P2 [110] were made public in 1993 and 1996 respectively; then biped walking performances of the robots amazed over the world. The development of humanoid robot by HONDA more moved ahead; in 1997, P3 [111] and in 2000, ASIMO [246] were made public respectively. Nowadays, ASIMO is often used as a benchmark of humanoid robots [246]. ASIMO can walk on flat terrain by controlling COG using predictive motion control, and also realized good movement ability: up-down of stairs, turning, and dance. Development of ASIMO was progressed; new ASIMO unveiled in November, 2011, intelligence, physical, task-performing capability were improved largely [116]. The combination of strengthened legs, an expanded range of leg movement and a newly developed control technology that enables ASIMO to change landing positions in the middle of a motion has enabled ASIMO to walk, run, run backward, hop on one leg or on two legs continuously. The running velocity of new ASIMO is 9 km/h. As a result of acquiring the ability to make such agile motions, ASIMO has become capable of more flexibly adapting to changing external situations so that it can, as an example, walk over an uneven surface while maintaining a stable posture.

1.2.6.3 Learning and Fuzzy Method

Bessonnet *et al.* proposed the dynamics-based optimization method of sagittal gait using Pontryagin maximum principal[28]. By expressing the dynamics in double-supporting phase as a closed-loop kinematic system, and by modeling the single-support phase as a open kinematic system, they optimize globally gait cycles. Moreover, in order to solve the problem that the necessary conditions for optimality stated by the Pontryagin maximum principal are complexified by the constraints set of phase transition time, they developed the optimization method, and proposed the method by use of spline-based parametric optimization technique[29].

Kawamura *et al.* expressed the robot as a linear system approximately, and applied the learning algorithm to the biped walking robot in order to generate bipedal locomotion pattern[143]. They showed the convergency of servo error after several trials, and realized biped walking with the robot. Moreover, it was suggested that an input pattern that generates other walking motion could be directly constructed from the input pattern obtained previously by learning.

Kwek *et al.* applied Active Force Control (proposed by Hewit *et al.*[107]) and its leaning method (proposed by Mailah[174]) to 5-link planar biped robot, and compared with the conventional PD controller in terms of performance and robustness against disturbance[161]. By numerical simulation, they showed that Active Force Control strategies could achieve better performance compared to PD control. They concluded that Active Force Control schemes successfully compensate the nonlinear terms and external disturbances acting on the biped system.

Arakawa *et al.* employed the genetic algorithm (GA) and evolutionary programming (EP) in order to find natural walking motion for a biped robot[94, 95]. They proposed the hierarchical evolutionary algorithm consisting of GA and EP through energy optimization and stability criterion based on ZMP. The GA aimed to minimize the total energy during walking, and EP intended to optimize interpolated configuration of the biped locomotion robot in light of ZMP position. They obtained the natural trajectory in numerical simulation, and verified the validity by experiment.

Many studies of biped walking control by combining the fuzzy-logic and learning algorithm were reported. Magdalena *et al.* introduced GA into the fuzzy-logic controllers, and applied the designed controller to a planar biped robot[173]. They employed the stability of walk as the evaluation function, which is the function of the time and the number of steps before falling. As a result of simulation, the learning system obtained new stable gait patterns. Jha *et al.* applied three types of fuzzy-logic controller (one based on the manually designed membership function and fuzzy-rule bases, one based on the membership function and fuzzy-rule bases whose parameters were optimized by GA, one based on the membership function and fuzzy-rule bases designed automatically by GA) to planar biped walking, and compared them in terms of the performance[127]. As a result of simulation, they demonstrated that the third approach could obtain the better performance. Zhou *et al.* incorporated different kinds of expert knowledge into the fuzzy reinforcement learning agent, based on a neuro-fuzzy network architecture, and applied to the biped robot starting walking[320]. In numerical simulation, they found that the incorporation of the expert knowledge and evaluation data could enhance the learning speed.

1.2.6.4 Neural Networks and CPG Approach

On the other hand, some work applying artificial neural networks to biped walking were reported. Miller presented the biped-walking controller using CMAC (Cerebellar Model Arithmetic Computer) neural networks[180]. By means of this method, a robot can walk without any detailed kinematics or dynamic models. He built respective learning architecture in the sagittal and lateral plane individually and achieved the short-step walking after 45-60 minutes learning. Furthermore, Miller *et al.* developed and designed the control method of variable-speed gaits for a biped robot based on the neural network[157]. In their work, the walking period is varied according to the desired walking velocity, and walking motion is adjusted by five modules of neural networks. After long-time learning, biped walking in various range of forward velocity was confirmed by the robot, called Toddler.

Benbrahim *et al.* studied biped dynamic walking using reinforcement learning [25]. They designed a new learning architecture based on modules that are composed of simple controllers and small neural networks. This modularizing dramatically increased the learning speed, and allowed easy incorporation of new modules that represent new knowledge, or new requirements for the desired task.

Reil *et al.* applied CPGs to the control of biped walking, and employed artificial evolution (genetic algorithm) to optimize the parameters of CPGs[236]. They showed that CPGs in conjunction with an appropriate mechanical implementation of the biped could generate stable walking patterns on planar surface, after about a learning. Furthermore, they developed and applied the proposed algorithm to three dimensional biped[235]. In a biped physical simulator, a six degree-of-freedom robot optimized evolutionary its recurrent neural networks, and achieved the three dimensional biped locomotion toward the target direction.

The control scheme proposed by Juang *et al.* was a fuzzy modeling neural network controller with the backpropagation through time algorithm, which could solve the problem of uncertainty of network size in the conventional neural network[130]. They gave desired walking parameters (such as the step-length, crossing clearance, and walking speed), and demonstrated that these parameters were obtained by numerical simulation.

Liu *et al.* proposed a new fuzzy neural networks (FNN) quadratic stabilization output feedback control, and applied it to the trajectory tracking of biped walking[168]. They designed a robust quadratic stabilization FNN nonlinear observer using H_∞ approach and variable structure control (VSC) in order to adapt external disturbance and parametric uncertainties, and proposed a quadratic stabilization FNN controller, a hybrid control system containing FNNs, inverse system method, H_∞ optimal control, and VSC, which guaranteed all signals in closed-loop system are bounded.

Salatian *et al.* designed the neural network mechanism to adapt the biped gait on the unknown sloping surface[247]. Their proposed algorithm allowed a robot to climb a sloping surface from a level ground with no a priori knowledge regarding the inclination of ground. The neural network was trained by the a reinforcement learning algorithm, while PI control was employed for position control of each joint. They finally tested the proposed control in experiment, and demonstrated the biped walking on the unknown surface.

1.2.6.5 Sliding Mode Control

Chang *et al.* proposed the novel approach of sliding mode control based on the second method of Lyapunov[34]. This approach modifies the tracking errors such that the system response starts on the sliding surface for arbitrary initial condition, which eliminates the reaching phase. They applied the proposed controller to a planar five-link biped model, and demonstrated the validity by numerical simulation.

In order to improve the robustness against modeling error, Tzafestas *et al.* presented a robust sliding-mode control, and applied it to nine-link biped robot which was assumed to involve large parametric uncertainty[284]. Employing sine function to switch dynamics at sliding surface, and saturation function to avoid chattering, they confirmed that the proposed controller achieved quite high robustness against estimation error including maximum 200% parametric inaccuracy.

Mu *et al.* divided the whole biped walking cycle into a single-support phase and a double-support phase, and formulated the dynamic model in double-support phase as the motion under holonomic constraints. Treating mechanical system of a biped robot in double-support phase as a redundant manipulator, they developed a sliding mode controller for motion regulation in double-support phase, and examined the stability and robustness of the proposed sliding mode controller[187].

Lum *et al.* applied both the computed torque method and the control method with robustness against modeling error proposed by Dawson *et al.*[50] to a planar five-link biped robot[172]. The robust controller switches the control structures according to the positional error for the robustness. They designed the desired trajectory of each joint including starting/stopping a motion by use of linear functions on the level ground and stairs. In the numerical simulation, they confirmed the robustness of their biped walking.

1.2.6.6 Impedance Control

Some researchers approached realization of active walking in terms of impedance control. Park *et al.* applied the concept of impedance control to the controller of bipedal locomotion[218, 220]. They proposed the hybrid control combining impedance control and computed-torque method. In the swing phase, impedance control was employed, and in stance phase, computed-torque method was used. Their proposed control scheme determines the torque according to the predetermined impedance between the actual posture and desired trajectory. The robustness against impact shock at foot-contact and irregularity of the ground was verified by numerical simulation.

Loffler *et al.* built the large-scaled biped robot (about 1.8[m] in height), and tested two control approach[169]. First one was feedback linearization calculating the desired joint torque according to the robot state, however this method has the problem of requirement of quite high-performance computer and accurate sensors with a high bandwidth. Hence, they employed the tracking controller of joint trajectory derived using an inverted pendulum model, and designed the foot-contact damping control similarly to the impedance control.

1.2.6.7 Adaptive Control

Yang presented the control law, which has the structure of the inverse dynamics servo, and the adaptation law to converge the system parameters on the actual values estimated by the tracking error[313]. They showed that the joint tracking error could be regulated to the acceptably small value and the performance was robust in spite of large deviations in the initial estimation of the system parameters.

Wieber *et al.* proposed the online adaptation method based on the trajectory tracking control[305]. In their method, a robot adaptably determines the reference trajectory according to the viability kernel, using the pre-designed well deliminated set of reference trajectories. This is one of online trajectory adjustment methods based on a strict trajectory tracking.

1.2.6.8 Other Active-Walking Methods

Meanwhile, many other control methods were proposed and brought results. There is the attempt to realize active walking based on passive walking. Asano *et al.* proposed Virtual Passive Dynamic Walking which is the control architecture to achieve natural dynamic gait based on the dynamics of passive walking with a full-actuated robot. They defined Virtual Gravity which is the gravitational field fixed virtually on a robot as if it was on the downslope and proposed various control methods such as Energy Constraint Control or Virtual Coupling Control[18, 17, 15]. Furthermore, they formulated the biped gait generation problem by deriving the indeterminate equation that determines the relation between the joint torques and its center-of-gravity[16]. By solving this indeterminate equation, a robust control method obtained from the property of mechanical energy behavior during 1 step was proposed.

Azevedo *et al.* applied the Non-linear Model Predictive Control to planar biped walking based on coherent physical inequality constraint (c.f. knee stopper etc.)[20]. In their method, on-line optimization was implemented, hence the reference trajectory was not necessary. Besides, constraints could be designed including physical limitations, geometry of the environment, and desired motion specifications. However, their algorithm required heavy processing in order to implement real-time optimization, and a small on-board processor necessitated long time to execute such process, hence it was difficult to implement their method into actual biped robots.

Furusho modeled a planar biped robot as 5-link robot and examined the eigenvalues of closed loop when all joints are controlled by local feedback[74]. As a result, they showed that there are two dominant modes that are identical with the modes of an inverted pendulum. The paper demonstrated that low-dimensional model derived by using these modes is attracted to stable cycle. Furthermore, utilizing the assumption that the 3-D motion can be decomposed in the lateral and sagittal planes, Furusho *et al.* designed the both motions[75]. The optimal regulator strategy was applied to the lateral motion control, and the sagittal motion control was designed based on the desired body speed determined by use of the dynamics of an inverted pendulum. Dividing the sagittal motion into four phases, single-support phase increasing the body speed, tip-toe rotation phase and double-support phase decreasing the body speed, the sagittal ankle torque was decided based on the body speed. By combining both motions, they realized 3-D dynamic walking.

Cotsaftis *et al.* applied the direct nonlinear decoupling method, which was developed in the context of canonical systems, to an n-link 2-D biped robot[48, 49]. They proposed the Direct Nonlinear Decoupling Method, which directly operated on the Lagrangian of mechanical systems, and solved the difficulty to control the large number of equations including intrinsic nonlinearity by using decoupling and linearization.

Jalics *et al.* proposed the control strategy of a planar biped robot powered by cable-like actuators that mimic human muscles, which can pull but not push[126]. Their proposed method assumed that the support forces were originated in not muscular action but primarily the gravity, and decided the desired joints state of support-leg. Meanwhile, joints of a swing-leg and torso was decided based on the desired initial point and end point of trajectory.

Gorce *et al.* designed a general control architecture for pneumatic legged robots, and proved the asymptotic stability by using Popov criteria[82]. Their proposed architecture consists of two layers, "Coordinator" and "Limb" layer. The coordinator layer maintains the robot stability by correcting the COG acceleration and by applying an appropriate distribution of the limb forces to the environment. Limb layer controls each limb according to the desired position and force trajectories from Coordinator layer. Although they demonstrated the asymptotic stability of trajectory tracking, the stability of biped locomotion was not discussed.

Some research focusing on double-support phase are reported. Goddard *et al.* worked on heel-off and toe-off motion of a double-support phase[80]. They supposed that heel-off and toe-off motion played an important roll of the human walking, and designed the algorithm of double-support phase including these motion. The point of the proposed controller was adaptability to an unknown surface of the ground.

1.2.6.9 Disadvantages of Active Walking

As mentioned above, active walking method realized stable walking in the various environment attained good results. However, in active walking, it is difficult to conform to the robot dynamics since all joints are basically actuated by high gain feedback. Thus, it has the following disadvantage: 1) Low energy-efficiency 2) Difficulty to achieve natural motion 3) Low adaptability to ground irregularity. The actual ground has small irregularities that cannot be observed by a vision sensor. In active walking, this kind of irregularity probably causes the detachment of a support-foot sole from the ground or the inaccuracy of ZMP measurement. Besides, the stabilizing controller based on the feedback using ZMP requires the specifically designed sole to obtain the precise measurement of ZMP, since the inadequate impedance of foot-sole results in the oscillating reaction force. That is, it is impossible to attenuate the foot-impact shock by wearing a pair of shoes like human.

Passive Walking

Passive walking is contrastive with active walking. In passive walking, passive-walking robots—it may be inappropriate to denominate it a robot—possess no active joint or energy source and perform walking on the downslope by use of solely the mechanical energy. The decline slope plays a role to compensate by use of potential energy for the mechanical energy lost during walking by joint friction or impact shock at foot-contact. Passive walking takes advantage of robot dynamics to the fullest extent, thus it can realize high-energy-efficient and natural dynamic walking.

1.2.6.10 Passive Dynamic Walker

McGeer initiated passive walking. He produced Passive Dynamic Walker and realized planar passive walking in the sagittal plane[177, 178, 179]. In addition, McGeer conducted the theoretical analysis regarding stability of passive walking and the valiation of gait patterns caused by that of robot structure.

After the McGeer's work, many researchers, who were fascinated with the concept that a robot performs locomotion according to solely its dynamics, investigated actively dynamics or stability of the passive dynamic walking by numerical simulation or robots.

1.2.6.11 2-D Passive Walking

Most of the research about passive walking was intended for 2-dimensional walking due to the complexity and difficulty to analyze 3-D dynamics. Thus, dynamics of 2-D passive walking was deeply pursued and many efficient facts were illuminated.

Schwab *et al.* derived the return map on the poincare section by means of the simple passive-walking-robot model and showed that its domain of attraction is quite small[253]. Based these results, they concluded that the domain of attraction is independent of the stability of periodic walking motion.

It is interesting to note that some work reported the chaos occurred under a certain condition in passive walking. Garcia *et al.* analyzed the energy efficiency, stability and kinetic property of passive walking by modeling Passive Dynamic Walker as a simple robot[77, 76]. Their numerical simulation showed that bifurcation in passive walking appeared by changing some parameters and that chaos occurred in a certain domain. In addition, they described the possibility of the control method to stabilize walking by confining robot dynamics in the stable bound. Thuilot *et al.* investigated the behavior of passive walking dynamics according to physical parameters in detail the by use of compass-like biped model which is the simplest model of passive walking robot[275]. In the work, they analyze and examine the bifurcation and chaos that occur in the dynamics of compass-like biped.

Hass *et al.* investigated the effects of mass distribution on walking speed and stability [100]. As a result of optimization, they found optimal manifolds in terms of speed, and showed the basin of attraction was high along these manifolds. Furthermore, it was estimated how speed and stability vary on different ground slope. In conclusion, they showed maximal robustness and highest walking speed was partly conflicting objectives of optimization.

1.2.6.12 3-D Passive Walking

On the other hand, some researchers studied 3-D passive walking analytically. Indeed, 3-D passive dynamic walking was realized in substance.

Coleman *et al.* analyzed the 3-D dynamics of passive biped aiming at expansion of planar passive walking, which was realized in the sagittal plane, to 3-D passive walking by applying the lateral sway motion like human[44]. They built the robot satisfying the desirable parameter for 3-D passive biped and achieved 3-D passive walking.

Collins *et al.* proposed the mechanism to utilize potential energy for not only the forward motion in the sagittal plane but lateral sway motion[47]. They adjusted the physical parameters of the robot experimentally and realized 3-D passive dynamic walking first in the world. It was said that its walking was quite similar to the human.

1.2.6.13 Disadvantages of Passive Walking

It was demonstrated by Collins *et al.* that passive walking can realize the energy-efficient, natural and dynamic motion since its source of power is dynamics itself. However, it has the following disadvantages: 1) Highly limited capability to walk—passive walking has limitations on types of gait in the specific environment such as the downslope— 2) Low robustness against disturbance 3) Disability to perform another tasks during walking.

Hybrid Walking

In hybrid walking which is in between passive walking and active walking, there are mainly two types of approaches, quasi-passive walking and quasi-active walking. Quasi-passive approach is the gait control method by use of small number of active joints to realize the passive-like walking on the non-decline ground or to enhance the stability. Meanwhile, in quasi-active walking, passive dynamics is introduced into active walking. That is, the energy efficiency or/and adaptability to ground irregularity are enhanced by making a few joints of active walking passive, such as point-contact or passive swing.

Fig. 1.4 shows the characteristics of active, passive, and hybrid walking. The passively, actives, and hybrid approach are compared from several aspects such as stability, adaptability, energy efficiency and so one. As far the number of underactuated joints and passivity, the passive approach is the highest. The hybrid approach is the seconds. In contrast, the order inverts as far the requirement of motor power. Thus, the order of energy efficiency becomes passive, hybrid, and active walking. As mentioned previously, a passive walking robot has no actuation, and the basin of attraction is very small, hence adaptability to environmental variation and walking stability are quite low. Meanwhile, Active walking and hybrid walking have the possibility to adapt the circumstances and to enhance stability. The notable point is that the adaptability to ground irregularity of hybrid walking is higher than passive and active walking. As described in the section of active walking, on the actual ground in human residential environment which includes small irregularity unable to be observed in advance, it becomes hard to measure the actual ZMP by means of force sensors, and a support-foot sole probably detaches from the ground. In other words, it is difficult for active walking to adapt to such ground when the environment is not propely measured. As a matter of course, passive walking cannot adapt to ground irregularity due to no actuation and no controller. Meanwhile, in most of methods in hybrid walking, point-contact condition between a robot and the ground is employed. The problems peculiar to active walking, i.e. the detachment of a support-foot sole from the ground and the inaccuracy of ZMP measurement, do not occur in hybrid walking. In addition, active joints make it possible to feedback the error arisen by ground irregularity and to converge the error.

Most of the research for hybrid walking deals with two-dimensional walking since the dynamics of three dimensional walking is complex and difficult to control. Since 2-D walkers are designed to possess the four legs or large supporting rod in the lateral direction in order to eliminate any lateral motion, it requires the large area on the ground to stand, and cannot change the walking direction. For the practical use of biped walking robots, we think that it is necessary to realize 3-D walking.

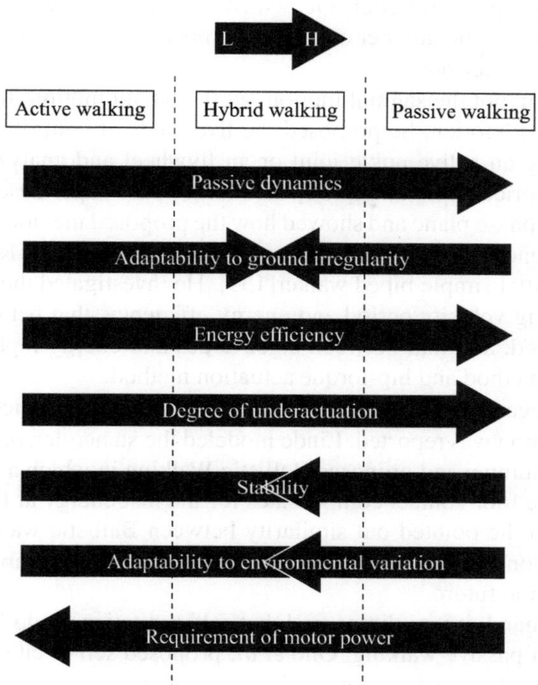

Fig. 1.4 Characteristics of active, passive, and hybrid walking

Hereinafter, we summarize the research of hybrid walking including our method in terms of dimension (planar of spatial, i.e. 2-D or 3-D) and the approach (quasi-passive or quasi-active walking).

2-D Hybrid Walking

Quasi-passive Approach

As a pioneering work of quasi-passive walking, Goswami *et al.* analyzed passive walking in terms of mechanical energy by means of compass-like biped and, by use of the simulation result of analysis, proposed control algorithm determining input torque based on the mechanical energy[84]. This was the pioneering research of quasi-passive walking. After his work, many researchers started to conduct work of quasi-passive walking.

Spong applied the passivity based control[272] to the compass gait biped[265]. His method augmented the Gravity term of dynamic equation by applying torque to the ankle or hip joint so that the energy balance of one walking cycle satisfied the desired value, and achieved the passive gait on flat ground, and up/down slope.

Furthermore, he proposed the controller which could have global stability by switching the control law or the number of passive joints according to the error of angular velocity around contact-point.

Kuo focused on not the sagittal motion but the lateral motion. In order to do the stable lateral sway motion, he proposed the five kinds of control methods such as the method using an active ankle joint or an flywheel and analyzed each control methods by numerical simulation[158]. In addition, the paper described the lateral dynamics on the phase plane and showed how the proposed method compensates for the mechanical energy lost by the impact shock at foot-contact. Kuo also pursued the powered sagittal simple biped walker[159]. He investigated the relationship between the walking velocity/period and energy efficiency, that between step-length and walking speed, and that between eigenvalues and energy input, regarding the toe-off impulse method and hip-torque actuation method.

Some work regarding the controller using spring that can achieve more energy-efficient locomotion was reported. Linde modeled the stance-leg of passive walking robot as spring-dumper and proposed Ballistic Walking in which a spring extension of rear leg before foot-contact compensates for the lost energy at foot-contact[165, 166]. In addition, he pointed out similarity between Ballistic walking and human actual gait and considered that this method would become valuable in the field of rehabilitation in the future.

On the other hand, it was discovered that self-excitation could be applied to the energy-supply in passive walking. Ono *et al.* proposed self-excited walking which generates biped walking with a four-link biped that possesses an actuated hip joint and passive knee joints with stoppers by means of self-excitation[214, 215]. They found the two types of self-excitation—Van del Pol type and the asymmetrical stiffness matrix type—and analyze them in detail. The controller to stabilize walking by means of the asymmetrical stiffness matrix type self-excitation was designed. Finally, the planar natural biped walking was achieved with four-link biped. Following Ono's work, Sangwan *et al.* investigated the self-excited walking in detail by using the complete dynamical system modifying the dimensionally incorrectness of the analyzed equations[248]. They detected the range of physical parameters such as leg-length and leg masses in which the stable walking could be achieved.

Ogino *et al.* designed the passive-dynamic-walking-based controller consisting of two layers[211]. The roll of lower layer is stabilization of walking, while that of upper layer is to minimize the energy consumption. In their walking termed ballistic walking, the lower layer applies the instantaneous torque to swing-leg immediately after the swing-foot is detached from the ground, and the upper layer controller adjusts parameters of the lower layer to minimize the energy cost.

Haruna *et al.* examined the dynamics of passive dynamic walking robot with a torso as the first step toward achievement of passive walking by a humanoid robot[93]. They employed PD control to stand the torso up, and showed the convergence of robot dynamics on a stable cycle in numerical simulation. Furthermore, three principals to reduce the torso torque: changing the desired posture of the torso, adding the soft leg tips, and changing the curvature of the sole, are proposed and effective in numerical simulation.

Hitomi *et al.* applied a reinforcement learning to stabilizing control of planar passive dynamic walking on the various gradient slope[115]. They employed the stochastic policy gradient reinforcement learning[148], and set the reward function making the robot dynamics attract the periodic stable cycle. The simulation results showed that their proposed learning framework achieved stable walking after about 1000 steps, and that the controller was robust against the variation of slope gradient and sudden perturbations.

In terms of the control of passive walking, expansion of the basin of attraction by the active actuation is important challenge. Wisse *et al.* investigated the basin of attraction of the passive walking by use of simple model, and showed that passive walker can only handle quite small disturbances[308]. Although there is no way to solve the problem of falling backward, they found the basin of attraction under the swing-leg control, and demonstrated that a simple controller can completely solve the problem of falling forward.

Quasi-active Approach

The objective of most of quasi-passive walking work is to realize passive walking not on the downslope but on the level ground, whereas point-contact method is intended for the practical use of bipedal walking robot. Point-contact method has the possibility to achieve the natural energy-efficient walking that is adaptable to not only flat ground but also the various ground condition. From those reasons, many studies of point-contact method are pursued actively in recent years.

Some researchers reported the controller to use telescopic-type leg supplying energy. Grishin *et al.* proposed the algorithm to control adaptively the robot with two telescopic legs of a passive ankle and a trunk in the sagittal plane[86]. The support leg length is adjusted so as to stabilize walking. They calculated some parameters in advance based on the mathematical model using linear approximation, and the robot determines the adjusting quantity. The sagittal dynamic walking was reported. Katoh *et al.* proposed the control method to use dynamical system possessing a stable limit cycle[142]. Dividing the walking motion into double-support phase and single support phase, they designed both controllers based on coupled Van der Pol's oscillator by varying the parameters of the oscillator. In the stance phase, hyperbolic curve generated by Van der Pol's oscillator was utilized, and in the swing phase, limit cycle was employed. Larin investigated the control of a biped robot with a torso and two telescopic legs[162]. He analyzed the impact at foot-contact within the framework of Carnot's theorem, and stabilized the periodic cycle by means of Riccati equation. However, his method was based on the linearization in the vicinity of equilibrium point of periodic dynamics to guarantee the linear stability.

Pratt achieved the smooth biped walking by means of the robot by applying the passive joint to the ankle and knee joint[228]. The ankle joint, which is the apparatus attached a spring to from toe to the link upper ankle joint, generates reaction force with a spring during support-leg phase and plays a roll to kick the ground at the shift from support-leg phase to swing-leg phase. As the walking velocity control, they design the controller based on the equilibrium concept of passive walking.

There are some research applying virtual holonomic constraint to 2-D point-contact robot, and pursued its dynamics. Grizzle and Westervelt *et al.* built the controller by use of the virtual holonomic constraint of joints named Virtual Constraint and realized stable dynamic walking by means of the planar point-contact biped robot with a torso[87, 224, 297, 298]. They described the walking dynamics as Hybrid Zero Dynamics and design the control framework. The stability was proved by means of Hybrid Zero Dynamics and the validity of controller was tested by experiment and simulation. Chevallereau presented the controller to converge robot dynamics on optical trajectory by introducing the virtual time[41]. Moreover, the control law to track not a reference joint trajectory but the geometrically associated path was proposed[42]. This is the similar concept to Virtual Constraint. In the paper, he found convergency of dynamics, conditions of existence and uniqueness of cyclic walking dynamics, and robustness against modeling error by numerical simulation. Kajita *et al.* proposed Linear Inverted Pendulum Mode that models a robot as an inverted pendulum and gives it the constraint of horizontal COG trajectory[136]. He derived the conserved quantity from the dynamics and built the controller by means of that value. The planar biped walking according to the ground shape was realized with Meltran II.

Chemori *et al.* proposed the finite-time stabilizing control law for a planar seven degree-of-freedom underactuated robot, inspired by previous results for mechanical systems under holonomic and unilateral constraints[38]. Dividing the whole planar biped motion into three fundamental phases, single-support phase, double-support phase, and impact instantaneous phase which is treated as external perturbation, they designed the controller of single-support phase with robustness in order to ensure the appropriate initial conditions of double-support phase despite of the impact. Furthermore, Chemori *et al.* proposed the dimension reduction scheme decomposing the vector of coordinates in actuated and unactuated variables, and applied a nonlinear predictive control to the low-dimensional system[37]. They detected both the sufficient conditions for the achievement of stable walking and the basin of attraction, and demonstrated the proposed controller's robustness against model uncertainties and ground irregularity, in numerical simulation.

Sabourin *et al.* proposed new control strategy for an underactuated robot, using the CMAC neural network[240, 241]. In their algorithm, the CMAC neural network is input the inclination of stance-leg into, and outputs the desired whole robot state. During learning phase, the controller gradually diminishes the error between desired output and actual one. In addition, they designed the high-level velocity controller of stance-leg that regulates the average walking velocity. The whole controller achieved robustness against external impulsive perturbation forces.

Geng *et al.* designed the sensor-driven controller, which is built with biologically inspired sensor- and motor-neuron models, and applied a policy gradient reinforcement learning algorithm[148] to tune the parameters of the sensor-driven controller in real-time during walking[78]. By means of the proposed method, they achieved the fastest planar biped walking, taking into consideration of leg-length.

Takuma *et al.* developed a planar biped robot with McKibben pneumatic artificial muscle, and estimated the relationship between the walking period and the control parameters of the pneumatic muscle[273]. By designing a step-by-step feedback controller to stabilize walking based on the estimated relation, they succeeded in stabilizing walking, and achieved stable walking over 9[mm] stair.

Vermeulen *et al.* proposed a real-time joint trajectory generator for planar walking bipeds[287, 286]. The trajectory planner they designed generates the dynamically stable motion patterns under the assumption of point-contact, by use of a set of objective locomotion parameters as its input. In their method, in the condition of no disturbance and perfect tracking, point-contact state is held. They examined the proposed method in two instances of perfect tracking and imperfect tracking by muscle-valves actuation system.

In this field, some controllers based on the periodic signals from oscillators were pursued. Nakanishi*et al.* proposed the new control framework for learning biped locomotion from demonstration and its adaptation through coupling between the pattern generator and the mechanical system[196, 195]. In their work, both the learning and encoding method of complex human movements, which they proposed previously[123], and phase adaptation based on phase resetting were applied to a planar underactuated biped as CPG, and they demonstrated the validity of the proposed controller by numerical simulation and experiment. The controller designed by Aoi *et al.* decided the motion based on nonlinear oscillator, and reset the phase of oscillator according to the feedback signals form touch sensors at the tips of the legsAoi. They demonstrated the stability of the walking motion analytically by means of Poincare return map. Furthermore, Aoi *et al.* pursued the dynamics of simple underactuated biped robot driven by a rhythmic signal from an oscillator, and found that the bifurcation of the motion which leads to chaotic motion is occurred depending on the model parameter values[9]. By use of approximation, they constructed discrete dynamics using Poincare map, and obtained both the period-doubling solutions and the critical value by means of Newton-Raphson method.

Agrawal found that the nonlinear terms in the centrifugal and Coriolis terms disappear, and that dynamics of biped robots can be nearly linear by designing the biped robot such that the inertia matrix is a constant and the gravity terms are simple[1]. He employed the counterweights to cancel the complicated non-linearity of biped robot dynamics, and designed the trajectory tracking control of underactuated biped by applying feedback linearization.

Miossee *et al.* designed the joint trajectory as the polynomial functions in the absolute orientation angle of the biped, and double-support motion stabilizing the walking cycle including non-stabilized single-support motion and discrete foot-impact phase[182]. In their method, underactuated single-support phase had no stabilization mechanism, and overactuated double-support phase played an all roll of stabilization. With numerical simulation, they demonstrated that the large one-step convergence zone exists, and that the proposed controller handle starting and stopping walking problem.

In order to solve the difficulties to control underactuated dynamics, some researchers employed the reinforcement learning to achieve hybrid walking. Morimoto *et al.* proposed the simple leaning framework of bipedal locomotion by use of approximated poincare map, and achieved stable biped walking by an underactuated planar biped robot[186, 175]. In their method, the reward was defined by use of the hip height at foot-contact, and the controller learned the adequate foot placement. Given the nominal walking pattern, after about 100 steps of learning, the controller could realize stable walking on the various surface of the ground such as carpet or metal.

3-D Hybrid Walking

Quasi-passive Approach

Most of research regarding quasi-passive walking were intended for 2-D motion, whereas the group of passive dynamic walking (Cornell University, Delft University of Technology, and Massachusetts Institute of Technology) developed 3-D passive walking robot and achieved 3-D passive-like walking on the level ground[45]. Collins *et al.* installed the electric motors with springs on the ankle joints that push off the ground, and realized the natural human-like biped walking on the level ground[46]. Their employed control was quite simple: When the swing-leg foot touched the ground, the stance-leg (opposite leg) ankle extends to compensate the energy lost by foot impact. The walking of their robot was highly energy efficient, and the energy consumption was about ten times lower than that of other humanoid robot. Wisse *et al.* analyzed 3-D passive walking in detail and achieved the stable 3-D passive walking in the simulation[307]. Besides, he designed the mechanism of skateboard-like ankle joint and actuated hip joint, and realized 3-D stable quasi-passive walking on the flat ground. This gait was quite dynamic and natural like human. Tedrake *et al.* built the simple 3-D passive walker with only six internal degree of freedom and four actuators, and realized 3-D quasi-passive walking by means of the learning algorithm they proposed[274]. Decomposing the control system in the lateral and sagittal plane, they formulated the learning problem on the discrete return map dynamics, and applied a stochastic policy gradient algorithm[148]. Their learning algorithm converged quite quickly, and well functioned for their underactuated biped.

As theoretical research, Spong *et al.* extended the passivity based control of compass gait biped[265] to n degree-of-freedom biped in three dimensional space[266]. They showed that variation of the ground slope defined a group action on the configuration manifold of the system, and that both the kinetic energy and impact dynamics were invariant under this group action. In order to ensure that closed-loop system is invariant under the slope variation, the energy shaping controller, i.e. the compensation control of potential energy, was designed.

Table 1.1 Achievement of hybrid walking method. Each item indicates the following: (Dynamics) 2-D and 3-D signify that the controller is based on 2-D and 3-D dynamics respectively. 2-D×2 indicates that 3-D dynamics is decomposed in the lateral and sagittal plane approximately, and the controller is designed based on the pair of 2-D dynamics. (QP or QA) QP and QA denote quasi-passive and quasi-active approach respectively. (DoU) DoU denotes the degree of underactuation. (Stabilization) Stabilization indicates whether the walking stability was confirmed by simulation or experiment. (Proof) Proof signifies whether the stability proof was conducted or not. (Realization) Realization indicates whether the controller actually realized biped walking by the robot.

	Dynamics	QP or QA	DoU	Stabilization	Proof	Realization
Goswami *et al.*[84]	2-D	QP	1	*		
Spong [265]	2-D	QP	1	*		
Kuo[158, 159]	2-D	QP	1	*		
Linde[166, 165]	2-D	QP	1	*		
Ono *et al.*[215, 214]	2-D	QP	1	*		*
Sangwan *et al.*[248]	2-D	QP	1	*		
Ogino *et al.*[211]	2-D	QP	1	*		
Haruna *et al.*[93]	2-D	QP	1	*		
Hitomi *et al.*[115]	2-D	QP	1	*		
Wisse *et al.*[308]	2-D	QP	1	*		*
Grishin *et al.*[86]	2-D	QA	1			*
Katoh *et al.*[142]	2-D	QA	1	*		
Larin *et al.*[162]	2-D	QA	1	*		
Pratt *et al.*[228]	2-D	QA	1or2			*
Grizzle *et al.*[87, 224]	2-D	QA	1	*		
Westervelt *et al.*[297, 298]	2-D	QA	1	*		*
Chevallereau *et al.*[41, 42]	2-D	QA	1	*		
Kajita *et al.*[136]	2-D	QA	1	*		*
Sabourin *et al.*[240, 241]	2-D	QA	1	*		*
Geng *et al.*[78]	2-D	QA	1	*		*
Takuma *et al.*[273]	2-D	QA	1	*		*
Vermeulen *et al.*[286, 287]	2-D	QA	1			
Nakanishi *et al.*[196]	2-D	QA	1	*		
Aoi *et al.*[8, 9]	2-D	QA	1	*		
Chemori *et al.*[38, 37]	2-D	QA	1	*		
Agrawal[1]	2-D	QA	1	*		
Miossee *et al.*[182]	2-D	QA	1	*		
Morimoto *et al.*[186, 175]	2-D	QA	1	*		*
Aoyama *et al.*[13]	2-D	QP	1	*	*	
Tedrake *et al.*[274]	2-D×2	QP	2	*		*
Miura *et al.*[183, 258]	2-D×2	QA	2	*		*
Aoyama *et al.*[10] and [13]	2-D×2	QA	2	*		*
Collins *et al.*[46]	3-D	QP	3			*
Wisse *et al.*[307]	3-D	QP	3			*
Spong *et al.*[265]	3-D	QP	3			
Song *et al.*[264]	3-D	QA	2			
Aoyama *et al.*[11]	3-D	QA	2	*		*

Quasi-active Approach

Miura *et al.* built stilt-like biped which contacts the ground at a point, and designed the controller by decomposing 3-D dynamics in the lateral and sagittal planes, and by linearizing system in the vicinity of origin [183, 258]. They proposed the stabilizing control to adjust the landing position according to the error between the desired state and actual one, and realized 3-D dynamic walking. However in their method, there is the limit that step-length can not be long due to linearization.

Song *et al.* found the transverse dynamics by applying the dynamics decomposition (proposed by Hauser and Chung[102]) to the dynamics of a three dimensional biped robot, and proposed the robust controller stabilizing the derived transverse dynamics by use of linearization and discretization[264]. They showed the stability of the proposed method in the vicinity of the desired nominal periodic orbit. It is conceivable, however, that their proposed method ensures only the convergence of dynamics toward the nominal trajectory, and there is no assurance of the stability of underactuated robot walking. That is, in their method, biped robot might fall down if the large error enough not to be converged during one step is occurred.

Bipedal Running

As recent advances of robotic hardware technology and performance enable biped robots to run, some researchers pursued the control of bipedal running. Raibert built the hopping robot and realized dynamic biped running by installing two hopping leg[232]. This robot has the time that the robot does not contact with the ground, however it performs stilt-like motion during support-leg phase which the robot contact with the ground. As for the controller, they stabilized the originally unstable motion by varying the landing position according to the state during flying phase.

Chevallereau *et al.* developed a time-invariant feedback control strategy, and applied Virtual Constraint and Hybrid Zero Dynamics approach to running control[43]. Differently from walking, running motion includes the flight phase. They developed their method, and applied same control algorithm to the flight phase. The simulation results showed the asymptotically stability of the bipedal running they proposed.

1.3 Bio-inspired System

1.3.1 Foundation of Neural Network

1.3.1.1 Artificial Neural Network

Research in artificial neural networks has recently been active as a new means of information processing. Artificial neural networks try to mimic the biological brain neural networks into mathematical model. The brain is a large-scale system connecting many neural cells called neurons; it is estimated that a human being has around 10^{10} brain cells. The brain has many excellent capabilities: parallel processing of information, a learning function, self-organization capabilities, and so forth [69, 153]. The brain can also provide an associative memory [153] and is good for information processing such as pattern recognition [301].

The neuron is a cell and so has a nucleus, which is contained in the soma or body of the cell (Fig. 1.5). One may think oh the dendrites as forming a very fine filamentary bush, each fiber being thinner than axon, and of the axon itself as a long, thin cylinder carrying impulses from the body to other cells. The axon split into a fine arborization, each branch of which finally terminates in a little end bulb almost touching the dendrites or other parts of a neuron. Such a place of near contact is called a synapse. Impulses reaching synapse set up graded electrical sings in the dendrites of the neuron on which the synapse impinges, the interneuronal transmission being sometimes electrical but usually by diffusion of chemicals called transmitters. A particular neuron only fire an electrical impulse along its axon if sufficient impulses reach the end bulbs impinging on its dendrites in a short period of time, called the period of latent summation. Actually, these impulses may either help or hider the firing of an impulse and are correspondingly called excitatory or inhibitory. The condition for the firing of a neuron is then that the excitation should exceed the inhibition by a critical amount called the threshold of the neuron.

Artificial neural network, a model of the brain, connects many linear or nonlinear neuron models and processes information in a parallel distributed manner [167]. In a Von Neumann computer with a single processor, the speed of computation is limited by the propagation delay of the transistors. Because of the massively parallel nature, neural networks can perform computations at much higher speed [69]. In addition, the neural network has many interesting and attractive features. Neural networks have learning and self-organizing capabilities. Therefore, neural networks

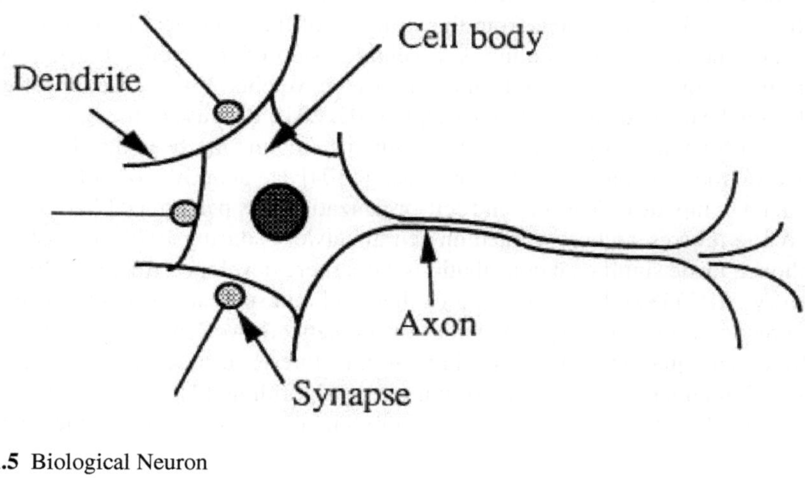

Fig. 1.5 Biological Neuron

can adapt to changes in data and learn characteristics of input signals. That is, neural networks can learn a mapping between an input and an output space and synthesize an associative memory that retrieves the appropriate output when presented with the input and generalizes when presented with new inputs [225]. Moreover, because of nonlinear nature, neural networks can perform functional approximation and signal filtering operations which are beyond optimal linear techniques [301].

1.3.1.2 History of Studies in Neural Networks

From 1946 to 1960, a movement attempted to carry out interdisciplinary research on the brain and computers to solve the basic principles for intelligent information processing. McCulloch and Pitts proposed an idea that a "mind-like machine" could be manufactured by interconnecting models based on behavior of biological neurons, the concept of "Neuro-logical Networks" [176]. They made a neuron model representing a basic component of the brain and showed its versatility as a logical operation system. In this era, cybernetics proposed by Wiener [306] was studied in depth. The principle of the cybernetics is the relationship between engineering principle, feedback, and brain function. Research was led to the basis of today's von Neumann-type computer [202].

With the progress of research on the brain and computers, the objective changed from the "mind-like machine" to "manufacturing a learning machine," for which Hebb's learning model was proposed [103]. In addition, by early 1960s, specific design guidelines for learning systems were given by Rosenblatt's perceptron [237], Widrow and Hoff's Adaline (AD Adaptive LINear Element) [302] and Steinbuch's Learning Matrix [268] which was a pattern recognition machine based on linear discriminate functions. The Perceptron received considerable excitement when it was first introduce because of its conceptual simplicity. However, Minsky and Papert (1969) proved mathematically that the Perceptron cannot be used for complex logic function [181]. Then, the excitement was terminated and research in Artificial Intelligence (AI) independent from the structure of the biological brain was started. On the other hand, the Adaline is a weighted sum of the inputs, together with a least mean square (LMS) algorithm to adjust the weights to minimize the difference between the desired signal and the output [302, 204]. Because of the rigorous mathematical foundation of the LMS algorithm, the Adaline has been developed in to a powerful tool for adaptive signal processing [304] and adaptive control [299]. Early work on competitive learning and self-organization was performed [267, 89].

While few researchers worked on neural networks during 1970s, Grossberg and Kohonen made significant contributions. Grossberg developed Adaptive Resonance Theory (ART) [88], based on the idea that the brain spontaneously organizes itself into recognition codes. The dynamics of the network were modeled by first order differential equations. There are three architectures and these are self-organizing neural implementations of pattern clustering algorithms [31]. On the other hand, Kohonen developed his work on self-organizing maps, based on the idea that neurons organize themselves to tune various and specific patterns [153, 152]. Moreover, Albus developed an adaptive Cerebellar Model Articulation Controller (CMAC) which is a distributed table-look-up system based on models of human memory[3]. More detail of the CMAC is explained in Section 1.3.4.

In the 1979s, Werbos originally developed a back-propagation algorithm [293], and its first practical application was for estimating a dynamic model to predict nationalism and social communications. However, Werbos' work remained almost unknown in the scientific community. In the mid-1980s, the back-propagation algorithm as the learning algorithm of the feed-forward neural network was also rediscovered by Parcker [221] and David Rumelhart et al. [238]. Moreover, the back-propagation through time is also powerful tool to deal with dynamical systems such as recurrent neural networks, feed-forward systems of equations, systems with time lag [295]. On the other hand, in the early 1980s, Hopfield introduced a recurrent-type neural network that is based on the interaction of neurons [117] and his approach was based on the Hebbian learning law [103]. The model consists of a set of first order (nonlinear) differential equations that minimize a certain energy function and this model is known as a Hopfield network. Furthermore, Kosko extended some of the ideas of Grossberg and Hopfield to develop his adaptive Bidirectional Associative Memory (BAM) [155]. Hinton, Sejnowski, and Ackley developed the Boltzmann Machine [109, 108] which is a kind of the Hopfield net that settles into solutions by a simulated annealing process [150] as a stochastic technique. These research works in the 1980s triggered the present boom in the scientific community. Recently, neural networks have found wide applications in many different fields.

Application of neural networks to pattern recognition has been widely studied. Neural Networks such as the Hopfield net and the feed-forward net with the back-propagation algorithm are applied and studied mainly for image and speech recognition. Research in image recognition includes initial vision (stereo vision of both eyes, outline extraction, etc.) close to the biological (particularly brain) function, manually-written character recognition by cognitron and neocognitron in the practical level [72], and cell recognition for mammalian cell cultivation by using the feed-forward neural network [65]. Speech recognition deals with time series, and it was reported that the Time Delay Neural Network (TDNN) with time delay input was effective as a buffer model of a feed-forward net [291].

Optimization is often required for the planning of actions, motion, and tasks, but many parameters causes the amount of calculation to be tremendous and the ordinary method cannot be applied. Hopfield network is an effective tool to find the optimal solution of the optimization problem [118], by defining an energy function.

Recently, in control field, there have been many cases, where automatic control theories and techniques have played an important role. With the progress of control theories, there are now many applications for automatic control with increased better performance. However, applicable systems become increasingly complicated and highly composite. It is therefore expected that control theories and techniques will make further progress. Adaptive control, such as the Model Reference Adaptive Control (MRAC) [54] and the Self Tuning Regulators (STR) [19, 154], has become available to those systems having more uncertainties [262]. Nevertheless, the traditional adaptive control has problems such as exponentially complicated calculation for the number of unknown parameters and limitations on the applicability to nonlinear systems [69, 200, 201]. Many attempts have been made to apply the neural network to control fields, where the neural network is used to deal with nonlinearities and uncertainties of the control system and to approximate functions [69], such as system identification [200]. Research in neural network application to control can be classified into some major methods depending on structure of the control systems [296], such as supervised control [303, 91], inverse control [230, 233, 203, 90, 144, 129, 62], neural adaptive control [200, 201, 276], back propagation of utility [296] which is an extended method of the back-propagation through time [295], and adaptive critics which include a reinforcement learning algorithm [300, 24, 61, 294].

1.3.1.3 Neural Network Models

Neuron Model

Each neuron building a network simulates a biological neuron as shown in Fig. 1.6. The neuron unit consists of multiple inputs, x, and one output, y, and its internal state is given as the weighted sum of input signals. The output of the neuron unit is as follows:

$$y(t) = f\left(\Sigma_{i=1}^{n} w_i \cdot x_i(t) - \theta\right), \quad (1.1)$$

where w_i is a weight of connection, θ is a bias of the neuron unit, t is time, and n is the number of inputs. The weight coefficient w_i that represents the strength of connection indicates the synapse load. It takes a positive value for excitation and a negative one for inhibition.

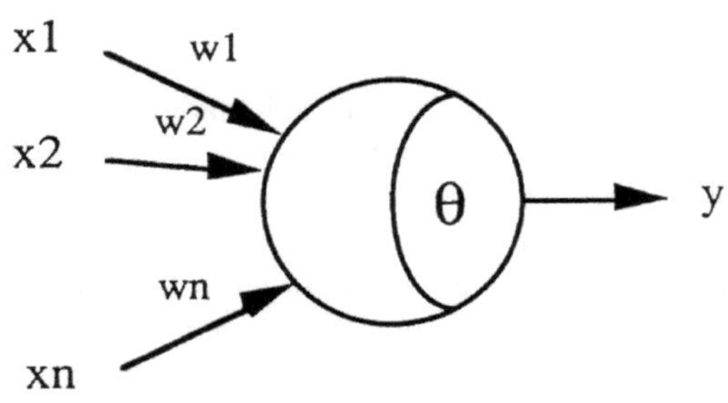

Fig. 1.6 Artificial neuron model

The neuron output function $f(x)$ often uses the two-valued function of 1 and 0 using threshold fend-limiter, or the sigmoid function that is a continuous and non-linear function (Fig. 1.7) [238]. A conventional sigmoid function is represented by following equations:

$$f(x) = \frac{1}{1+e^{-ax}} \quad (0 < f(x) < 1). \tag{1.2}$$

A tanh function is often used instead of the conventional sigmoid function since the output value of the sigmoid function is positive while that of the tanh function being both positive and negative. The tanh function is as follows:

$$f(x) = \frac{1-e^{-ax}}{1+e^{-ax}} \quad (-1 < f(x) < 1). \tag{1.3}$$

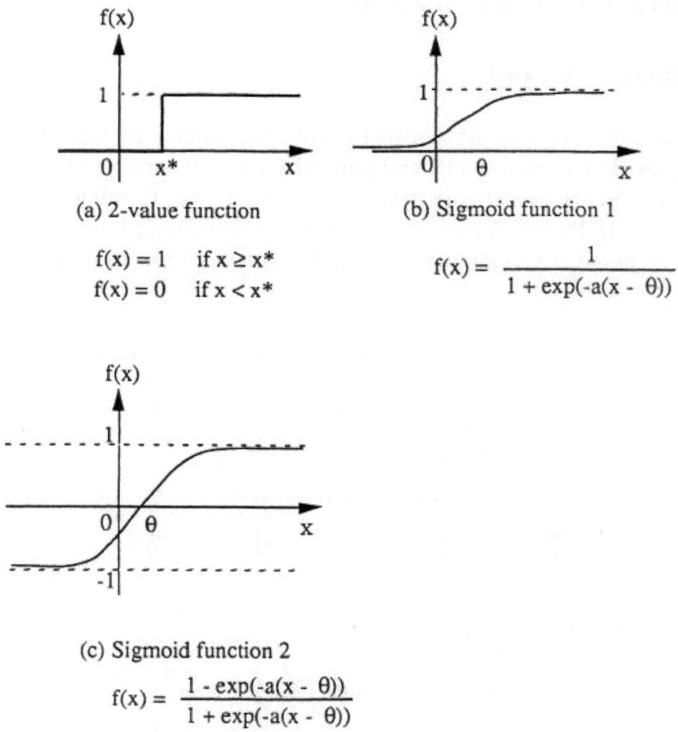

Fig. 1.7 Examples of non-linear functions in neurons

Network Model

In the biological brain, a tremendous number of neurons are interconnected to form the network is built by these inter-connection neuron models. The artificial neural network is built by these inter-connection neuron models. The organization of the neural network is classified largely into two types: a recurrent network and a feed-forward network. In the recurrent network, multiple neurons are interconnected organize the network (Fig. 1.8). There are feedback loops of signal flow. Therefore, the recurrent network has dynamic characteristics.

The feed-forward network has a hierarchical structure which consists of some layers without interconnections between neurons in each layer, and signals flow from input layer to output layer in one direction (Fig. fig:Feed-forwardNet). The following sections describe typical examples of the recurrent net and the feed-forward net, respectively.

1.3.2 Recurrent Neural Network

1.3.2.1 Hopfield Network

Recurrent neural networks are networks with interconnected neurons. Recurrent networks provide feedback paths and generally have dynamics. In the 1980s, the theory of the recurrent network has been often called Hopfield network (Fig. fig:RecurrentNet). Research on recurrent nets has been carried out, for example, to implement an associative memory [117] and to solve optimization problems [118].

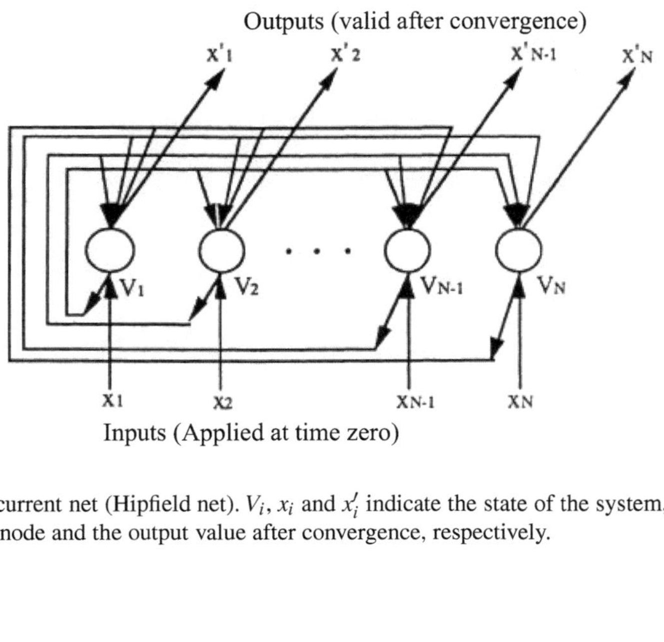

Fig. 1.8 Recurrent net (Hipfield net). V_i, x_i and x_i' indicate the state of the system, the initial value of the node and the output value after convergence, respectively.

The theory of the recurrent network basically consists of an entrainment effect to the stable equilibrium of a potential field. Although discrete and continuous time systems were discussed [6, 217, 259, 5], only the continuous time system is surveyed in this section.

Let the state of the system be $V_i(i = 1,2,\cdots,n)$ and related potential be $E(V_1,V_2,\cdots,V_n)$. The potential E given by Hopfield is as follows:

$$E = -\frac{1}{2}\Sigma_i\Sigma_j W_{ij}V_iVj + \Sigma_i\frac{1}{R_i}\int_0^{V_i} f^{-1}(V)dV + \Sigma_i I_i V_i, \quad (1.4)$$

where W_{ij} is the interconnection weight of neurons. The derivative of E with respect to time can be expressed:

$$\frac{dE}{dt} = \Sigma_i\left(\frac{\partial E}{\partial V_i}\right)\left(\frac{dV_i}{dt}\right) = -\Sigma_i\left\{f^{-1}(V_i)\right\}'\left(\frac{dV_i}{dt}\right)^2, \quad (1.5)$$

where f_i is the sigmoid function in Fig. 1.7 (b). The variable u_i is defined by the gradient system in the following equation:

$$V_i = f_i(u_i) \quad \text{and} \quad \frac{du_i}{dt} = -\frac{\partial E}{\partial V_i} \quad (1.6)$$

As if f_i is a monotonously increasing function, f_i^{-1} is positive. E is a Liapunov function and then, $\frac{dE}{dt} < 0$. Therefore, when the system is stable, the state will approach

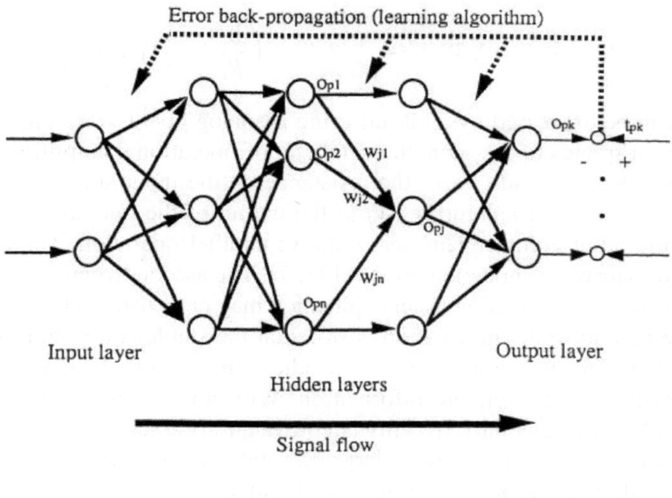

Fig. 1.9 Feed-forward net. The solid line indicates the physical cignal flow, while the dotted line does the back propagation.

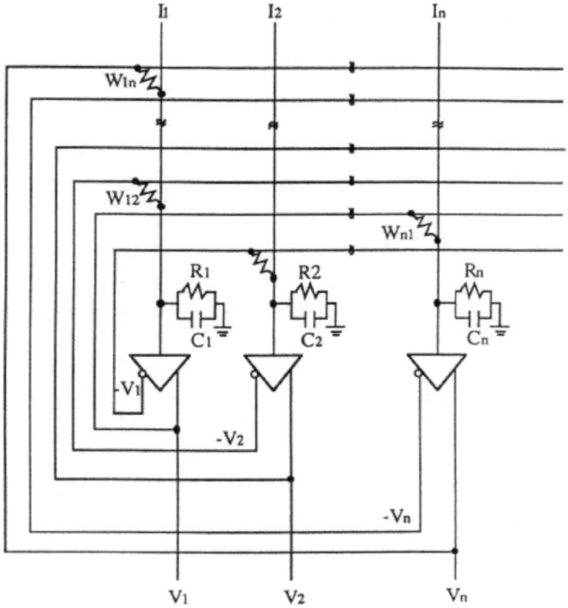

Fig. 1.10 A structure of the Hopfield network for hardware implementation.

an equilibrium point as time goes to infinity. However, note that there may be not only one but many equilibrium points. Eq. (1.6) given by Hopfield is as follows:

$$C_i = \frac{du_i}{dt} = \Sigma_j W_{ij} V_j - \frac{u_i}{R_i} + I_i, \quad \left(\frac{1}{R_i} = \frac{1}{\rho_i} + \Sigma W_{ij} \right) \tag{1.7}$$

Eq. (1.7) can be expressed by a circuit using n analog operational amplifiers (Fig. 1.10). f_i corresponds to the gain characteristic of operational amplifires, C_i to the capacitor on the input side, R_i to the resistance on the input side, W_{ij} to the conductance, I_i to the external current, V_i to the output of the operational amplifiers, and u_i to the input voltage. This setup allows parallel operation. There are other examples of hardware implementations [188, 7]. The associative memory (content-addressable memory) function as an application may be considered as follows:

Assume that accurate information is stored at the stable equilibrium state S_1 in Fig. 1.11. This depends on W_{ij}. The state (indicate by x) shifted away slightly from the equilibrium has incomplete information. With this as the starting point, Eq. (eq:Hopfield) converges into S_1. This is equivalent to associative memory [153]. If Shift from the equilibrium state is large, the state may lead to other equilibriums and correct information cannot possibly be obtained.

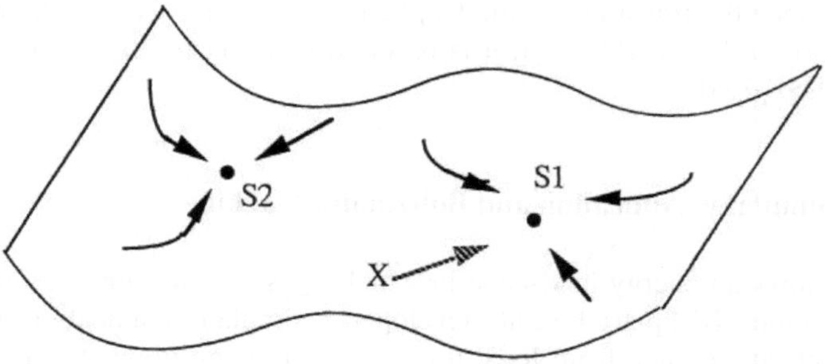

(a) Energy curve surface and entrainment effect of the Hopfield neural network

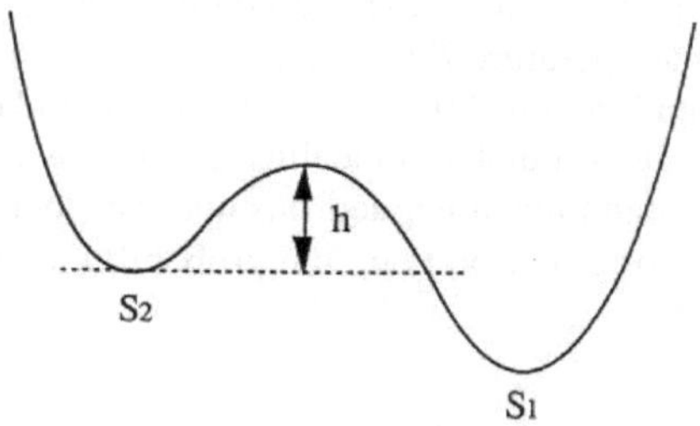

(b) energy increment h is required to escape from the valley of the local minimum S_2 to enter the attractor of the global minimum S_1.

Fig. 1.11 An energy landscape.

1.3.2.2 Application to Optimization Problem

Hopfield et al. applied their neural network to the Traveling Salesman Problem [291]. This finds a pattern minimizing the total distance when a salesman pays sequential visit n cities. A solution can be obtained, but the amount of calculations explosively increases as the number of cities increases. Hopfield et al. showed [291] that the solution to this problem can be obtained theoretically by parallel calculation. A matrix is assumed to represent cities and sequences of visits as rows and columns, respectively. Neurons are applied to these elements to make the distance between cities correspond to Eq. (1.4), as well as conditions for the solution. The convergence solution is obtained from the Eq. (1.7). This method does not guarantee the optimal solution and the state of the network is sometimes stuck in a local minimum, since the transition of the Hopfield network is sometimes stuck in a local minimum, since the transition of the Hopfield network is based on the least mean square algorithm [302]. However, it is reported that quasi-optimal solutions were often obtained [204].

1.3.2.3 Simulated Annealing and Boltzmann Machine

Fig. 1.11 shows an energy landscape in which S_2 is a local minimum and S_1 is a global minimum. Kirkpatrick et al. developed a simulation annealing method for making likely the escape from local minima by allowing jumps to higher energy states [150]. In the statistical theory of gases, the gas is distributed not by a deterministic dynamics, but rather by the probability that it will be in different states. The 19th century physicist L. Boltzmann developed a theory that included a probability distribution for the states of a gas when it had reached a uniform distribution of temperature. Hinton, Sejnowski, and Ackley used this distribution to describe neural interactions, where low temperature T is replaced by a small noise term. Then, they developed the Boltzmann Machine [109, 108] which is a kind of recurrent networks that settle solutions by the simulated annealing process as a stochastic technique (Fig. 1.12). In the Boltzmann machine, neurons operate stochastically; and assuming that an internal state of a neuron is u_i, the probability of its excitation is given by the following equation:

$$P_i = \frac{1}{1+exp(1-u_i/T)}, \tag{1.8}$$

where T is parameter corresponding to the temperature of the statistical dynamics and changes the probability of the neuron's excitation. This technique is based on the simulated annealing method [108].

As the Boltzmann machine operators in probability, it can reduce the energy function like the Hopfield network. The parameter T determines the operational fluctuation of each neuron, and simulated annealing can be performed easily by adjusting T [108].

1.3.3 *Feed-forward Neural Network*

1.3.3.1 Learning Algorithm

Fig. fig:Feed-forwardNet illustrates an example of a feed-forward network with only forward connections. For the input/output of neurons in the hidden and output layers, Eq. (1.1) is used. Since a form of the feed-forward network indicates a multi-layering of perceptrons, it is also called a perceptron type. In the case of the original perceptron, the hidden layer and the output layer [268]. In the 1980s, Rumelhart et al. proposed a learning algorithm with the teacher of the feed-forward

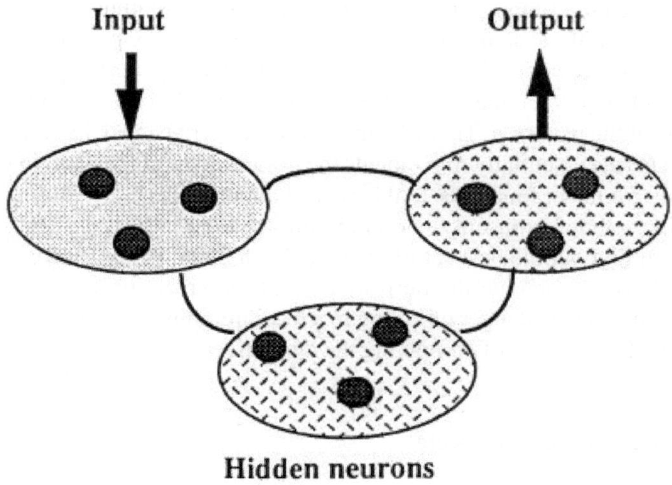

Fig. 1.12 Boltzmann machine network.

network to solve this problem by multi-layering using a back-propagation (BP) algorithm [238]. In the 1970s, Werbos originally developed a back-propagation algorithm [293], and its first practical application was for estimating a dynamic model to predict nationalism and social communications. However, Werbo's work remained almost unknown in the scientific community. The back-propagation is a generalization of the Least Mean Squares algorithm [302]. In this algorithm, an effort function is defined and equal to the mean square difference between the desired output and the actual output of the feed-forward network. In order to minimize this error function, the back-propagation algorithm uses a gradient search technique. It is called the generalized delta rule [238].

1.3.3.2 Standard Delta Rule

The learning procedure requires a set of pairs of input and output patterns. The system first uses the input vector to produce its own output vector and then compares this with the desired output, or the target vector. If there is no difference, no learning takes a place. Otherwise, the weights are updated to reduce the difference. This procedure with no hidden units gives the standard delta rule. The rule for updating weights using an input/output pair p is given by

$$\Delta_p w_{ij} = \eta (t_{pj} - o_{pj}) i_{pi} = \eta \delta_{pj} i_{pj} \tag{1.9}$$

where t_{pj} is the target input for the jth component of the output pattern for p, o_{pj} is the jth element of the actual output pattern produced by the presentation of input pattern p, i_{pi} is the value of the ith element of the input pattern, δ_{pj}=t_{pj}-o_{pj}, and $\Delta_p w_{ji}$ is the change to be made to the weight from the ith to the jth unit following presentation of the pattern p. There are many ways to derive the delta rule [221]. The standard delta rule essentially implements the gradient descent method in the sum of squared error for linear functions. In this case, without hidden units, the error surface is shaped like a bowl with only one minimum, so the gradient descent method is guaranteed to find out the best set of weights. With hidden units, however, it is not so obvious how to compute the derivatives, and the error surface is not concave upwards, so there is the danger of falling in a local minimum. The following section describes an efficient way of computing the derivatives.

1.3.3.3 Generalized Delta Rule

Here, we consider the multi-layered feed-forward networks. In these networks, the input units are the bottom layer and the output units are the top layer. There can be many layers of hidden units in between, but every unit must send its output to higher layers than its own and must receive its input from lower layers than its own. Given an input vector, the output vector is computed by a forward pass which computers the activity levels in the earlier layers.

The back-propagation algorithm [3, 221] performs the steepest descent method on a surface in a weight space whose height at any point in the weight space is equal to the error measure. In order to show the algorithm, let

$$E_p = \frac{1}{2}\Sigma_j(t_{pj} - o_{pj})^2 \tag{1.10}$$

be the measure of the error on input/output pattern p and let $E = \Sigma E_p$ be the overall measure of the error.

Here, we define the weighted sum of the output in the previous layer,

$$S_{pj} = \Sigma_i w_{ij} o_{pi} \tag{1.11}$$

as the state of the unit. The output,

$$o_{pj} = f_j(s_{pj}) \tag{1.12}$$

uses the sigmoid function which is a non-decreasing and differentiable function described in Section 1.3.1.3. To obtain the correct generalization of the delta rule, w_{ij} is set as

$$\Delta_p w_{ij} \infty - \frac{\partial E_p}{\partial w_{ij}} \tag{1.13}$$

It is useful to see this derivative as resulting from the product of two parts: one part reflecting the change in error as a function of the change in the net input to the unit and one part representing the effect of changing a particular weight on the net input. Thus the derivative in Eq. (1.13) can be written by

$$\frac{\partial E_p}{\partial w_{ij}} = \frac{\partial E_p \partial s_{pj}}{\partial s_{ij} \partial w_{ij}} \tag{1.14}$$

By Eq. (1.11), the second factor is expressed by

$$\frac{\partial s_{pj}}{\partial w_{ij}} = \frac{\partial}{\partial w_{ij}} \Sigma_k w_{jk} o_{pk} = o_{pi} \tag{1.15}$$

Now let us define

$$\delta_{pj} = -\frac{\partial E_p}{\partial s_{pj}} \tag{1.16}$$

Equation (1.14) thus has the equivalent form

$$-\frac{\partial s_{pj}}{\partial w_{ij}} = \delta_{pj} o_{pj} \tag{1.17}$$

To implement gradient descent in E, the weight must be updated according to

$$\Delta_p w_{ij} = \eta \delta_{pj} o_{pj} \tag{1.18}$$

just as in the standard delta rule. The interesting result is that there is a simple recursive computation of these δ's which can be implemented by propagating an error signal backward from the output to the input through the network.

To compute Eq. (1.16), the chain rule is applied to write this partial derivative as the product of the unit, and one reflecting the change in the output as a function of changes in the input,

$$\delta_{pj} = -\frac{\partial E_p}{\partial s_{pj}} = -\frac{\partial E_p \partial o_{pj}}{\partial o_{pj} \partial s_{pj}} \tag{1.19}$$

By Eq. (1.12) , we see that

$$\frac{\partial o_{pj}}{\partial s_{pj}} = f'(s_{pj}), \tag{1.20}$$

which is simply the derivative of the compressing function f_j for the jth unit, evaluated at the net input s_{pj} to that unit. To compute the first factor, there are two cases. First, assume that unit u_i is an output unit of the network. In this case, it follows from the definition of E_p that

$$\frac{\partial E_p}{\partial o_{pj}} - (t_{pj} - o_{pj}). \tag{1.21}$$

substituting Eqs. (1.20) and (1.21) into Eq. (1.19), we can obtain

$$\delta_{pj} = (t_{pj} - o_{pj}) f_j'(s_{pj}) \tag{1.22}$$

for any output unit u_j. If u_j is not an output unit, the chain rule is used to write

$$\begin{aligned} \Sigma_k \frac{\partial E_p \partial s_{pk}}{\partial s_{pk} \partial o_{pj}} &= \Sigma_k \frac{\partial E_p \partial}{\partial s_{pk} \partial o_{pj}} \Sigma_i w_{ki} o_{pi} \\ &= \Sigma_k \frac{\partial E_p}{\partial s_{pk}} w_{kj} = -\Sigma_k \delta_{pk} w_{kj}. \end{aligned} \tag{1.23}$$

In this case, substituting Eqs. (1.20) and (1.21) into Eq. (1.19) yields

$$\delta_{pj} = f_j'(s_{pj}) \Sigma \delta_{pk} w_{kj}, \tag{1.24}$$

whenever u_j is not an output unit. Eqs. (1.22) and (1.24) give a recursive procedure for computing the δ's for all units in the network, which are then used to compute the weight changes in the network according to Eq. (1.18). This procedure constitutes the generalized delta rule for a feed-forward network.

1.3.3.4 Back-Propagation Algorithm

This section describes the back-propagation algorithm for training. The back-propagation training algorithm is an interactive gradient algorithm designed to minimize the mean square error between the actual output of a feed-forward network and the desired output. It requires continuous differentiable non-linearities [293, 238]. Fig. 1.13 shows a flow chart of the back-propagation training algorithm. Assumptions are made that a sigmoid logistic non-linearity in Fig. 1.7 is used.

Step 1. Initialize Weights and Offsets

Set all weight and unit offsets to small random values.

Step 2. Give Input and Desired Output

Give a continuous function of input vectors and specify the desired outputs. If the network is used as a classifier, then all desired outputs are typically set to zero except for the output corresponding to the class. The input could be new at each trial or samples from a training set can be given cyclically until weights are stabilized.

Step 3. Calculate Actual Outputs

Use the nonlinearity of the sigmoid function as shown in Fig. 1.7 and formulas as shown in Fig. 1.9 to calculate outputs.

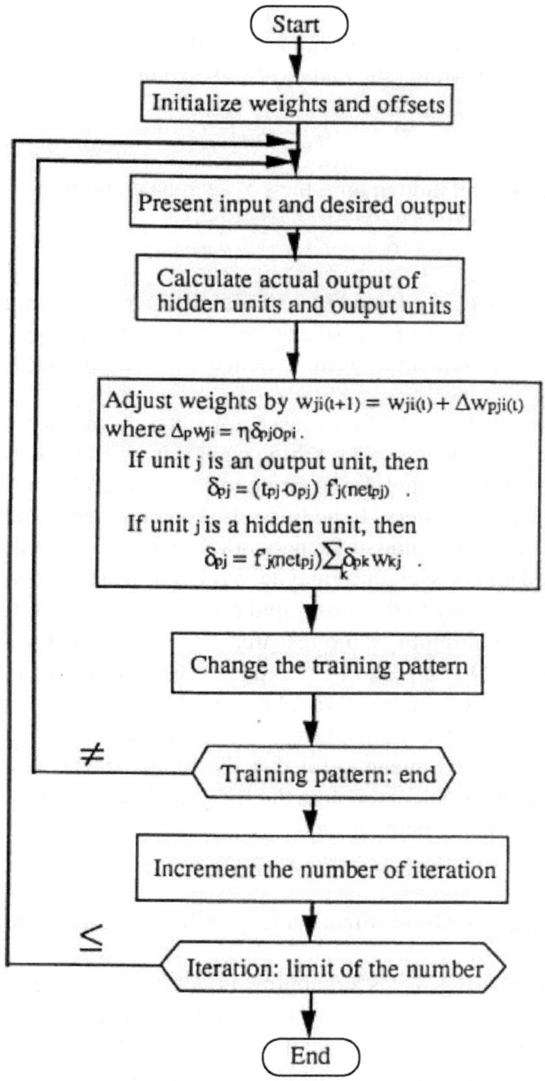

Fig. 1.13 Flowchart of the back-propagation training algorithm.

Step 4. Update Weights

Use recursive algorithm starting from the output units to the first hidden layer to update weights in each layer level. Adjust weights by

$$w_{ij}(t+1) = w_{ij}(t) + \Delta w_{ij}(t), \tag{1.25}$$

where $\Delta_p w_{ji}=\eta \delta_{pj} o_{pi}$ as shown in Eq. (1.9). In this equation, $w_{ij}(t)$ is the weight, η is a gain term of a learning rate, and δ_{pj} can be computed by Eq. (1.22),

$$\delta_{pj} = (t_{pj} - o_{pj}) f'_j(s_{pj}). \tag{1.26}$$

If the unit j is an internal hidden unit, then δ_{pj} can be computed by Eq. (1.24),

$$\delta_{pj} = f'_j(s_{pj}) \Sigma_k \delta_{pk} w_{kj}. \tag{1.27}$$

Internal unit thresholds are adapted in a similar manner by assuming they are connection weights on links from auxiliary constant-valued inputs. Convergence is sometimes faster if a momentum term is added and weight changes are smoothed [238, 292].

The Eqs. From (1.9) to (1.25) are used for a single teacher signal. In fact, a pattern of several input signals and teaching signals is given for iterative learning. If the error becomes small enough for learning to be successful, correct output can be obtained for the learned pattern. In addition, if the unlearned pattern is close to the learned pattern, the output value of the network. However, successful learning is not always guaranteed and convergence may take time, or learning may be trapped at a local minimum on the way to the most rapid drop. In order to improve convergence and find the global minimum, some research has been done (A. R. Jacobs, 1998 [128]), (N. Baba, 1989 [22]), and (T. Fukuda, T. Shibata et al., 1990 [70]). There still remain problems how to code variables and how to determine the number of neurons (particularly for the hidden layer) in the network.

Nevertheless, it is very meaningful that the hidden layer was introduced and multi-layered to allow problems such as the EX-OR classification to be solved. This cannot be implemented by perceptrons.

1.3.3.5 Mapping and Generalization Capabilities of Networks

The network shown in Fig. 1.9 is a mapping model from m inputs to n outputs variables. Funahashi [73] indicates that any multi-variable function can be approximated by using a sufficient number of units at the hidden/middle layer in the three-layered network, which is very important to neural network applications. If the neural network derives a rule from a specific example, it will be desirable that the rule can be applied to slightly different cases and obtain effective results. This capability is called generalization. However, the generalization capability has not yet been well formulated.

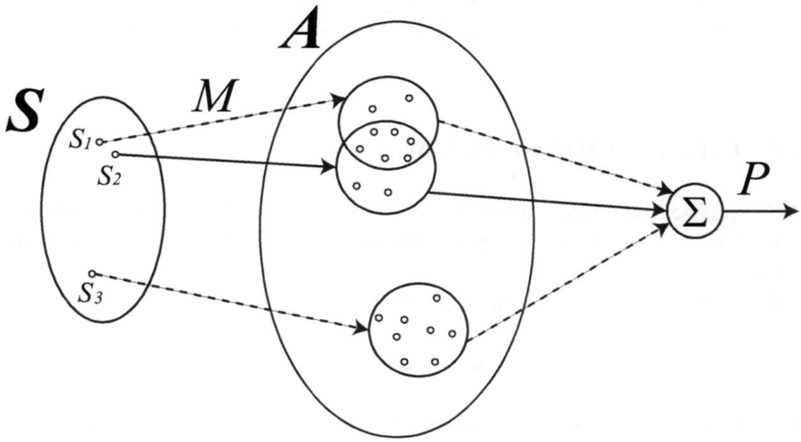

Fig. 1.14 CMAC.

1.3.4 *Cerebellar Model Arithmetic Computer (CMAC)*

1.3.4.1 Characteristics of CMAC

Cerebellar Model Arithmetic Computer (CMAC) is a neural network that models the structure of information processing in cerebellar cortex proposed by Albus [3].

CMAC expresses nonlinear function of multi-input single-output system as linear sum of loading as shown in Fig. 1.14. Mapping space of multi-input multi-output can be derived by using multi CMAC that share input. The characteristic of CMAC that can mimic complex nonlinear function of multi-input multi-output system is same as multi-layer neural network. Although CMAC requires large size memory for loading memory, nowadays large size memory can be used easily; thus application of CMAC is expanding. The advantage of CMAC is clear mapping space compare to neural networks. In CMAC, generalization range is set every input valuable; thus input pattern over the range is independent and interference does not occur. In contrast, discrete learning data of multi-layer neural network affects all of parameters. Thus, it is difficult to require high accuracy of each relationship between input and output relation. For instance, it is difficult to express mapping space that output changes intricately due to sequential alteration of single input such as time-dependent function.

Although multi-layer neural network has ability of self-organized generalization, generalization learning for distribution of learning data needs to be given beforehand for CMAC in order to generalize successfully. CMAC does not output in the case that input data is not in the generalization range of learning data even if input data is less important dimension. This is different from multi-layer neural network that output any value from input data. This characteristic of CMAC is advantage for some applications because output of multi-layer neural network is not always suitable.

1.3.4.2 Calculation Method of CMAC

CMAC expresses a complex nonlinear function $P = f(s)$ that is from N dimensional input vector $s \in S$ to one dimensional output P by referring to numerical table as shown in Fig. 1.14.

At first, CMAC is defined following mapping.

$$\mathbf{S} \xrightarrow{M} \mathbf{A} \xrightarrow{\Sigma} P, \tag{1.28}$$

where $\mathbf{S}$ is input vector space, $\mathbf{A}$ is set of load unit, M is mapping function, P is output value of CMAC. That is,

$$P = \Sigma_{i=1}^{C} a_i. \tag{1.29}$$

Entire mapping is denoted by multi-input single output $\mathbf{S} \to P$; L dimensional CMAC standardizes input and denotes N-input L-output $\mathbf{S} \to \mathbf{P}$ mapping relationship.

The mapping from input vector space $\mathbf{S}$ to set of load unit $\mathbf{A}$ satisfy following property: the common mappings C-load unit increase as the distance between two input vector $S \in s_1, s_2$ closer, and same load unit does not include if the distance is longer than a constant value. Thus, output of CMAC, that is sum load, is similar value if input vector is similar; then generalization is conducted.

Learning is conducted by adding amount of correction Δ calculated from learning data $\hat{P}$ and output of CMAC P to all load $a_i (i = 1, 2, \cdots, C)$ used in calculation of P. That is,

$$a_i \leftarrow a_i + \Delta \qquad (i = 1, 2, \cdots, C), \tag{1.30}$$

$$\Delta = g \frac{\hat{P} - P}{C}, \tag{1.31}$$

where g is learning coefficient, the value is $0 < g \leq 1$.

1.3.5 Fuzzy Neural Network

In this section, we consider synthesis of fuzzy logic and neural networks. Fuzzy Neural Networks (FNN) have been proposed synthesizing the fuzzy logic and neural networks in order to solve drawbacks of the neural networks and the fuzzy logic [92, 199, 122, 261]. FNN is believed to have considerable potentials in the areas of expert systems, medical diagnosis, control systems, pattern recognition, system modeling, and so forth. Fuzzy Associative Memory (FAM) is also a powerful tool.

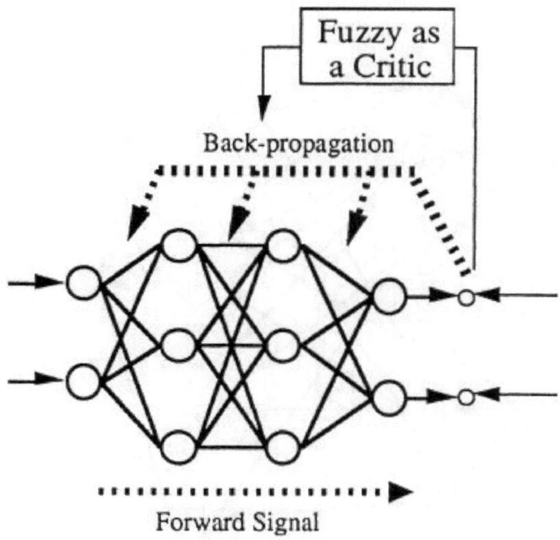

Fig. 1.15 Fuzzy + Neural network: improvement of convergence through learnig by using the fuzzy logic as a critic.

FAM (Kosko, 1987 [156]) is a two-layer feed-forward, heteroassociative, and fuzzy classifier that stores an arbitrary fuzzy spatial pattern pair using fuzzy Hebbian learning.

1.3.5.1 Neural Network and Fuzzy

The fuzzy logic and the neural network are used independently in a system or either one serves as a preprocessor for the other. For example, Fukuda, Shibata et al. uses the fuzzy logic as a supervise or a critic for the neural network in order to improve convergence of learning (Fig. 1.15) [70, 68]. There, the learning rate is adaptively determined by using fuzzy rules.

1.3.5.2 Neural-Like Fuzzy

In the case, the network uses the fuzzy neurons which are described by fuzzy sets, instead of the non-fuzzy meuron [92, 122]. In knowledge based systems, one often uses a set of conditional statements, 'If-then' rules to represent human knowledge extracted from human experts. Very often this knowledge is associated with uncertain and fuzzy terms, such as "many", "few", "often", and "sometimes". Therefore, antecedents and consequents in the 'If-then' rules are treated as fuzzy sets.

In this book, we survey a FNN proposed by Ichihashi [122] (Fig. 1.16). The FNN clusters the input parameters into the fuzzy sub-space and identified input-output

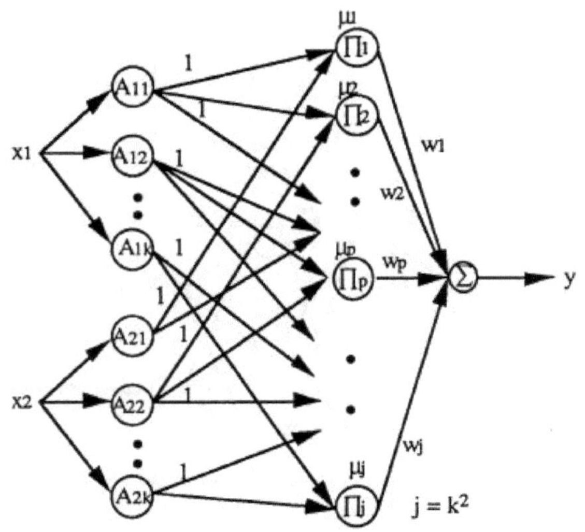

Fig. 1.16 A structure of Fuzzy Neural Network (FNN).

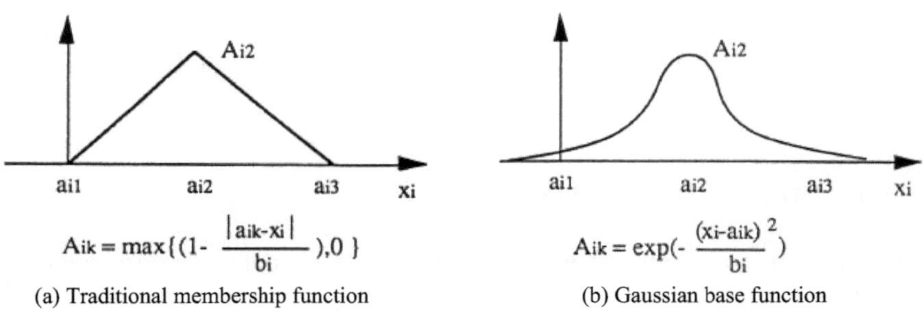

Fig. 1.17 Membership function for FNN.

relationships by using a weighted network. The FNN is based on the simplified fuzzy inference [21]. A_{ip} is the membership function for ith input variable $x_i(i = 1\cdots n)$ in pth rule. w_p is the consequence of the Pth rule. Individual rules' results are given by μ_p as follows:

$$\mu_p = \Pi A_{ip}(X_i) \tag{1.32}$$
$$= A_{1p}(x_1) \times A_{2p}(x_2) \cdots \times A_{np}(x_n). \tag{1.33}$$

then, the output of the FNN is as follows:

$$y = \Sigma \mu_p \cdot w_p. \tag{1.34}$$

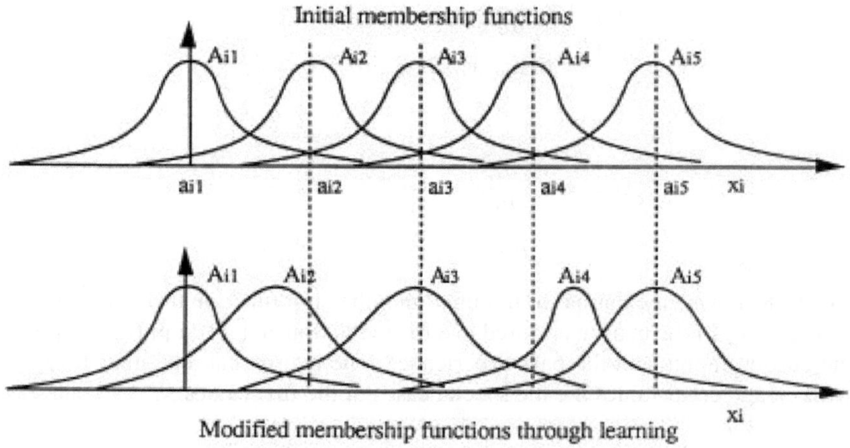

Fig. 1.18 Modified membership functions through learning FNN.

In order to obtain the inference rule in the FNN, the back-propagation algorithm is used for learning to minimize the difference δ between the desired output y_d obtained from human experts and the actual output y of the FNN. An energy function to be minimized is defined as follows:

$$E = \frac{(y_d - y)^2}{2}. \tag{1.35}$$

Then, the change of weights in the network based on the back-propagation is as follows:

$$\begin{aligned} \delta &= \tau \cdot \frac{\partial E}{\partial w_p} \\ &= \tau \cdot \mu_p \cdot \delta, \end{aligned} \tag{1.36}$$

where τ is a learning rate.

Ichihashi used a Gaussian base function (Fig. 1.17)as a membership function in the FNN as follows:

$$A_{ip}(x_i) = exp\left(-(x_i - a_{ip})^2/b_{ip}\right). \tag{1.37}$$

The parameters of a_{ip} and b_{ip} are defined depending on the rules. The Gaussian base function is both continuous and differentiable (i.e. possess derivatives) function. In the back-propagation algorithm, there characteristics are required [295]. In order to modify or tune membership functions in the input layer through learning (Fig. 1.18), the following equations are used based on the back-propagation:

$$\Delta a_{ip} = \tau \cdot \frac{\partial E}{\partial a_{ip}} \\ = \tau \cdot \mu_p \cdot w_p \cdot 2(x_i - a_{ip})/b_{ip} \cdot \delta, \quad (1.38)$$

and

$$\Delta b_{ip} = \tau \cdot \frac{\partial E}{\partial b_{ip}} \\ = \tau \cdot \mu_p \cdot w_p \cdot 2(x_i - a_{ip})/b_{ip} \cdot \delta. \quad (1.39)$$

Figure 1.18 shows an example of modified membership function through learning. The experience of the neuron is stored in a fuzzy relation and its output is composed of the current inputs and the past experiences. The neuron can be either fuzzy or non-fuzzy, and crisp values are the special cases of the fuzzy ones.

1.3.6 Genetic Algorithms

This section will briefly describe the general approach to some applications of Genetic Algorithms (GA). An excellent introduction and survey of GA is given in Goldberg's book [81].

1.3.6.1 Characteristics of Genetic Algorithm

GAs are search algorithms based on the mechanics of natural selection and natural genetics. They combine survival of the fittest among string structures with a structured yet randomized information exchange to form a search algorithm with some of the innovative flair of human search. An occasional new part is tried for a good measure avoiding local minima. While randomized, GAs are no simple random walk. They efficiently exploit historical information to speculate on new search points with the expected improved performance.

In their purest form, GAs can be seen as a technique for solving optimization problems in which the elements of the solution space are coded as binary strings and in which there is a scalar objective function that can be used to compute the "fitness" of the solution represented by any strings. The GA maintains a "population" of strings, which are initially chosen randomly. The fitness of each member with high fitness values are reproduced in order to keep the population at a constant size. After the reproduction phase, operations are applied to introduce variation in the population. Common operators are crossover and mutation. In crossover, two population elements are chosen, at random, as operands. They are recombined by randomly choosing an index into the string and making two new strings, one that

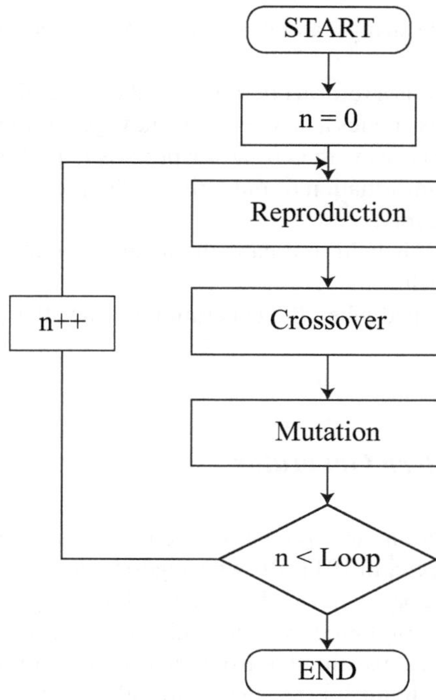

Fig. 1.19 Flow chart of Genetic Algorithms

consists of the first part of the first string and the second part the second string and one that consists of the first part of the second string and the second part of the first string. Mutation simply changes bits in population elements, with very low probability.

1.3.6.2 Flow Chart of Genetic algorithm

Genetic algorithms have three operations to abstract and rigorously explain the adaptive process of natural systems as follows:

1. Reproduction operation,
2. Crossover operation,
3. Mutation operation.

Figure 1.19 shows a flow chart of GA. The reproduction process is an operation to select the survival in a set of candidate strings. In this process, the fitness value is calculated for each candidate string by using the fitness function which depends on a goal for searching problems. According to the value, the selection rate is determined for the present candidate strings, and the survival is selected in any rate depending on the selection rate.

The crossover process is a reform operation for the survival candidates. In natural system, a set of creatures creates a new set of the next generation by crossing among the creatures. In the same way, the crossover process is performed by exchanging pieces of strings using information of old strings. The pieces are crossed in couple of strings selected randomly.

The mutation process is held to escape the local minima in search space in the artificial genetic approach.

The algorithm is stopped when the generation reaches the last one which is defined beforehand.

1.3.7 Central Pattern Generator

Rhythmic motions are found in basic animal behaviors, such as in walking, swimming, flying, breathing, etc. In 1911, Brown suggested the presence of neural circuit of generating an oscillatory behavior from the walking experiment of decerebrate cat [85]. However, such rhythmic motions had been regarded as a result of reflex interaction due to exogenous stimulus from environment, rather than the intrinsic behavior of the neural circuit for decades. Currently, numerous experimental evidences have proven the existence of neural circuits to generate the rhythmic motions, what is called Central Pattern Generator (CPG). Although CPG produces an intrinsic rhythmic pattern without an external sensory input or descending signals from high-level brain functions, it exhibits modulation property interacting with musculoskeletal systems. Experiments on decerebrate cat walking on the treadmill observed the transition of locomotion gait due to frequency modulation of CPG according to the moving speed of the treadmill. CPG is thus considered to take a fundamental role in animal locomotion including human, and it is applied to walking rehabilitation program for the patients suffering from spinal cord injuries[23]. Some advanced features in CPG-based locomotion control are summarized as follows;

1. Locomotion pattern is self-organized through interaction between CPG and musculoskeletal systems, that is physical body dynamics.
2. CPG-based control does not require the concrete trajectory design of each joint, which is particularly suitable for multi-legged robot, swimming robot, and modular robotic systems control.
3. CPG controller has both natures of hierarchical and distributed structure where multiple CPGs are reciprocally coordinated for unified motion.
4. Frequency modulation of CPG due to the external signal input can exhibit spontaneous gait transitions.
5. Natural gait transition may enhance adaptability to unknown environment and improve energy efficiency.

Taking advantage of these features, CPG has been extensively applied in the robotics field since early '90s. Related works were introduced in Section 1.2.1.4 and 1.2.6.4.

Although CPG-based method is promising, we should mention about some remaining challenges as follows;

1. There need to be a number of parameter setting in CPG model, and they have to be determined often relying on a heuristic search.
2. Interactions between CPGs and musculoskeletal systems are case-sensitive and difficult to analyze except for a simple model.
3. Stability analysis is not sufficient; hence implementation to the robot is still limited to the primitive locomotion.
4. Flexible adaptation to the wide range of time is still a challenge. Actual robot motions are integration of rhythmic and non-rhythmic motions depending on the environmental condition.

1.4 Multi-Locomotion Robot

1.4.1 Bio-inspired Robot

In the background of robotics development, there are two explicit expectancies toward robots as follows.

1. Executing repetitive tasks human get tired to do.
2. Executing dangerous or hard work human cannot do on behalf of humans.

From such background, industrial robots were developed as robots to conduct tasks that was hard physically for humans such as conveyance heavy materials or to implement oppressive work such as marathon uncomplicated operation. Besides, in construction site, nuclear power plant or outer space, it is dangerous for workers to put tasks in practice, hence it is desired to develop the robot substitute for human workers.

On the other hand, in the age of an aging society, the prospective role of robots is turning gradually from just working machines to do monotonous work in a factories to partners who support human life. In order to realize this kind of function, it is necessary to accomplish the ability to recognize the circumstances and also to achieve the ability to move toward the objective point autonomously in the various environment. However much research about bio-inspired robotics were specific to just one type of locomotion mode. In addition, the robots developed in these work were designed to be specialized to perform such locomotion.

Here, focusing on animals in the nature and considering the intrinsic difference between the previous bio-inspired robot and them, it is seen that diversity of locomotion is important. They are capable to perform several kinds of locomotion by stand-alone and to accommodate the alteration of environment by choosing the adequate locomotion from multiple locomotion modes. Such diversity of locomotion, which was not pursued in the previous work regarding bio-inspired robotics, is significant challenge to achieve the above-mentioned goal. In the next section, we describe the diversity of locomotion in animals at first, then propose the novel concept of bio-inspired robotics, Multi-Locomotion Robot. Multi-Locomotion Robot is designed to be able to perform several kinds of locomotion such as biped walking, quadruped walking, and brachiation.

1.4.2 Diversity of Locomotion in Animals

There are great many types of living creatures on the earth. The number of species that have been identified so far is about 1.5 million. It is said that it becomes no fewer than 30 million if unidentified living creatures in the place that has not been explored yet such as the rain forest are included. They have adapted to their environment and built a variety of unique society in the 40 billion-year evolutionary history from birth of the earth.

Here, let the environment on the earth be classified into three factors (air, ground, hydrospace) and let locomotion pattern of animals in each circumstances be described.

At first, in order to move in the air, it is necessary to generate the higher lift force than its own weight. As for the living creatures, flying animals like birds and

Fig. 1.20 Biological diversity

a part of insects (snake doctor or butterfly, cicada, bee) acquire the lifting force by flapping their own wings. Fast free movement irrelevant to the environmental condition of the ground is enabled by moving in the air, hence it has possibility to collect nourishments effectively. On the other hand, ambulatory movement on the ground is performed by actuating joints to move limbs and a body to take advantage of the friction force and gravity. As for ambulation on the ground, differently from movement in the air or water, suitable locomotion pattern is dependent on the environmental condition since the ground includes desert, grassland, forest, and mountain terrain and the ground condition is of infinite variety. A part of insects (grasshopper or calicoback, ant, mantis), quadruped animals (horse or camel, dog) and primates (human or monkey) walk to ambulate by use of their own limbs. The living creatures without limbs such as serpent or earthworm can move by creeping. As just described, these living creatures acquire the locomotion suitable for the environmental condition and adapt to the circumstances. Lastly, in the hydrospace such as sea or river, natant animals (fish, calamary, octopus, and medusoid) move freely in water with their fin. Additionally, there are some animals which ambulate by walking on the bottom of the ocean such as crustacea such as shrimp or crab.

As mentioned above, the living creatures acquire a variety of locomotion modes according to the environment. These locomotion patterns are significantly related to physical structures. The dominant dynamics of system during movement is dependent on both the locomotion pattern and the physical structure [4]. It is conceivable that the locomotion modes and physical structure are co-evolved to accommodate to the circumstances in evolutionary history.

In addition, it should be noticed that the locomotion pattern that many of living creatures acquired is not single. For instance, birds, which basically possess not only wings but also legs. Snakes moves by creeping, however some of them are able to swim in water. Many of insects, which make up 75 percent of animals, have not only six legs but also wings and fly in air. By selecting multiple locomotion patterns, these animals have the high ability to ambulate or escape from their enemy.

As described above, it can be said that animals possess the diversity of locomotion and that such diversity is one of important characteristics in order to live in nature intelligently.

1.4.3 Multi-Locomotion Robot

Diversity of locomotion, which has not been pursued in previous work of bio-inspired robotics, is one of significant points to realize so as to put life-supporting robots like a human partner into practical use. Considering both the mechanism necessary to realize tasks that human conduct on a daily life and locomotion mode to accomplish, in evolutionary term, the physical structure which is situated in between primates such as chimpanzee and orangutan, human seem to be most appropriate [56, 260].

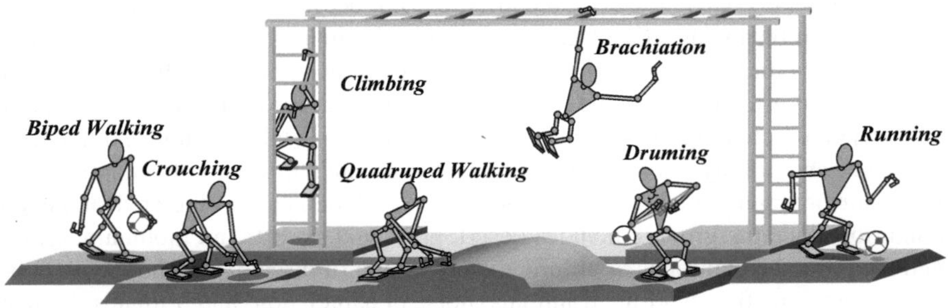

Fig. 1.21 Concept of the multi-locomotion robot

Thus, as the new concept of bio-inspired robot, we propose multi-locomotion robot shown in Fig. 1.21. The multi-locomotion robot is the novel bio-inspired robot which can perform several kinds of locomotion such as biped walking, quadruped walking, and brachiation.

1.4.3.1 Biped Walking

In all of living creature with limbs in nature, the animals that acquire the biped walking are part of mammal such as the anthropoid closely related to human (human, chimpanzee, orangutan, gibbon, and gorilla). These creatures acquire biped locomotion by accommodating to the circumstances in the evolutionary process. Acquirement of biped walking made their hands free, hence it became possible to conduct some tasks by hands. Consequently, it is thought that their intelligence is superior to other animals and they accomplish quite high ability to adapt to the dynamic environment.

Biped walking is normally modeled as an inverted pendulum [53]. The dynamics of an inverted pendulum is intrinsically unstable, hence in order to achieve biped walking, it is necessary to control the body dynamics skillfully so as to compensate for the walking stability. However, the height of COM (Center of Mass) is kept higher position than quadruped locomotion or creeping, it has merit that the energy efficiency to ambulate can be quite high by taking advantage of the gravity effects. Also, since, as for biped walking, the view point is kept in high position, it is possible to observe larger range of environment than quadruped walking.

Since most of human routine tasks are conducted in bipedal posture, the necessity for a robot to use biped walking is natural. In addition, the residential environment is designed to fit human body structure. That is reason why biped locomotion is suited for life supporting robots to work there. From this background, recently, lots of researches aiming to realize biped walking with humanoid robots have been conducted actively. It can be said that it is one of the most developed area in recent years. Besides, accomplishment of faster locomotion by realizing running is anticipated [189] .

1.4.3.2 Quadruped Walking

Animals performing quadruped locomotion are familiar to us [189]. Quadruped walking is locomotion mode that many of animals to live on the ground employ.

It is known that there are several gait patterns in quadruped walking such as crawl gait (Fig. 1.22 (1)), pace gait (Fig. 1.22 (2)), trot gait(Fig. 1.22 (3)), bounce gait(Fig. 1.22 (4)), and gallop gait (Fig. 1.22 (5)) [59]. Crawl gait is the walking pattern employed by most of animals ambulating at low velocity such as reptile and amphibian. In crawl gait, more than 3 feet contact the ground constantly and there is the period of four-feet-contact. Thus, the walking stability is quite high, however it does not suit for fast locomotion in terms of energy efficiency.

Most of mammals such as cat, dog, and horse at low velocity, perform gait that a body is supported by 2 legs constantly and each leg moves forward in the following order: right rear-leg, right front-leg, left rear-leg, left front-leg. As the walking velocity increases, the number of supporting legs is decreases in trot gait [106] and pace gait. In trot gait, legs of opposing corner—e.g. right rear-leg and left front-leg—contact the ground simultaneously by turns at foot-contact. Meanwhile, in pace gait, legs in the same side—e.g. right rear-leg and right front-leg—do at foot-contact. At higher velocity, the period of no-feet-contact is shown, namely gallop gait [105]. It is known that animals realize the high energy efficient locomotion by switching these gait pattern according to walking velocity or state [119].

Quadruped locomotion is more stable than biped walking, hence it can be said that quadruped locomotion is suited for ambulation in the environment with many obstacles or the rough and rugged terrain. In order to put partner robots into practical use, robots need to be able to ambulate on not only flat ground but the irregular ground. Hence, in the environment where biped locomotion is unstable, robots should choose quadruped walking. Also, in the circumstances where the degree of spatial freedom is limited such as a narrow cave or tunnel, quadruped walking is effective.

1.4.3.3 Brachiation

Brachiation, which is one of motions in tree, is locomotion mode employed by arboreal animals such as gibbon and orangutan. It is the locomotion pattern that animals dangle on the branch and swing from branch to branch like pendulum as shown in Fig. 1.23. It is know that dynamics of brachiation is modeled as a pendulum [26, 27, 58]. As for locomotion on the ground, COM is basically in upper position than supporting point of its weight, whereas as to brachiation, motion is performed dangling on the branch and COM is in lower position. Chang *et al.* built the system to observe the gibbon brachiation, and estimated the COM trajectory and alteration of mechanical energy [35].

Brachiation enhances the ambulating ability of the multi-locomotion robot. It is difficult to perform biped or quadruped locomotion without the rigid ground.

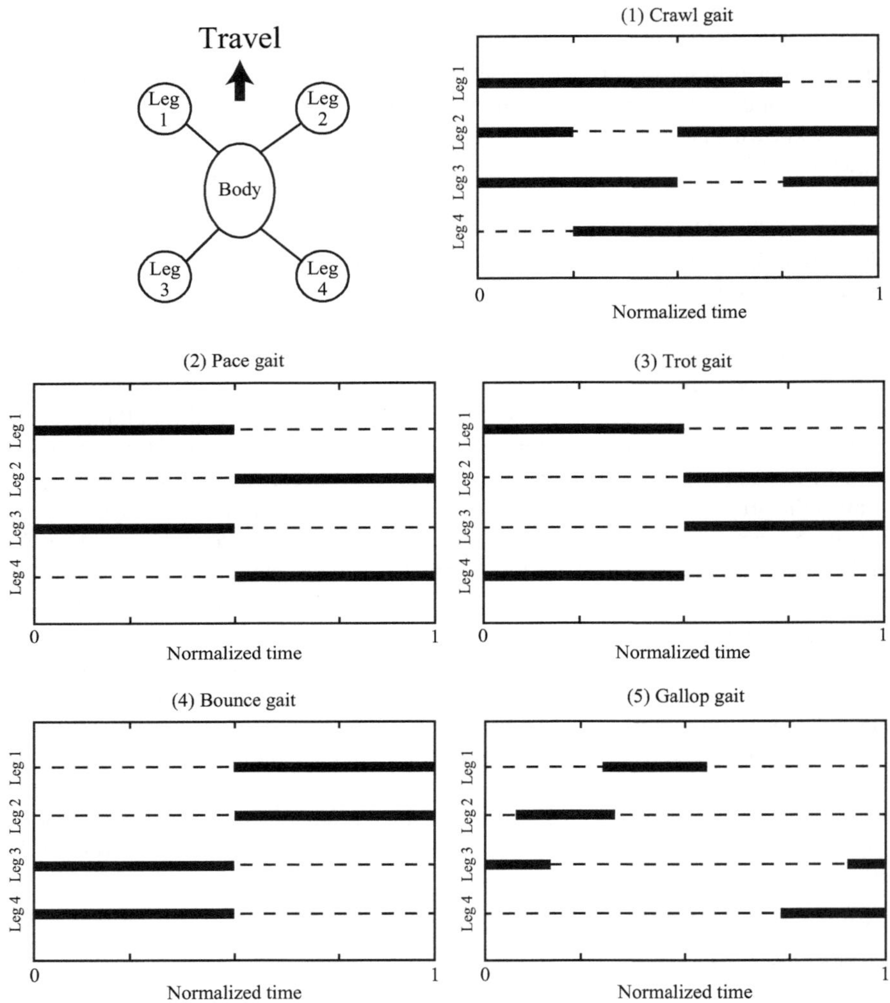

Fig. 1.22 Quadruped gait graphs. Stance phases are drawn as bars.

However, if there is something to hold and dangle such as branch or pipe, it is possible to ambulate by use of it.

1.5 Organization of This Book

In this chapter, we introduced the conventional studies with regard to control algorithm for several types of locomotion, and the novel concept of bio-inspired robotics

Fig. 1.23 Brachiation of a gibbon

named Multi-Locomotion Robot. In order to realize a life-supporting robot in the future, it is necessary that robots can ambulate autonomously in the various environments. Multi-Locomotion Robot has the high ability to ambulate by achieving several kinds of locomotion in stand-alone. This diversity of locomotion is inspired by living creatures on earth. We mentioned about diversity of locomotion in animals and described biped walking, quadruped walking, and brachiation as main locomotion mode of Multi-Locomotion Robot.

This book continues as follows. In Chapter 2, basics of robotic motion control are explained. Especially, the Passive Dynamic Autonomous Control (PDAC) is novel control method that we have proposed. Section 2.3 is based on [51]. Chapter 3 describes the link structure and control system including sensors and actuators of Gorilla Robot–designed as Multi-Locomotion Robot–that is employed in the experiment. In Chapter 4, multiple brachiating controllers (learning-based method and analytical method) are described. Section 4.2, 4.3,and 4.4 are based on [132], [131], and [63] respectively. In Chapter 5, static quadruped walking called crawl gait is realized so that Multi-Locomotion Robot can move stably. In addition, the structure of the Gorilla Robot is analyzed as quadruped hardware. Section 5.2 is based on [12]. Chapter 6 describes three climbing ladder motions of the Gorilla Robot. This Chapter is based on [316]. In Chapter 7, a load-allocation algorithm is introduced to balance the loads of the joint motors during transition from a ladder to another. By applying the load-allocation algorithm, the transition motion from ladder climbing to brachiation is achieved. This chapter is based on [171]. In Chapter 8, we propose a method to choose a suitable locomotion mode by estimating the falling down risk. The falling down risk is estimated from internal conditions of the robot using Bayesian Network. This chapter is based on [251]. In Chapter 9, the PDAC is applied to multiple motions. First applications are the 3-D biped walking of 2-D dynamics such as lateral and sagittal motions. Second one is the heel-off biped walking that makes it possible to avoid the problem for the impact shock at a foot landing. Third one is the 3-D biped walking based on 3-D dynamics without dividing into 2-D plane. Fourth one is the 3-D biped walking on uneven terrain. Fifth one is the quadruped walking, and the final application is the brachiation. Section 9.1, 9.2, 9.3, 9.4, 9.5, and 9.7 are based on [10], [13], [52], [11], [14], and [66] respectively. Finally, we summarize and discuss perspective of these studies in Chapter 10.

Fig. 1.25

Chapter 2
Basics

2.1 Trajectory Generation Method of Robots

2.1.1 Generation of a Desired Trajectory

Figure 2.1 shows a framework of motion control system including the trajectory generation. Motion Planning is sometimes called Trajectory Planning. Motion Planning is the function in order to decide how to move based on intent of an operator or decision of a robot. If a control system is the position control system, the information that Motion Planning output is the position information; if a control system is the force control system, the force information is given. In the following, we give explanation on the premise of the position control.

Generally, target trajectory is desired to be smooth function of time. To decide the "Specification of motion" is setting following valuables.

1. Position and posture
2. Maximum velocity
3. Maximum acceleration or acceleration time

There are PTP (Point to Point) motion and CP (Contineous Path) motion. PTP is the motion that path way is not considered if the position and posture of beginning and end point are correct. CP put importance on path way too.

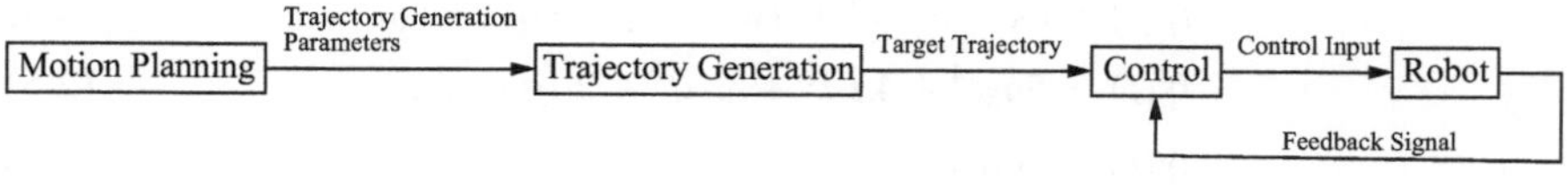

Fig. 2.1 Framework of Motion Control System.

T. Fukuda et al.: Multi-Locomotion Robotic Systems, STAR 81, pp. 61–73.
springerlink.com © Springer-Verlag Berlin Heidelberg 2012

2.1.2 *Basic Orbital Function*

In order to follow the target trajectory, a function of time form basis of the desired motion needs to be designed. A lot of methods to design a function of time are proposed. Here, we give following two examples.

1. PTP motion: each drive axis is controlled so that it could reach a desired position and posture at the beginning and end of the motion.
2. CP motion: each drive axis is controlled so that it tracks the target trajectory designed to satisfy the specification of acceleration and deceleration. Specification of motion is designed from position and posture; the position is based on movement distance and the posture is based on equivalent angle.

2.1.3 *Design of Basic Orbital Function Using n-Dimensional Polynomial*

The planning method of position, velocity, and acceleration by using n-dimensional polynomial is introduced. In this subsection, the polynomial means a function of time expressed as Eq. (2.1).

$$\eta\,(t) = a_n t^n + a_{n-1} t^{n-1} + \cdots + a_2 t^2 + a_1 t + a_0, \tag{2.1}$$

where n is the dimension of the polynomial, normally counting number. Coefficients $a_j (j = 0, 1, 2, \cdots, n)$ are decided by initial condition or termination condition. Assuming that Eq. 2.1 is the function of position, velocity and acceleration are first and second-order differential respectively. By differentiating Eq. 2.1, following equations are derived.

$$\dot{\eta}\,(t) = n a_n t^{n-1} + (n-1) a_{n-1} t^{n-2} + \cdots + 2a_2 t + a_1. \tag{2.2}$$

$$\ddot{\eta}\,(t) = n(n-1) a_n t^{n-2} + (n-1)(n-2) a_{n-1} t^{n-2} + \cdots + 2a_2. \tag{2.3}$$

2.1.3.1 Design of Orbital Function Using Five-Dimensional Prenominal

Design procedure when n=5 is shown here. Eq. (2.4) is the orbital function of position, and Eq. (2.5) and Eq. (2.6) are velocity and acceleration function respectively.

$$\eta\,(t) = a_5 t^5 + a_4 t^4 + a_3 t^3 + a_2 t^2 + a_1 t + a_0, \tag{2.4}$$

$$\dot{\eta}\,(t) = 5a_5 t^4 + 4a_4 t^3 + 3a_3 t^2 + 2a_2 t + a_1, \tag{2.5}$$

$$\ddot{\eta}\,(t) = 20a_5 t^3 + 12a_4 t^2 + 6a_3 t + 2a_2. \tag{2.6}$$

$$\tag{2.7}$$

The coefficients are decided if 6 independent conditions such as initial and termination position, velocity, and acceleration, because unknown variables are 6. By using initial and termination time t_0 and t_1, initial and termination position, velocity and acceleration $\eta_0, \dot{\eta}_0, \ddot{\eta}_0, \eta_1, \dot{\eta}_1$, and $\ddot{\eta}_1$, the following equation is derived:

$$\begin{bmatrix} \eta_0 \\ \dot{\eta}_0 \\ \ddot{\eta}_0 \\ \eta_1 \\ \dot{\eta}_1 \\ \ddot{\eta}_1 \end{bmatrix} = \begin{bmatrix} t_0^5 & t_0^4 & t_0^3 & t_0^2 & t_0 & 1 \\ 5t_0^4 & 4t_0^3 & 3t_0^2 & 2t_0 & 1 & 0 \\ 20t_0^3 & 12t_0^2 & 6t_0 & 2 & 0 & 0 \\ t_1^5 & t_1^4 & t_1^3 & t_1^2 & t_1 & 1 \\ 5t_1^4 & 4t_1^3 & 3t_1^2 & 2t_1 & 1 & 0 \\ 20t_1^3 & 12t_1^2 & 6t_1 & 2 & 0 & 0 \end{bmatrix} \begin{bmatrix} a_5 \\ a_4 \\ a_3 \\ a_2 \\ a_1 \\ a_0 \end{bmatrix}. \tag{2.8}$$

From Eq. (2.8), if coefficients matrix is holomorphic, a_1-a_5 are decided uniquely:

$$\begin{bmatrix} a_5 \\ a_4 \\ a_3 \\ a_2 \\ a_1 \\ a_0 \end{bmatrix} = \begin{bmatrix} t_0^5 & t_0^4 & t_0^3 & t_0^2 & t_0 & 1 \\ 5t_0^4 & 4t_0^3 & 3t_0^2 & 2t_0 & 1 & 0 \\ 20t_0^3 & 12t_0^2 & 6t_0 & 2 & 0 & 0 \\ t_1^5 & t_1^4 & t_1^3 & t_1^2 & t_1 & 1 \\ 5t_1^4 & 4t_1^3 & 3t_1^2 & 2t_1 & 1 & 0 \\ 20t_1^3 & 12t_1^2 & 6t_1 & 2 & 0 & 0 \end{bmatrix}^{-1} \begin{bmatrix} \eta_0 \\ \dot{\eta}_0 \\ \ddot{\eta}_0 \\ \eta_1 \\ \dot{\eta}_1 \\ \ddot{\eta}_1 \end{bmatrix}. \tag{2.9}$$

2.2 Limit Cycle

Normally, the robot locomotion is expressed as a nonlinear system. Limit cycle, which is a phenomenon of nonlinear system, is often used for stability analysis. Limit cycle is the phenomenon that periodical solution with a constant period, amplitude, and angular frequency is generated out of relation to and initial value. Representative example is the van der Pol oscillator (Fig. 2.2). The dynamic characteristic is expressed as follows:

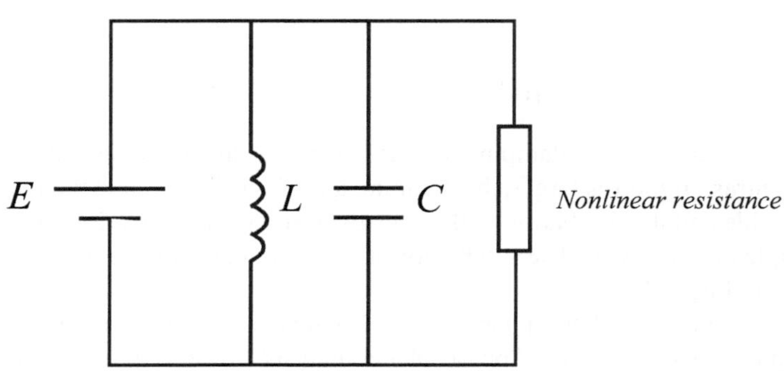

Fig. 2.2 van der Pol oscillating circuit.

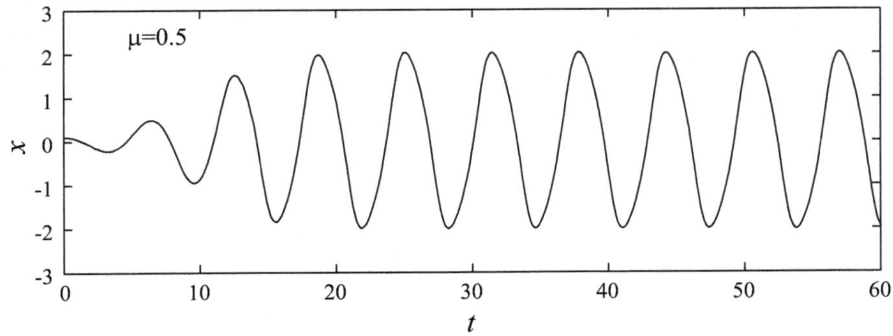

Fig. 2.3 Limit cycle.

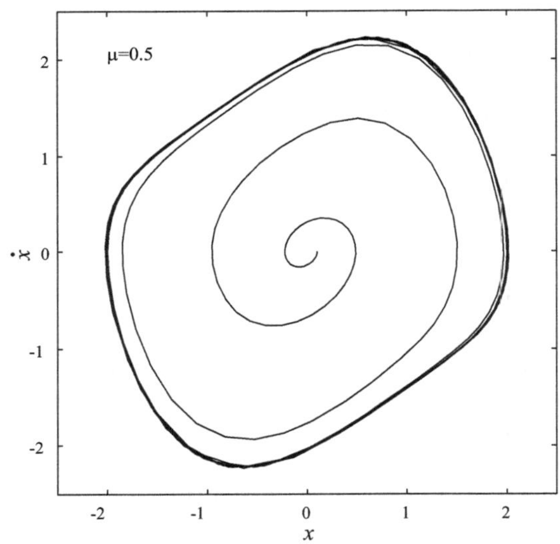

Fig. 2.4 x behavior of Fig. 2.3 in the $x-\dot{x}$ plain. (Limit cycle.)

$$\ddot{x}-\mu(1-x^2)\dot{x}+x=0 \quad (\mu>0) \tag{2.10}$$

When x is small value, the damping term is negative value; thus the system is unstable. In contrast, if the x is larger, the damping term is positive value; then the system will be stable and the oscillation will be steady at some value of x. Figure 2.3 shows an example of x behavior. The x behavior of this example depicted in the $x-\dot{x}$ plain as shown in Fig. 2.4.

The $x-\dot{x}$ plain is called "phase plane". To analyze behavior of solution trajectory in phase plane is called "phase plane analysis". The system needs to be an autonomous system when the phase plane is conducted.

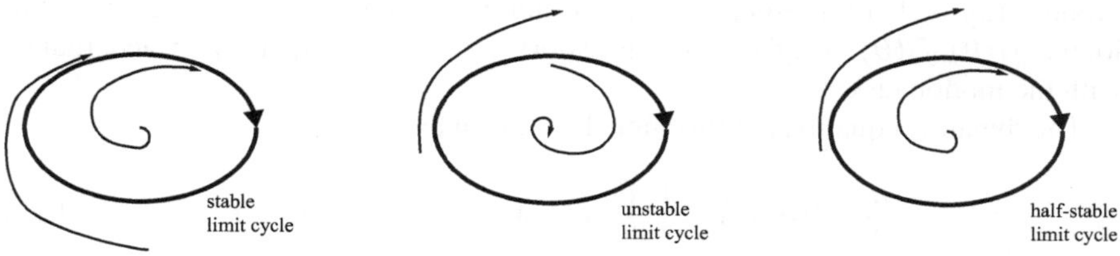

Fig. 2.5 Limit cycle stability.

If all phase plane trajectories starting from neighborhood of the limit cycle approach the limit cycle, the limit cycle is stable. If all or some neighboring trajectories get away from the limit cycle, the limit cycle is unstable or half-stable (Fig. 2.5).

2.3 Passive Dynamic Autonomous Control (PDAC)

In this section, we introduce Passive Dynamic Autonomous Control (PDAC) that is the novel control method based on point-contact and virtual holonomic constraint [51]. Point-contact denotes that a robot contacts the ground at a point (i.e. the first joint is passive), and makes it possible to achieve adaptability to ground irregularity and energy efficiency. The concept of the virtual holonomic constraint is proposed as *Virtual constraint* by Grizzle and Westervelt *et al.*[87, 297], and defined as a set of holonomic constraints on the robot's actuated DoF parameterized by the robot's unactuated DoF. The virtual holonomic constraint enables a robot to satisfy the desired path of postural motion.

2.3.1 Dynamics of PDAC

2.3.1.1 Converged Dynamics

Assuming that PDAC is applied to the serial n-link rigid robot shown in Fig. 2.6, above-mentioned two premises are expressed as follows:

$$\tau_1 = 0 \tag{2.11}$$

$$\Theta = [\theta_1, \theta_2, \cdots, \theta_n]^T = [f_1(\theta), f_2(\theta), \cdots, f_n(\theta)]^T := f(\theta) \tag{2.12}$$

where θ is the angle around the contact point in the absolute coordinate system. That is, letting the angle of ground slope be ζ (ascent is positive, and decent is negative), $\theta = \theta_1 + \zeta$. On the flat and level ground, $\theta_1 = f_1(\theta) = \theta$ is held.

Point-contact is realized by zero-torque applying to the ground around contact-point, Eq. (2.11) indicates the point-contact condition between a robot and the ground. Eq. (2.12) represents the virtual holonomic constraint interlocking robot joints. $f_1(\theta), f_2(\theta), \cdots, f_n(\theta)$ are necessary to design adequately in accordance with the motion task.

The dynamic equations of this model are given by

$$\frac{d}{dt}\left(M(\Theta)\dot{\Theta}\right) - \frac{1}{2}\frac{\partial}{\partial\Theta}\left(\dot{\Theta}^T M(\Theta)\dot{\Theta}\right) - G(\Theta) = \tau \tag{2.13}$$

$M(\Theta)$ denoting the inertia and centrifugal terms is defined as follows:

$$M(\Theta) := \begin{bmatrix} m_{11}(\Theta) & m_{12}(\Theta) & \cdots & m_{1n}(\Theta) \\ m_{21}(\Theta) & m_{22}(\Theta) & \cdots & m_{2n}(\Theta) \\ \vdots & \vdots & \ddots & \vdots \\ m_{n1}(\Theta) & m_{n2}(\Theta) & \cdots & m_{nn}(\Theta) \end{bmatrix} := \begin{bmatrix} m_1(\Theta) \\ m_2(\Theta) \\ \vdots \\ m_n(\Theta) \end{bmatrix}. \tag{2.14}$$

That is, $m_i(\Theta) = [m_{i1}(\Theta), m_{i2}(\Theta), \cdots, m_{in}(\Theta)]^T \quad (i = 1, 2, \cdots, n)$. The second term indicates the Corioli's forces, where

$$\frac{\partial}{\partial\Theta} = \left[\frac{\partial}{\partial\theta_1}, \frac{\partial}{\partial\theta_2}, \cdots, \frac{\partial}{\partial\theta_n}\right]^T. \tag{2.15}$$

$G(\Theta)$ and τ denote the gravity terms and torques, defined as below respectively,

$$G(\Theta) := [G_1(\Theta), G_2(\Theta), \cdots, G_n(\Theta)]^T \tag{2.16}$$

$$\tau := [\tau_1, \tau_2, \cdots, \tau_n]^T. \tag{2.17}$$

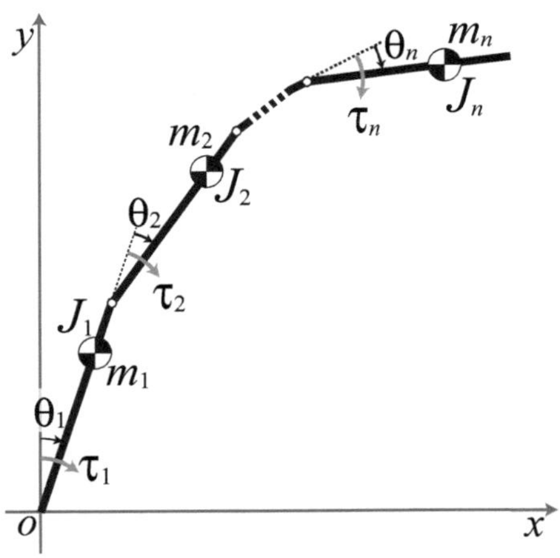

Fig. 2.6 Mechanical model of the serial n-link rigid robot. θ_i and τ_i are the angle and the torque of ith joint respectively. m_i and J_i are the mass and the moment of inertia of ith link respectively.

Since in this model, the dynamic equation around the contact point (the first equation) has no term of the Coriolis force, it is given as

$$\frac{d}{dt}\left(m_1(\Theta)^T\dot{\Theta}\right) - G_1(\Theta) = \tau_1. \tag{2.18}$$

By differentiating Eq. (2.12) with respect to time, the following equation is acquired,

$$\dot{\Theta} = \frac{\partial f(\theta)}{\partial \theta}\dot{\theta} = \left[\frac{\partial f_1(\theta)}{\partial \theta}, \frac{\partial f_2(\theta)}{\partial \theta}, \cdots, \frac{\partial f_n(\theta)}{\partial \theta}\right]^T \dot{\theta}. \tag{2.19}$$

Substituting Eqs. (2.11), (2.12) and (2.19) into Eq. (2.13) yields the following dynamic equation,

$$\frac{d}{dt}\left(M(\theta)\dot{\theta}\right) = G(\theta) \tag{2.20}$$

where

$$M(\theta) := m_1\left(f(\theta)\right)^T \frac{df(\theta)}{d\theta} \tag{2.21}$$

$$G(\theta) := G_1\left(f(\theta)\right). \tag{2.22}$$

By multiplying both sides of Eq. (2.20) by $M(\theta)\dot{\theta}$ and integrating with respect to time, the dynamics around the contact point is obtained as follows:

$$\int \left(M(\theta)\dot{\theta}\right)\frac{d}{dt}\left(M(\theta)\dot{\theta}\right) dt = \int M(\theta)G(\theta)\dot{\theta}\, dt \tag{2.23}$$

$$\Longleftrightarrow \frac{1}{2}\left(M(\theta)\dot{\theta}\right)^2 = \int M(\theta)G(\theta)\, d\theta. \tag{2.24}$$

Therefore, the whole robot dynamics is expressed as the following 1-dimensional autonomous system (that is, the phase around contact point),

$$\dot{\theta} = \frac{1}{M(\theta)}\sqrt{2\int M(\theta)G(\theta)\, d\theta} \tag{2.25}$$

$$:= \frac{1}{M(\theta)}\sqrt{2\left(D(\theta)+C\right)} \tag{2.26}$$

$$:= F(\theta). \tag{2.27}$$

In this book, we term Eqs. (2.26) and (2.27) Converged dynamics.

Additionally, by setting desired dynamics to Eq. (2.27), a robot achieves desired dynamic motion. However, it is quite difficult to solve this inverse problem, which is to find virtual holonomic constraint, Eq. (2.12), satisfying desired dynamics. Thus, in this book, the Virtual constraint is designed empirically based on the condition of desired posture at foot-contact or desired COG trajectory etc. without dealing with this problem.

Here, we have to describe about an unsolved problem of PDAC. To obtain the Converged dynamics by utilizing PDAC, Eqs. (2.23) and (2.24) should be integrable. However, the condition of Virtual constraint to make Eq. (2.23) integrable has not been clarified so far. Thus, in order to make PDAC more effective, it is required to find such condition, or to propose an approximate calculation method or a deriving algorithm of Converged dynamics without integration. This integrability problem is one of future work regarding PDAC.

2.3.1.2 PDAC Constant

Since the Converged dynamics is autonomous, and, independent of time, it is considered as a conservative system. The integral constant in right hand side of Eq. (2.24), C, is a conserved quantity, which is termed PDAC Constant. Its value is decided according to the initial condition (as for biped walking, the state immediately after foot-contact), and kept constant during a cycle of motion.

The dimension of PDAC Constant is equal to the square of angular momentum and has relevance to it. As is well know, assuming that the robot shown in Fig. 2.6 resides in the horizontal plane, the angular momentum around contact point is conserved since there is no effect of gravitational force on the robot dynamics. In this condition, it is clear that $M(\theta)\dot{\theta}$(angular momentum)$= \sqrt{2C}$ from Eq. (2.26), since $G(\Theta) = 0$ in Eq. (2.13) hence $D(\theta) = 0$. Note that, although angular momentum is not conserved in the condition which robot dynamics is affected by the gravitational force, PDAC Constant is conserved since it includes the effects of the gravitation. In conclusion, PDAC Constant is conserved quantity derived by embedding gravity term into angular momentum around the contact-point.

2.3.1.3 Dynamics Interlocking

As mentioned previously, PDAC is based on the two premises: passivity and virtual holonomic constraint. These premises make it possible to describe the whole robot dynamics as the 1 dimensional autonomous system, owing to which the simple and valid controller based on the robot dynamics can be composed. However, holonomic constraint of joint angles has possibility to cause the problem that the robot vibrates and the controller loses its stability during locomotion, especially at the moment of foot-contact, since if the passive joint vibrates, all of other active joints also do. As one of control technique to solve this problem, dynamics interlocking is proposed. In this method, all of robot joints are controlled according to the desired dynamics of each joint derived from the interlocking function Eq. (2.12) and the target dynamics Eq. (2.27) as follows:

$$\dot{\theta}_i = \frac{\partial f_i}{\partial \theta} F\left(f_i^{-1}(\theta_i)\right) \qquad (i = 1,2,3,\cdots). \tag{2.28}$$

These designed dynamics generate tracking error from the target dynamics, because the desired dynamics can be independently designed, and are not necessary and sufficient subset of the target dynamics. Hence, we add a feedback term that calculate the tracking error between the reference angle and the current angle. The desired angular velocity of each joint is calculated as described below,

$$\dot{\theta}_1^d = F\left(f_1^{-1}(\theta_1)\right) \tag{2.29}$$

$$\dot{\theta}_i^d = \frac{\partial f_i}{\partial \theta} F\left(f_i^{-1}(\theta_i)\right) + k_i\left(f_i(\theta) - \theta_i\right) \qquad (i = 2, 3, \cdots) \tag{2.30}$$

$$\Longleftrightarrow \dot{\Theta}^d := F_D(\Theta) \tag{2.31}$$

where k_i is the strength of connection determined experimentally since its value has little effect on the robot dynamics. As for humanoid robots, the ground slope at the contact point is estimated based on the angle of the ankle joint of the swing leg at foot-contact, and θ is calculated from θ_1 and the ground slope. The remarkable point is that if there is no error such as the model error or disturbance, the second term of Eq. (2.30) is constantly zero and the actual dynamics of θ is identical with the target dynamics definitely.

2.3.1.4 Virtual Passive Joint

In order to keep standing, most of humanoid robots are fully actuated i.e. ankle joints have actuators. In many instances, these robots which support human life are basically large and heavy, hence reduction gear ratio of their joints are relatively high e.g. as for our Gorilla Robot, their value are from 100 to 200. High reduction gear ratio generates high viscous force.

To apply PDAC to these kinds of robots, it is necessary to diminish the joint viscosity to simulate the point-contact state. As one of approaches to reduce the viscous forced, ankle joint is actuated to be virtually passive by applying adequate torque.

Generally, it is known that viscosity is associated with the angular velocity. That is, viscous torque of joints can be expressed as the function of the angular velocity. Thus, we use the technique to make the joint vertually passive based on PDAC by applying the following torque to ankle joints,

$$\tau_\beta = \beta(\dot{\theta}^d) = \beta \circ F(\theta). \tag{2.32}$$

Note that the viscous torque is estimated based on reference angular velocity. Providing the actual angular velocity is larger than the desired one, the angular velocity is decreased since the actual viscous torque is higher than the applied torque. Meanwhile, if the actual angular velocity is slower than the desired one, ankle torque for viscosity compensation works to increase the angular velocity. Therefore, the actual dynamics is weakly converged on the desired dynamics.

Additionally, breaking slightly the assumption of point-contact, it is possible to enhance the convergency toward the desired dynamics. For instance, the following torque is applied to the ankle joint of a support-leg,

$$\tau_\beta = \beta \circ F(\theta) + \tau_{Fb}. \tag{2.33}$$

where

$$\tau_{Fb} = k_\beta (F(\theta) - \dot{\theta}). \tag{2.34}$$

The second term in right side of Eq. 2.33 is feedback term, and various forms of τ_{Fb} can be employed such as PID, PI, or PD. Note that all gains in the feedback term are set at quite small value, since the objective of this term is to attenuate the slight error arisen from modeling error. Breaking the assumption of point-contact gives the probability that ZMP goes outside the support region, and that the support-foot sole detaches from the ground unexpectedly. Thus, it is necessary to observe the information from some sensors such as ZMP position or the error of upper body inclination, and to make the switch to the controller of falling avoidance if a robot has probability to fall down. Furthermore, if the controller modifies the model gradually according to the error of dynamics, the controller performance may be enhanced more.

The falling avoidance controller and model adjustment method are future work, and we do not discussed them in this book.

2.3.2 Control System

Figure 2.7 shows the block diagram of PDAC in bipedal locomotion without dynamics interlocking. The control loop including a robot (enclosed by the gray line in Fig. 2.7) has no input from outside the controller, thus it can be considered that the control system is autonomous. This autonomy makes it possible to achieve the natural dynamic motion based on the inherent dynamics of a robot. The loop described by the broken line is executed only at the moment of foot-contact. In this loop, virtual holonomic constraint and Converged dynamics of the next step are updated according to both the robot status and the desired parameters such as walking velocity. This update process makes it possible to stabilize walking or to vary the walking parameters.

Figure 2.8 depicts the block diagram of PDAC in bipedal locomotion with dynamics interlocking. As shown in Fig. 2.8, the dynamics interlocking has no effect on autonomy of the controller since the dynamics interlocking affects only the lower layer controller—virtual passive joint, Virtual constraint. When the dynamics interlocking is employed, basically the angular velocity of active joints is controlled by PD control.

Additionally, the controller with the weak feedback for modeling error is shown in Fig. 2.9. As shown in this figure, the controller measures the actual angular velocity around contact-point, and output the compensation torque to attenuate the error between the desired dynamics and actual one. Note that, as described previously, the feedback gain is quite small hence this feedback has the weak effects. When there a robot to start falling down, the robot controller is switched to the falling avoidance controller as depicted in Fig. 2.9.

2.3.3 Advantage of PDAC

Finally, we describe advantage of PDAC method briefly. PDAC has the several advantages due to the point-contact and the virtual holonomic constraint mentioned above.

The first advantage is PDAC Constant and Converged dynamics. Although robot dynamics is a high degree-of-freedom complex system, it is possible to express robot dynamics as a one-dimensional autonomous system, i.e. the phase around contact-point. This 1-D dynamics named Converged dynamics makes it easy to control robot dynamics and to take a desired motion. In addition, Converged dynamics includes the conserved quantity that is maintained constant theoretically during walking even if conservation of angular momentum and energy conservation law are broken. This conserved quantity termed PDAC Constant makes it possible to stabilize, analize, and prove the stability of the walking dynamics easily.

The second one is to be able to realize the dynamic motion which is based on the robot dynamics and follows the desired postural motion given by a operator or the circumstances so that a robot can avoid obstacles or maintain the upper body at the certain inclination (c.f. in the upright position) in order to perform some tasks.

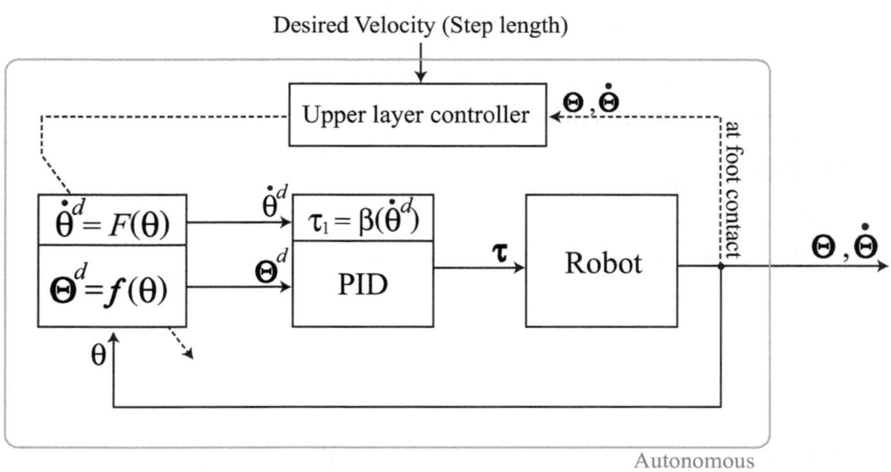

Fig. 2.7 Block diagram of PDAC in bipedal locomotion without dynamics interlocking

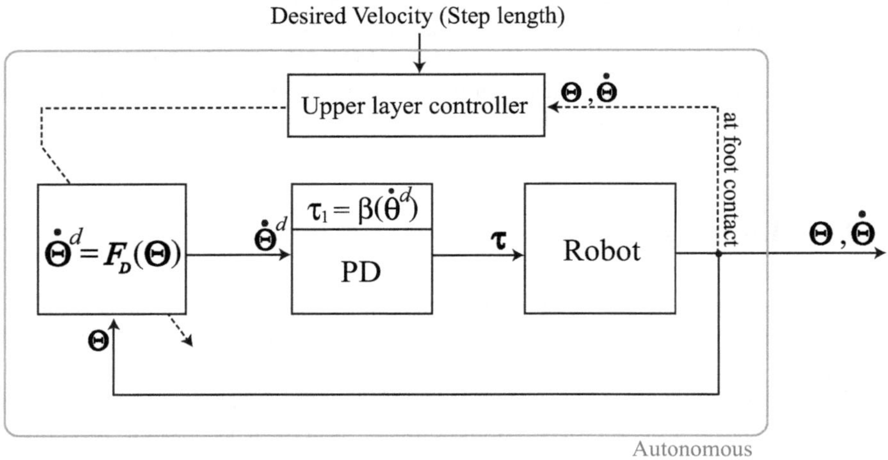

Fig. 2.8 Block diagram of PDAC in bipedal locomotion with dynamics interlocking

In most of hybrid walking methods, it is difficult to design the rigorous postural motion due to the torque control of joints, whereas in PDAC method, it is possible to give the desired path of postural motion by designing the virtual holonomic constraint adequately. This ability to follow the designed path is considered another to make advantage a robot quite practicable in the actual environment.

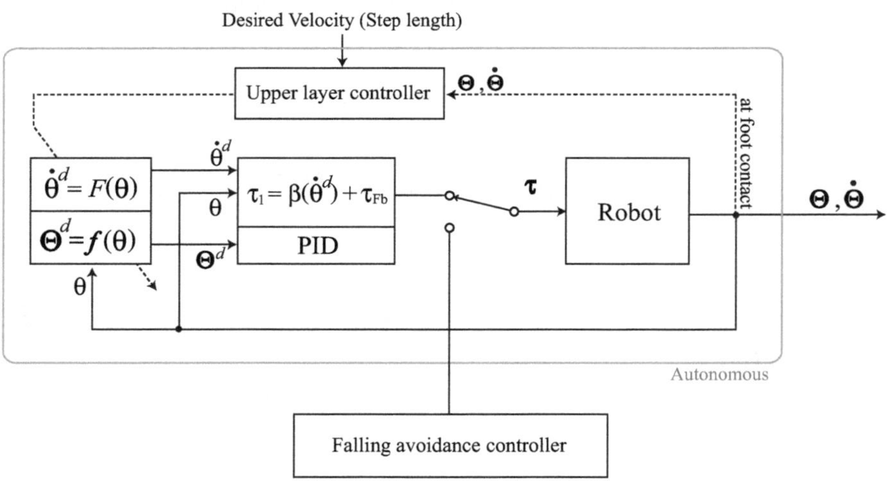

Fig. 2.9 Block diagram of PDAC in bipedal locomotion with the weak feedback for modeling error compensation

The third is capability to obtain the advantage of hybrid walking and point-contact method. Basically, it is difficult for active walking to utilize the natural dynamics of the robot since it controls joints by high-gain feedback, hence active walking has the following disadvantages compared with the passive walking: 1) Unnatural walking motion 2) Vulnerability to the ground irregularity 3) Low energy efficiency. Meanwhile, passive walking has the following disadvantages: 1) Highly limited capability to walk 2) Low robustness against disturbance 3) Disability to perform another tasks during walking.

Finally, the fourth advantage is safety. Safety means that walking motion is suspended without applying the excessive torque to the joints when a robot conflict with a wall or human. In order to realize this function in the active walking, it is necessary that a robot is equipped with enormous touch sensors all over the body. As for PDAC, whole robot motion is dependent on the passive joint one and the controller is independent of time, thus the robot motion is stopped if the passive joint is stopped by external force.

Chapter 3
Hardware of Multi-Locomotion Robot

3.1 Brachiation Robot (Conventional Bio-inspired Robot)

This section introduces brachiation robots that are monkey-type robots developed previously in our laboratory as one of bio-inspired robots. In pioneering research, the dynamics of locomotion were analyzed and locomotion types proposed using a six-link brachiation robot [64]. Following this research, the second one is a two-link brachiation robot "Brachiator II" shown in Fig. 3.1. This robot has one actuator at the elbow connecting two links, each of which has a gripper. Because the gripper cannot impose torque on the handhold directly, this is an underactuated mechanical system. We developed a two-link brachiation robot and proposed a heuristic method to find feasible motions [67, 244, 245]. We also proposed a self-scaling reinforcement learning algorithm to achieve the robustness against some disturbances [96]. As an

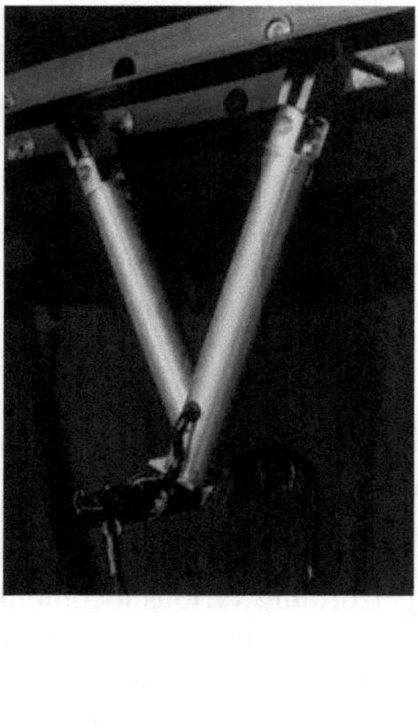

Fig. 3.1 Brachiator II

T. Fukuda et al.: Multi-Locomotion Robotic Systems, STAR 81, pp. 75–81.
springerlink.com © Springer-Verlag Berlin Heidelberg 2012

Fig. 3.2 Brachiator III

analytical approach, We proposed a target dynamics method to solve the"ladder", "swing up", "rope" and "leap" problems, where they encoded the task as the output of the target dynamical system inspired by the pendulum-like motion of an ape's brachiation [193, 192, 191]. In those works, a brachiating robot with multi-dynamic-body has not been discussed. Then, we consider the control of the higher degrees of freedom robot, and develop the third brachiation robot "Brachiator III" shown in Fig. 3.2. Brachiator III consists of 13 links, 12 degrees of freedom, and 14 motors including two motors for a gripper control. The dimension and location of the joints are designed based on a real long-armed ape. We proposed the adaptation method that measures effects of the each local behavior to the total behavior, and determines the activation level of each behavior controller [97]. The controller can adjust the amplitude of the behavior controller. This method is effective for the online learning by real robot because learning iteration of this method is less than that of other learning algorithms.

The motion of these brachiation robots are designed to use specific locomotion form such as brachiation. In the next subsection, gorilla robots are introduced based on the concept of the multi-locomotion robot described in Chapter 1.

3.2 Gorilla Robot (Multi-Locomotion Robot)

In order to build the robot performing various locomotion mode mentioned above, various devices of design is needed such as link structure, mass distribution, actuators, gear ratio, sensors selection, and range of joint motion. For instance, to make biped and quadruped walking—locomotion on the ground with legs—and

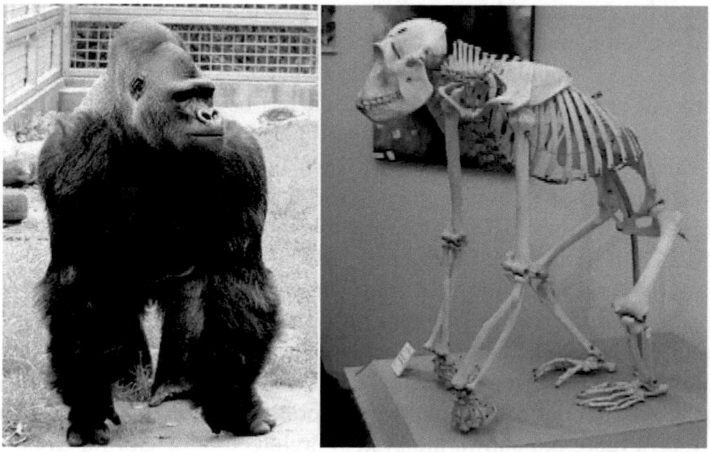

Fig. 3.3 Western Lowland Gorilla. Left: shape. Right: skeletal preparations.

brachiation—locomotion with arms—compatible, both arms and legs that can generate high enough torque to be able to support its own weight. Besides, since swing leg and swing arm is needed to move forward in short period, knee joints and elbow joints are demanded high angular velocity. Therefore, so as to realize multi-locomotion robot shown in Fig. 1.21, it is necessary to develop a balanced robot capable to satisfy high torque and high angular velocity. However, in previous work, design guideline of such type of robots has not clarified yet.

Hence, as an animal close to the concept of multi-locomotion robot, we select male gorilla (order: Primates, family: Pongidae, genus: Gorilla). Gorillas normally ambulate in particular type of quadruped locomotion named knuckle walk. It often employs biped walking to carry something by hand and sometimes performs brachiation. In addition, since body size of mature male gorilla is about 1.7-1.8[m], it is conceivable that gorilla is suited for the model of multi-locomotion robot in the human residential environment.

We built and developed three gorilla robots as the prototype of multi-locomotion robot. Controllers of biped walking, quadruped walking, and brachiation have been composed and whole control system of multi-locomotion robot has been developed from both sides of software and hardware.

In the following, hardware architecture of 3 gorilla robots and control system is explained.

3.2.1 Gorilla Robot I

The overview of first prototype of multi-locomotion robot, Gorilla Robot I, is shown in Fig. 3.4. This robot was designed based on skeletal preparations of mature male gorilla by taking into consideration degree of freedom necessary to perform biped, quadruped walking, and brachiation. Robot link structure and joints placement are shown in Fig. 3.5. Shoulder joints and hip joints are primarily classified into ball joint. In recent years, realization of ball joint has begun to be investigated [213], however its development is on the way and there are many problems to solve in order to realize such as endurance and performance. Hence, in this research, these joints are composed of 2 joints—pitch and roll axis and the mechanical structure was designed as follows: 5 DOF leg (2 DOF hip, 1 DOF knee, 2 DOF ankle), 5 DOF arm (2 DOF shoulder, 2 DOF elbow, 1 DOF wrist).

Motor output and reduction gear ratio was decided according to the numerical simulation and motion analysis of each locomotion. Each joint is actuated by DC servo motor (DC servo actuator, RH series, Harmonic Drive Systems Inc.) and as driver amplifier, TITech Driver Version 2 (Okazaki Sangyo Co.) is employed. The operating system of controller is real-time operating system, VxWorks (Wind River Systems Inc.) Figure 3.6 shows the system architecture of Gorilla Robot I. Computer to control, AD/DA board, counter board, and power is set outside robot and connected with wires.

Since the counter board (Interface PCI-6201) counts the pulse from encoder embedded in each actuator, joint angle is obtained at all time. Force sensors (wrists: IFS-67M25A50 ankles:IFS-100M40A100, NITTA CORPORATION) are installed at wrists and ankles of robot. This makes it possible to measure floor reaction force and ZMP (Zero Moment Point) during biped or quadruped locomotion and to observe reaction force from bar and impact force at hand-contact in brachiation.

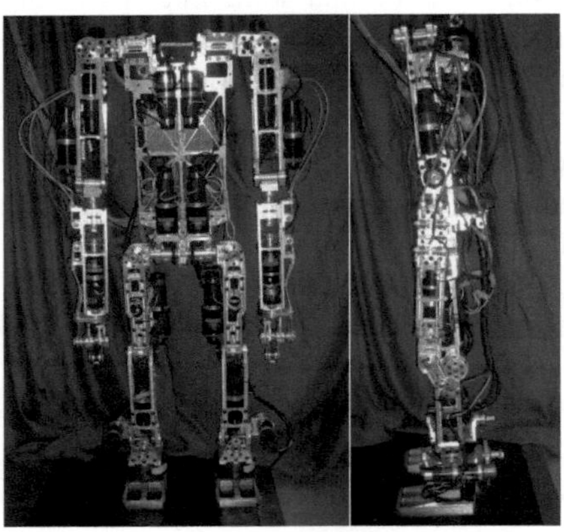

Fig. 3.4 Overview of Gorilla Robot I

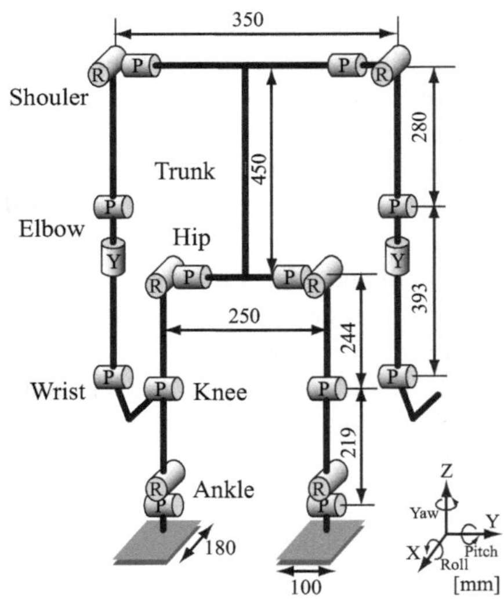

Fig. 3.5 Link structure of Gorilla Robot I

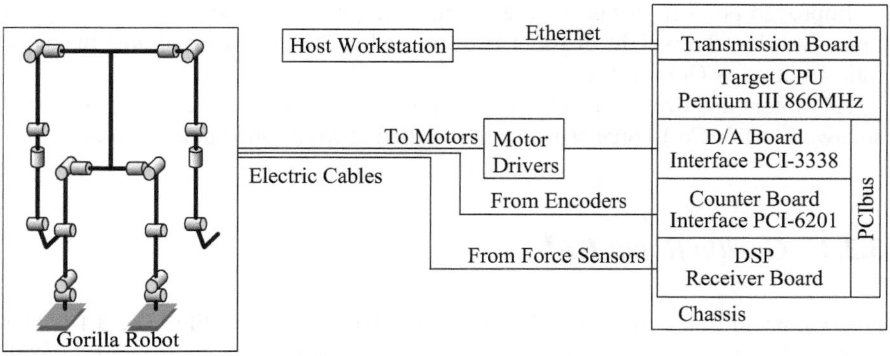

Fig. 3.6 Control system of Gorilla Robot I & II

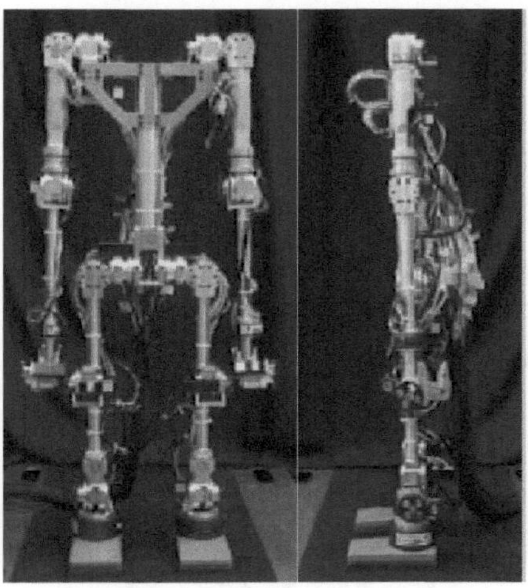

Fig. 3.7 Overview of Gorilla Robot II

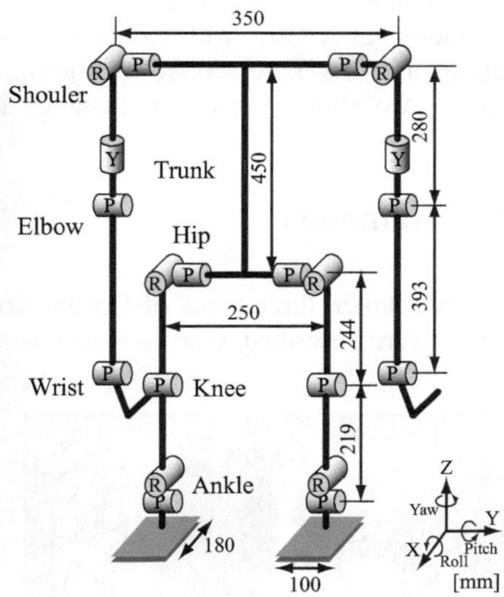

Fig. 3.8 Link structure of Gorilla Robot II

Consequently, this robot could not perform brachiation since its weight is too heavy, over 30.0[kg]. In order to solve this problem, we built Gorilla Robot II, which is lighter than Gorilla Robot I.

3.2.2 Gorilla Robot II

Overview and link structure of Gorilla Robot II is shown in Fig. 3.7, 3.8. The control system of this robot is same as that of Gorilla Robot I shown in Fig. 3.6.

Improved point in design is to exchange the pitch axis and yaw axis of elbow joint. Besides, the weight of robot is decreased at 20.0[kg] with almost the same link structure as Gorilla Robot I.

As motors and drivers to actuate each joint, AC servo drive (Σ-mini series, Yasukawa Electric Co.) Force sensors are installed at wrists and ankles of robot.

3.2.3 Gorilla Robot III

Overview and link structure of Gorilla Robot III is shown in Fig. 3.9, 3.10. This robot is designed to add the following 4 DOF to 20 DOF of Gorilla Robot II: 2 DOF lumbar (roll and yaw axis), 1 DOF hip joints (yaw axis). Gorilla Robot III became able to perform more smooth motion by use of the added DOF.

Its height is about 1.0[m] and weight is about 22.0[kg]. This robot is driven by 24 AC motors of 20-30W with 100-200 times of speed reduction by harmonic gears. As motors and drivers to actuate each joint, AC servo drive (Σ-mini series, Yasukawa Electric Co.) Force sensors are installed at wrists and ankles of robot.

3.3 Summary

In this chapter, three types of Gorilla Robots developed as the Multi-Locomotion Robot were described. Gorilla Robot was designed to be able to perform biped lo-

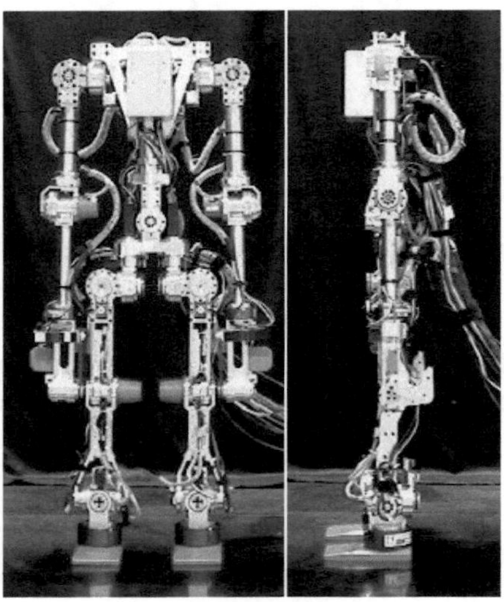

Fig. 3.9 Overview of Gorilla Robot III

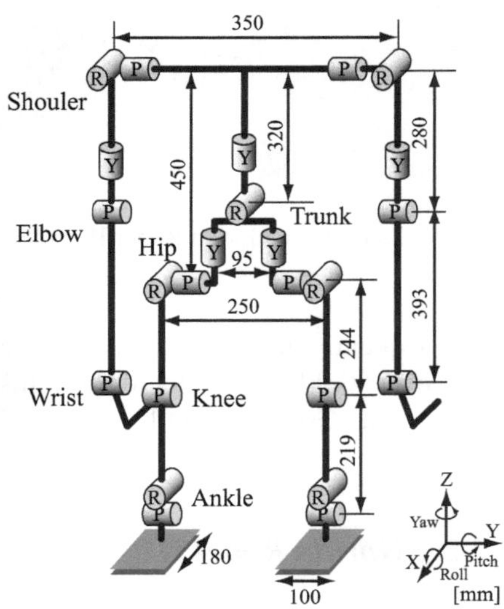

Fig. 3.10 Link structure of Gorilla Robot III

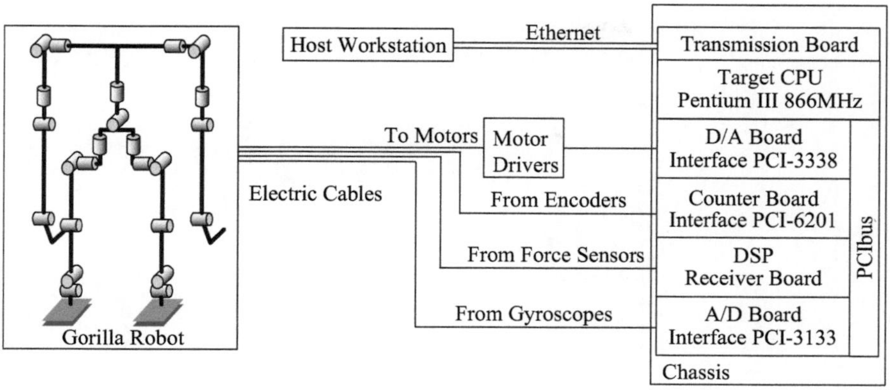

Fig. 3.11 Control system of Gorilla Robot III

comotion, quadruped locomotion, and brachiation etc. The hardware configuration of each Gorilla Robot and control system architecture were presented. In the subsequent chapters, Gorilla Robot is employed in some experiments.

Chapter 4
Brachiation

4.1 What Is Brachiation?

Brachiation—arboreal locomotion via arms swinging hand over hand through the trees—is an interesting form of locomotion unique to long armed apes. In primatology [198], apes are in the anthropoid families, which is common to man, and they are classified into two families. One of them is *Pongidae* including the gorilla, chimpanzee and orang-utan, and the other is *Hylobatidae* including the gibbon and siamang which are arboreal in habitat in tropical rain forest of Southeast Asia [198]. In fact, apes and monkeys are quite different anatomically. One of the basic differences between apes and monkeys is that apes have much greater flexibility of movement in their long forelimbs than monkeys—with their free swinging arms that rotate at shoulders, apes can travel in arm swinging locomotion while monkeys are basically quadrupeds [56].

Historically, Keith introduced the term "brachiation" in 1899 [145] and its participial adjective "brachiating" as well according to [280]. However, the agent-noun "brachiator" had been applied by Owen at least forty years earlier (e.g. in [216]) to the "long-armed" gibbons in contrast to the orang-utan, chimpanzee and gorilla [280]. Napier reviews the historical usage of these terms "brachiation" and "brachiator," and defines the terminology based on his classification of arm-swinging locomotion of anthropoid apes and monkeys [197]. Only the gibbons and the siamangs can be regarded as "full-time" brachiators since they perform efficient brachiation by means of the suspensory activities of the forelimbs alone at almost all the time in locomotion [197]. Carpenter estimates that approximately 90% of locomotion of a gibbon is brachiation by his observation of the behavior of gibbons in his field study in Thailand [30]. While orang-utans are entirely arboreal in habitat, they climb and move cautiously with the aid by the hind limbs because of their body weight. Gorillas are essentially forest floor animals, and chimpanzees appear to take an intermediate position between gibbons and orangs in terms of brachiating forms [197]. Among all four apes, the gibbons are the best at brachiating whose slight body, elongated arms and fingers are morphologically suited to this form of locomotion

T. Fukuda et al.: Multi-Locomotion Robotic Systems, STAR 81, pp. 83–116.
springerlink.com © Springer-Verlag Berlin Heidelberg 2012

[56]. Most commonly, gibbons engage in "slow brachiation," traveling at about the speed of the average human walk. But when excited or frightened, they can plunge through the forest canopy at astonishing speeds, sometimes covering 30 feet or more in a single jump without a break in "stride" (fast brachiation, ricocheting) [56].

In the studies of biomechanics of brachiation [58, 222, 229], specifically Preuschoft et al. identified a close correspondence between slow brachiation and the motion of a simplified pendulum [229]. Although the ape's moment of inertia varies during the swing locomotion according to its change of posture, the motion of a simplified pendulum gives a fairly good approximation.

4.2 Learning Algorithm for a Gorilla Robot Brachiation

4.2.1 Motion Learning

We focus on behavior generation architecture and its adaptation algorithm for a dexterous motion like an animal. The model-based approach such as the target dynamics can be one of the feasible approaches for designing of the motion controller, but there are two major problems. One is modeling problem of the robot's dynamics. It is hard to acquire a precise dynamical model of the multi-body robot. As a result we spend much time to model and adjust its parameters. The other is design problem of the target dynamics. The target motion design of some representative parts in a multi-body robot becomes complex because the motion is constrained by a robot mechanism and its kinematics. On the contrary, some learning approaches are introduced for motion control such as evolutionary algorithm, reinforcement learning algorithm, back-propagation method for a neural networks and so on. They can generate a robot controller only from an evaluation index of the robot motion through a n iterative searching process, but they have two limitations: learning time and controller size. They take many iteration processes for controller design and apply only to a simple robot dynamics with a few DOF. Therefore we proposed an on-line learning algorithm for the hierarchical behavior controller so that a multi-body robot can adjust dynamically dexterous motion to the desired motion based only on some evaluation indices. The hierarchical behavior controller with two layers is designed based on the behavior-based approach. A control output for the dexterous motion is generated on the upper layer coordinating some simple behaviors on the lower layer. The structure is explained in the next subsection.

4.2.1.1 Structure of Hierarchical Behavior Controller

We discuss a brachiation as an objective task of Gorilla Robot II. The hierarchical behavior controller is designed based on behavior-based approach since the robot has multiple degrees of freedom and the objective task is complex. In a designing process, the brachiation behavior is firstly divided into two actions: "Preliminary Swing" that stores a sufficient energy prior to Locomotion and "Locomotion" that is a movement toward a next branch. After that, these actions are decomposed into several local behaviors. "Preliminary Swing" consists of two behavior controllers: "Leg Swing" and "Body Rotation 1." "Leg Swing" is a swing up and down motion of the legs. This behavior controller drives all of joints included in both legs. "Body Rotation 1" is a rotational motion of the body to take suitable posture for the swing action, especially to control the direction of the movement, according to control yaw axis joints of both arms. Additionally, "Locomotion" consists of four behavior controllers: "Leg Stretch", "Body Rotation 2", "Body Lift" and "Arm Reaching." "Leg Stretch" is a motion of extending both legs. This behavior controller drives legs' joints as well as Leg Swing. "Body Rotation 2" is a rotational motion of the body controlling yaw axis joint of the supporting arm. "Body Lift" is a lift-up motion of the supporting arm controlling shoulder and elbow joints of the supporting arm. "Arm Reaching" is a swing motion of the free arm to get at the target branch and grasp it.

The behavior controllers are feedforward controllers which output the desired trajectories expressed by the cubic spline function to the feedback controllers. The feedback controller is a PD-controller to make a corresponding actuator follows the desired trajectory. The video tracking system measures some representative points in order to evaluate a performance of the robot. Then the evaluator adjusts desired trajectories according to a learning algorithm as following subsection. The hierarchical behavior controller for Gorilla Robot II is shown in Fig. 4.1.

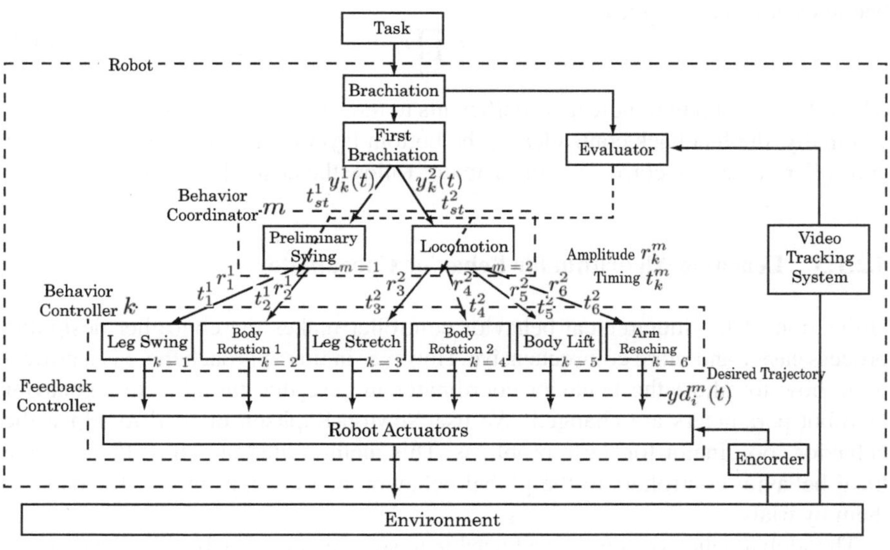

Fig. 4.1 Hierarchical behavior controller for Gorilla Robot II

4.2.1.2 Desired Trajectory Adjustment

Firstly, we design the trajectory $y_k^m(t)$ and timing parameter t_{st}^m as a primitive input of the behavior coordinator m according to a target motion. The behavior coordinator m outputs the coefficients r_k^m and t_k^m to the behavior controller k. In the lower layer, the behavior controller k outputs the desired trajectory $yd_i^m(t)$ modified by the coefficients r_k^m and t_k^m to the actuator i. The desired trajectory $yd_i^m(t)$ for the actuator i included in the behavior controller k is rescaled by two kinds of coefficients r_k^m and t_k^m from the behavior coordinator m as follows:

$$yd_i^m(t) = r_k^m(\tilde{y}_k^m(t) - b_k^m(t)) \tag{4.1}$$

$$\tilde{y}_k^m(t) = y_k^m \left\{ \frac{(t_{st}^{m+1} + t_k^{m+1}) - (t_{st}^m + t_k^m)}{t_{st}^{m+1} - t_{st}^m} t \right\} \tag{4.2}$$

$$b_k^m(t) = y_k^m(t_{st}^m) \frac{(t_{st}^{m+1} + t_k^{m+1}) - t}{(t_{st}^{m+1} + t_k^{m+1}) - (t_{st}^m + t_k^m)} + y_k^m(t_{st}^{m+1}) \frac{t - (t_{st}^m + t_k^m)}{(t_{st}^{m+1} + t_k^{m+1}) - (t_{st}^m + t_k^m)} \tag{4.3}$$

where $y_k^m(t)$ is the primitive trajectory from the behavior coordinator m to the behavior controller k. $\tilde{y}_k^m(t)$ is the trajectory modified just by the timing t_k^m. $b_k^m(t)$ is the base line connecting an initial point with an end point of $\tilde{y}_k^m(t)$. t_{st}^m is the primitive timing parameter of the behavior coordinator m that represents when it gets the motion started. These modifications shown in Fig. 4.2. If multiple behavior coordinators indicate coefficients to one behavior controller, multiplication of these values becomes new one as follows:

$$r_k = \prod_{i \in I} r_i \tag{4.4}$$

where I is a group that indicates coefficients to the behavior controller k.

Finally, the feedback controller on the bottom layer of the hierarchical behavior controller makes the corresponding actuator follow the desired trajectory.

4.2.1.3 Learning Algorithm for Behavior Coordinator

Utilization of the hierarchical behavior controller makes the controller designing process easier and more comprehensive. However in this approach there is a critical issue how to adjust the behavior coordinator in case that the objective behavior or robot parameters are changed. We use Newton-Raphson method to adjust the behavior coordinator for these problems. This method measures the effects of the local behavior controllers to the global behavior, and changes the coefficients for them by trials.

The global behavior could be adjustable to some extent only by adjusting the activation coefficients. The relation between change of the activation coefficient and the resultant behavior is strongly nonlinear. However, we assume that the relations could be expressed as multiplication of the degree of contributions and activation

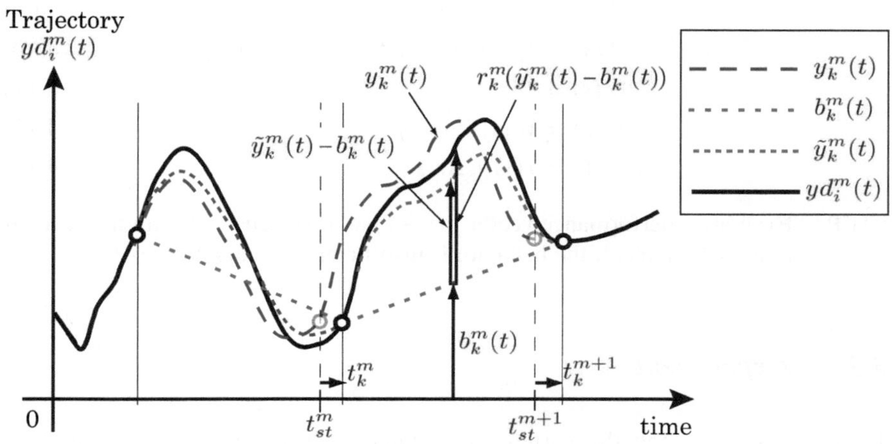

Fig. 4.2 Adaptation of desired trajectory

coefficients only in limited neighborhood of the current state. Therefore a performance vector $p(s)$ at step s which has elements as indices of evaluation for the total behavior is written by

$$p(s) = W(s) \cdot r(s) \tag{4.5}$$

where $W(s)$ is a gradient matrix and $r(s)$ is an activation vector.

The error vector $e(s)$ from the desired performance p^* and a current performance vector $p(s)$ are written by

$$e(s) = p^* - p(s) \tag{4.6}$$
$$= W(s) \cdot r^*(s) - W(s) \cdot r(s) \tag{4.7}$$
$$= W(s) \cdot (r^*(s) - r(s)). \tag{4.8}$$

This calculated activation coefficient r^* is not the desired ones, because linearized equation is adopted for the nonlinear system. Therefore the target activation vector are searched iteratively using by Newton-Raphson method. The procedure is as follows:

STEP1: At first, evaluate total behavior performance $p(s)$ by means of tuning with the activation vector $r(s)$.

STEP2: Explore the performance $p'(s), p''(s)$ and $p^{n-1}(s)$around neighborhoods area, $r'(s)$, $r''(s)$ and $r^{n-1}(s)$. These $r(s)$ are linear independent.

STEP3: Update gradient matrix $W(s)$ using Eq. (4.11), and calculate new activation coefficients using Eq. (4.12) and Eq. (4.13).

$$R(s) = (r(s), r'(s), r''(s), \ldots, r^{n-1}(s)) \tag{4.9}$$
$$P(s) = (p(s), p'(s), p''(s), \ldots, p^{n-1}(s)) \tag{4.10}$$
$$W(s) = P(s) \cdot R(s)^{-1} \tag{4.11}$$
$$\Delta r(s) = W(s)^{-1} \cdot e(s) \tag{4.12}$$
$$r(s+1) = r(s) + \Delta r(s) \tag{4.13}$$

STEP4: Evaluate a performance vector $p(s+1)$ at next step $s+1$ with new activation coefficients. If the behavior is insufficient, go back to step 2.

4.2.2 *Experiment*

In this section, we apply the learning algorithm explained in the previous section to the control problem of a three-dimensional brachiation motion using the new Gorilla Robot.

Generally, apes does "slow brachiation": moving by almost same speed of human walk. Moreover, when excited or frightened, apes can plunge through the forest canopy at astonishing speeds, sometimes covering 30 feet or more in a single jump without a break in "stride" (fast brachiation, ricocheting)[56]. In this section, we adapt the proposed learning algorithm to two types of show brachiation: overhand brachiation and side-hand brachiation.

4.2.2.1 Motion Measurement Using Real-Time Tracking System

A vision sensor is very useful to measure a dynamical motion without constant constrains because the constrained points are switched in accordance with the body posture. In the brachiation motion, it is almost impossible to measure the body position, for example the tip of the free arm, the center of gravity of the robot and so on, because the slip angle at the catching grip is not directly measurable using a potentiometer or rotary encoder. We therefore use the real-time tracking system, "Quick MAG System IV", which measures the three-dimensional locations of the eight points at 60Hz sampling frequency, using two CCD cameras and colored markers.

The seven measuring positions shown in Fig. 4.4 are chosen to calculate the center of gravity of the robot approximately based on the assumptions as follows:

1. Elbow of the catching arm keeps straight.
2. Both legs are controlled synchronously.
3. Two joints on the shoulder are adjoining and attached on the almost same position.

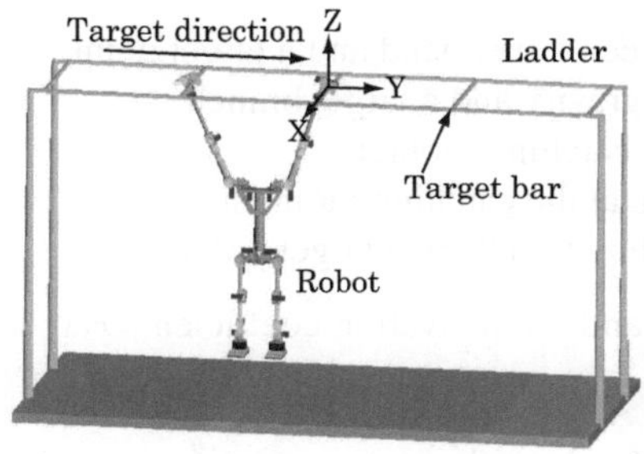

Fig. 4.3 Coordinate system

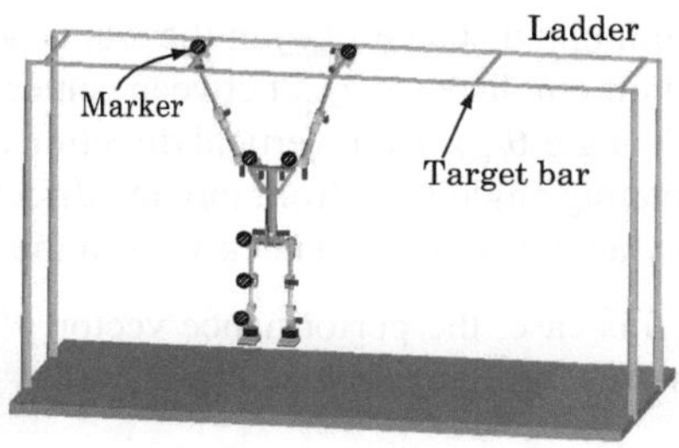

Fig. 4.4 Measuring points

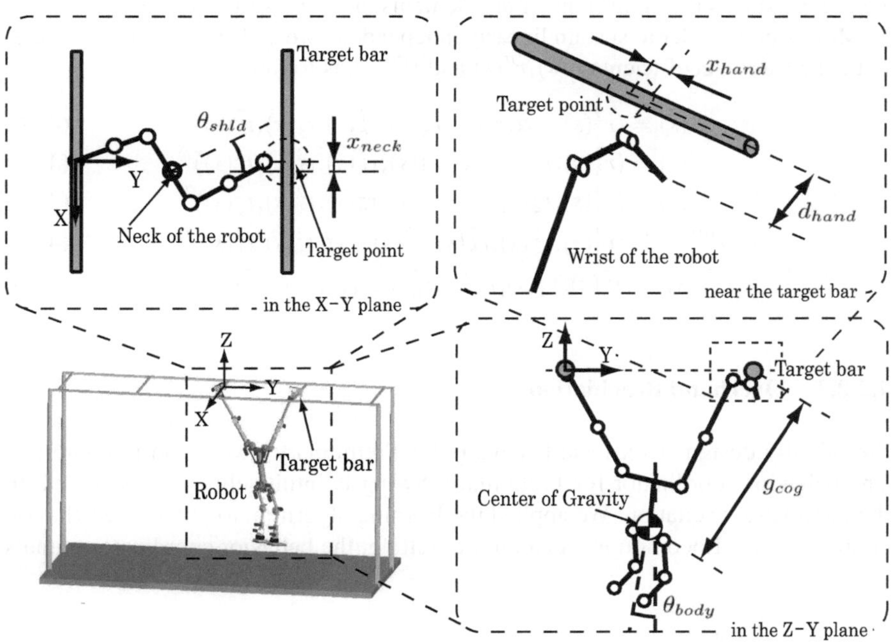

Fig. 4.5 Elements of the performance vector $p(s)$

4.2.2.2 Experiment Setting

We applied the adaptation algorithm to adjust Locomotion, which indicates six activation coefficients $(r_3^2, r_4^2, r_5^2, r_6^2, t_3^2, t_4^2)$ to the corresponding four local behavior controllers: Leg Stretch, Body Rotation 2, Body Lift and Arm Reaching. In order to apply the adaptation algorithm to these six activation coefficient, we chose six performance indices to represent performance of the total behavior as follows:

- Minimum distance d_{hand} in the y-z plane
- Minimum distance x_{hand} in the x axis between a free hand and a target point
- Minimum distance g_{cog} between center of gravity and a target branch
- Tilt angle θ_{body} from vertical direction at a catching moment
- Rotating angle θ_{shld} from moving direction at the catching moment
- Deviated position of a neck x_{neck} in the x direction from a target point

In this case, the performance vector $p(s)$ and the activation coefficients $r(s)$ at step s are therefore expressed by

$$p(s) = (d_{hand}(s), x_{hand}(s), g_{cog}(s), \theta_{body}(s), \theta_{shld}(s), x_{neck}(s))^T \tag{4.14}$$
$$r(s) = (r_3^2(s), r_4^2(s), r_5^2(s), r_6^2(s), t_3^2(s), t_4^2(s))^T. \tag{4.15}$$

Figure 4.5 shows the definitions of the elements of the performance vector.

Moreover, in order to sustain linearly independent, we add small perturbations α to the activation coefficients $r'(s), r''(s)$ and $r^{(5)}(s)$ as follows:

$$r'(s) = (r_3^2(s)+\alpha, r_4^2(s), r_5^2(s), r_6^2(s), t_3^2(s), t_4^2(s))^T \tag{4.16}$$
$$r''(s) = (r_3^2(s), r_4^2(s)+\alpha, r_5^2(s), r_6^2(s), t_3^2(s), t_4^2(s))^T \tag{4.17}$$
$$r'''(s) = (r_3^2(s), r_4^2(s), r_5^2(s)+\alpha, r_6^2(s), t_3^2(s), t_4^2(s))^T \tag{4.18}$$
$$r^{(4)}(s) = (r_3^2(s), r_4^2(s), r_5^2(s), r_6^2(s)+\alpha, t_3^2(s), t_4^2(s))^T \tag{4.19}$$
$$r^{(5)}(s) = (r_3^2(s), r_4^2(s), r_5^2(s), r_6^2(s), t_3^2(s)+\alpha, t_4^2(s))^T. \tag{4.20}$$

4.2.2.3 Overhand Brachiation

We initially designed six local behaviors for Preliminary Swing and Locomotion. The behavior coordinator for Preliminary Swing is intuitively designed based on the parametric excitation. We applied the learning algorithm to obtain the behavior coordinator for Locomotion. As an initial setting, the behavior coordinator outputs

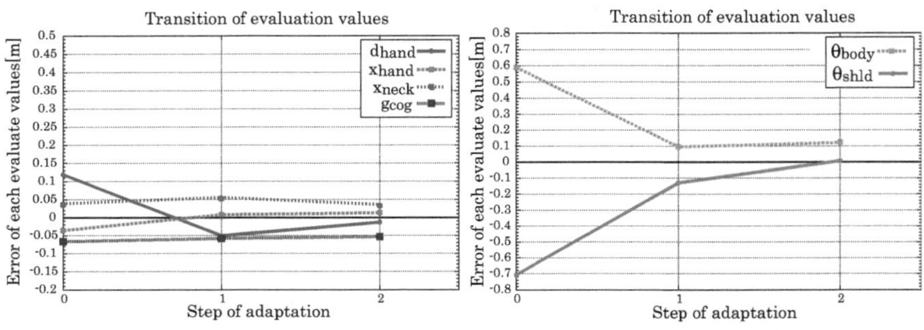

Fig. 4.6 Transitions of evaluation values (Overhand brachiation)

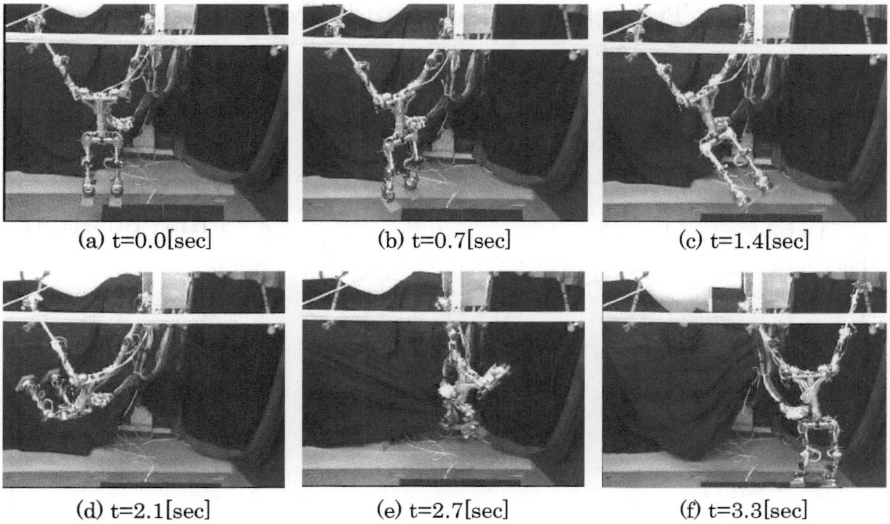

Fig. 4.7 Stroboscopic photography of overhand brachiation

Table 4.1 Transitions of adjustment values of overhand brachiation

Trial No.	d_{hand}	x_{hand}	x_{neck}	g_{cog}	θ_{shld}	θ_{body}
0	0.1184	-0.036	0.035	-0.067	-0.708	0.589
1	-0.051	0.008	0.052	-0.058	-0.129	0.095
2	-0.013	0.013	0.032	-0.053	0.008	0.122

Table 4.2 Transitions of evaluation values of overhand brachiation

Trial No.	r_3^2	r_4^2	r_5^2	r_6^2	t_3^2	t_4^2
0	1.0	1.0	1.0	1.0	0.0	0.0
1	0.733	1.047	1.031	1.313	-0.072	0.052
2	0.894	1.213	1.273	1.048	-0.121	0.024

one as an activation level to each local behavior. In this experiment, we set the perturbation $\alpha = 0.05$. In each step the robot makes six trials with different activation coefficients. Through the learning process, each performance error goes to zero as shown in Fig. 4.6, which shows transitions of each element in error vector "e". The overhand brachiation is achieved at the second step. In each step the robot makes six trials with different activation coefficients for local search. Figure 4.7 shows the obtained overhand brachiation and Fig. 4.8 shows a trajectory of a free hand. The locomotion time from one branch to another is $1.20[sec]$.

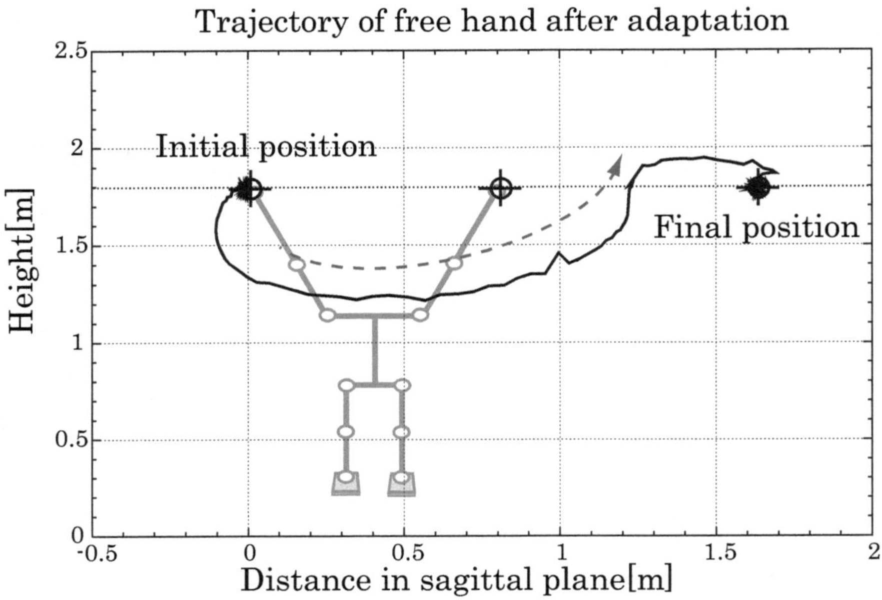

Fig. 4.8 Free hand trajectories of overhand brachiation

4.2.2.4 Side-Hand Brachiation

At first, in order to achieve the side-hand brachiation, we redesigned two local behaviors: Body Rotation 2 and Arm Reaching. The we applied the learning algorithm to obtain the behavior coordinator for Locomotion as well as overhand brachiation.

Through four trials, the minimum distance x_{hand} and θ_{body} is especially decreasing. At the forth step, robot caught the target branch. The target motion is generated in four steps by the same learning algorithm. Transitions of each element in error vector e and a trajectory of a free hand are shown in Figures 4.9 and 4.11 respectively. Figure 4.10 shows the obtained side-hand brachiation. It takes a little longer time $1.53[sec]$ for this locomotion.

Table 4.3 Transitions of adjustment values of side-hand brachiation

Trial No.	d_{hand}	x_{hand}	x_{neck}	g_{cog}	θ_{shld}	θ_{body}
0	0.026	0.208	0.146	-0.042	-0.428	-0.367
1	0.073	0.421	0.003	-0.149	-0.723	-0.488
2	0.121	0.361	0.043	-0.121	-0.274	-0.521
3	0.034	-0.066	0.128	-0.008	-0.548	-0.332
4	0.035	0.121	0.156	-0.044	-0547	-0.267

Table 4.4 Transitions of evaluation values of side-hand brachiation

Trial No.	r_3^2	r_4^2	r_5^2	r_6^2	t_3^2	t_4^2
0	1.0	1.0	1.0	1.0	0.0	0.0
1	1.167	1.467	1.116	0.613	-0.035	0.204
2	0.081	1.612	0.339	1.163	0.004	0.050
3	0.904	1.265	0.259	0.494	0.003	0.118
4	0.748	0.920	1.256	0.873	-0.050	-0.036

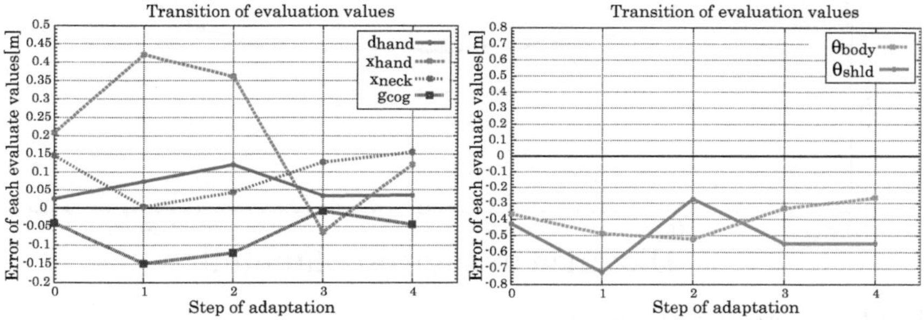

Fig. 4.9 Transitions of evaluation values (Side-hand brachiation)

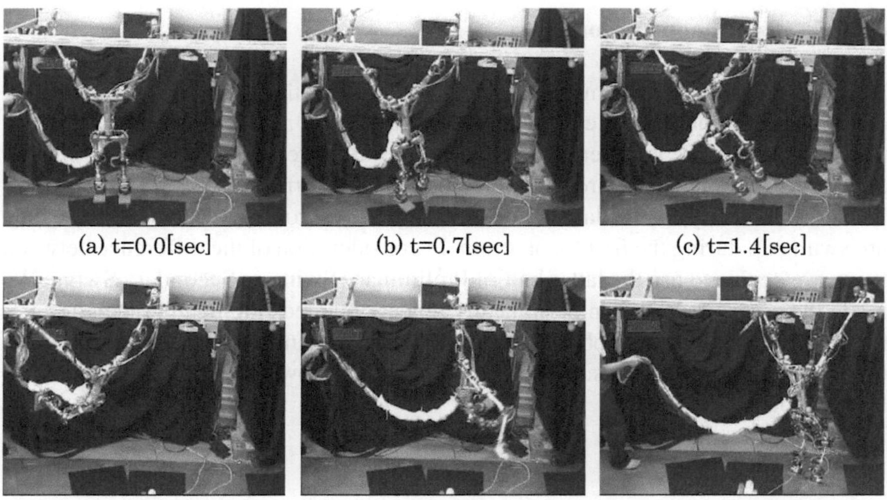

(a) t=0.0[sec] (b) t=0.7[sec] (c) t=1.4[sec]

(d) t=2.1[sec] (e) t=2.9[sec] (f) t=3.6[sec]

Fig. 4.10 Stroboscopic photography of side-hand brachiation

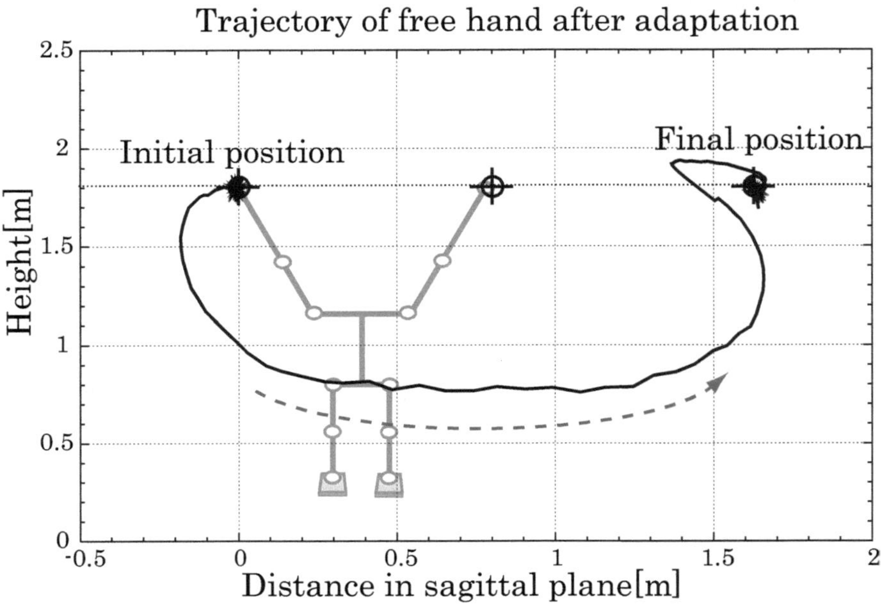

Fig. 4.11 Free hand trajectories of side-hand brachiation

4.2.2.5 Continuous Locomotion

Now we consider achievement of continuous locomotion without suspending. A continuous locomotion generally consists of a periodic repetition of one cycle motion. Although there is difference between the start state of Preliminary Swing and Secondary Swing with regard to the brachiation. The former state is that the legs are just under the body, but the latter state is the legs are in front of the body because they are swung up during the first locomotion. In consideration of the difference between them, we decide to use the latter half of Preliminary Swing as Secondary Swing. We made the controller that can periodically repeat obtained side-hand brachiation periodically. Then Gorilla Robot II could achieve a dynamic continuous locomotion that moved twice without suspending. Figure 4.12 shows the obtained continuous locomotion.

4.2.3 Summary of This Section

In this section, we showed the algorithm for generation of dynamically dexterous behavior, which is including the architecture for behavior coordination and adjustment. We improved the learning algorithm that timing of each behavior as well as

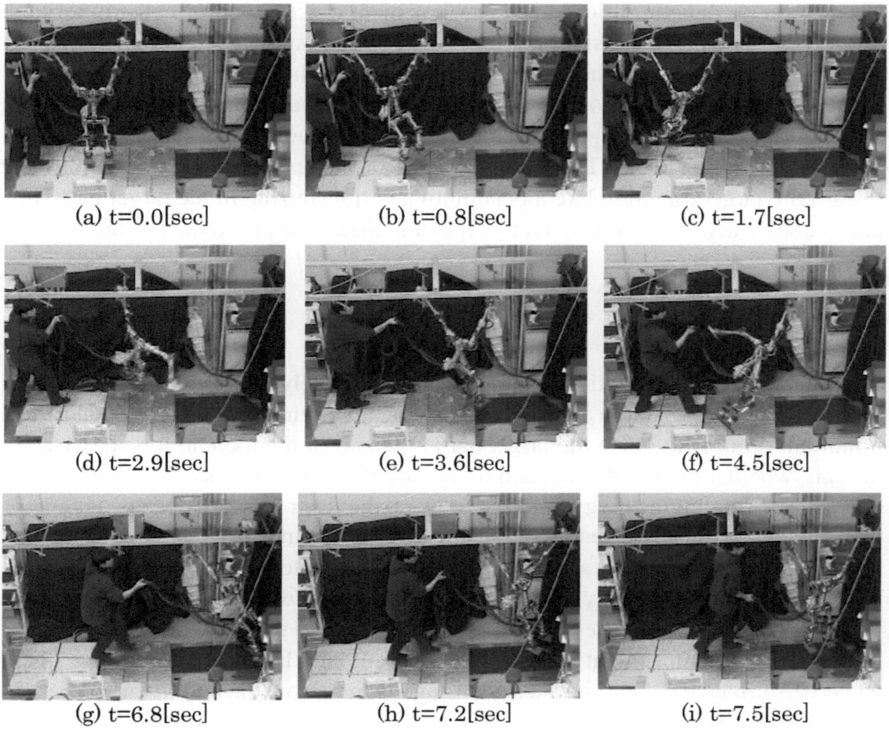

Fig. 4.12 Stroboscopic photography of continuous locomotion

those activation levels could be properly adjusted. We applied the learning algorithm to achieve two types of brachiation: overhand brachiation and side-hand brachiation by Gorilla Robot II. Finally, we made a controller that can periodically repeat the obtained side-hand brachiation and Gorilla Robot II was able to achieve a continuous locomotion that moved twice without suspending.

4.3 Continuous Brachiation Using the Gorilla Robot

4.3.1 Smooth, Continuous Brachiation

Some apes, such as chimpanzees, orangutans, and bonobos, generally do "slow brachiation": moving at almost the same speed of human walk. Moreover, when excited or frightened, apes can plunge through the forest canopy at astonishing speeds, sometimes covering 30 feet or more in a single jump without a break in "stride" (fast brachiation, ricocheting)[56]. Apes can brachiate quickly and effectively through the forest canopy. We consider the essential control strategies to achieve such dynamic and effective motions by the multilink robot. The target brachiation is basically divided into two actions: a swing action and a locomotion action. The swing action injects enough energy to enable the robot to reach the target bar in the successive locomotion phase. The locomotion action is the one responsible for moving toward the target direction like a pendulum and transferring to the target bar. Then continuous brachiation is realized by repeating these two phases alternately.

4.3.1.1 Excess Energy

In order to realize the continuous brachiation effectively and smoothly, it is necessary to start the next action as soon as the robot grasps the front target bar. The most important thing is to grasp the target bar at the end of the locomotion action without fail; whether the robot can reach to the target bar or not depends on mechanical energy. The minimum energy required to grasp the front target bar can be geometrically estimated due to the length of the arm and the distance between the suspending bar and the target bar. Thus an effective brachiation strategy is that which injects the minimum amount of energy into the robot during the swing phase and approaches to the target bar in the subsequent locomotion phase.

However, any movement towards a target generally has an error. If some unexpected disturbance reduces its energy through locomotion, the loss of energy results in failing to grasp the target bar.

The remarkable point is that it is difficult to recover from the unexpected error completely during the locomotion phase because the angle around the contact bar, i.e. the pivot point of the pendulum, is unactuated. Therefore it is necessary for the robot to inject excess mechanical energy during the swing phase in order to improve its stability. Actually, in biology, Usherwood *et al.* reported in [285] that the experimental results obtained by means of a brachiating gibbon implied that mechanical energy of the swing phase is considerably greater than the minimum required to allow its hand to contact with the next target bar.

4.3.1.2 Gripping Problem

A gripping problem occurs due to injecting excess mechanical energy in order to improve stability at the end of the locomotion action. A collision, which occurs between the hand and the target bar at the grasping moment, affects the pendulum motion of the robot. The collision causes uncontrollable vibrations that disturb the pendulum motion of the robot. Though one of the simplest solutions of this vibration problem is just to fix all joint angles until the vibration has disappeared, the motion is not effective or smooth motion. Then, utilizing redundant degrees of freedom of the mechanical system is proposed to solve the gripping problem as follows. By bending the elbow joint of the swinging arm, the robot can safely grasp the front target bar at the end of the locomotion phase. In this case, there is no impact force because the angular velocity of the pendulum is zero. It is possible to start the subsequent swing action as soon as the robot grasps the target bar. During the swing phase, the robot can use the excess mechanical energy at the end of the locomotion phase. The elbow-bending action makes it possible for the robot to decrease the collisional energy loss and the waste of time during the subsequent swing phase. The elbow-bending action actually causes secondary effects: the robot can get larger amplitude from the pendulum motion due to transposing the position of the pivot point from the shoulder joint to the grasped bar.

Figure 4.13 shows the overview of smooth continuous brachiation described above. The swing-up motion from *Sa* to *Sb*, the preliminary swing, is performed only once in order to inject initial potential energy. After the preliminary swing, keeping the mechanical energy constant, the robot just repeats the swing-back action from the state *Sb* to *Sd* after the locomotion action. The control strategies of the swing action and the locomotion action are described in detail in the following section.

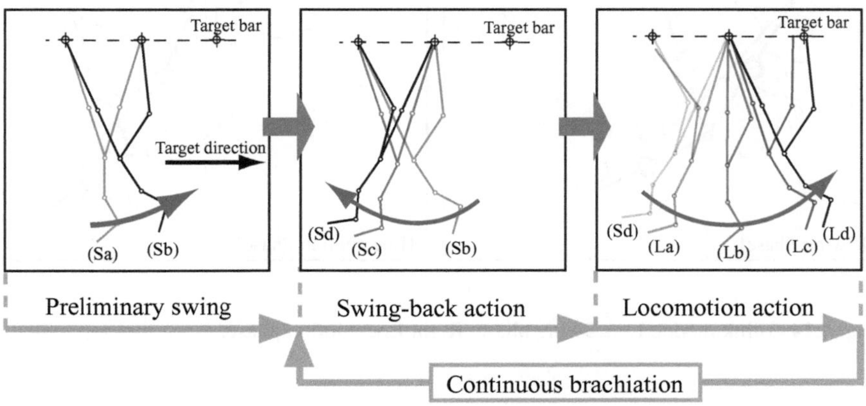

Fig. 4.13 Smooth, continuous brachiation

4.3.2 Controller Design

This section introduces a brachiating model which consists of a torso, shoulders and two arms. The length of each arm is variable due to its elbow-bending action. Further details about the controller of the swing action and the locomotion action are described in the following subsection respectively.

4.3.2.1 3-Link Model

We consider the 3-link model shown in Fig. 4.14. This model consists of a torso, shoulders and two arms. Link 1 and link 2 are a stretched arm and a bended arm respectively, and link 3 is the torso and legs. O and Q denote the hand position of the stretched arm and that of the bending arm respectively. P is the position of the shoulder, G is the position of the center of gravity. m_i, l_i and θ_i are the mass, the length and the angle of link i ($i = 1,2,3$), respectively. d_{bar} is the distance between bars. l_0 is the length of a stretched arm. θ_p is the pendulum angle. The angle around the contact bar can not be actuated, i.e. $\tau_1 = 0$. τ_2 and τ_3 are the torques applied between the support arm and the torso, and the support arm and the swinging arm, respectively. The length of the swinging arm l_2 can be changed by bending its elbow. The subscript "d" refers to the desired value of each parameter.

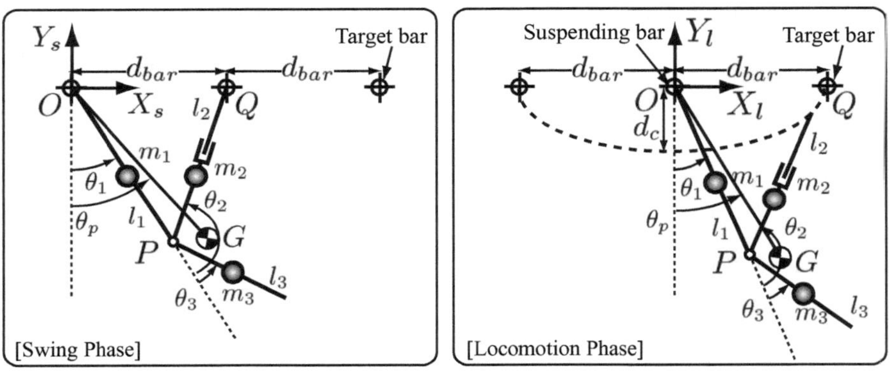

Fig. 4.14 3-link model. Left:Swing phase. Right:Locomotion phase.

4.3.2.2 Swing Phase

In this subsection, the controller of the swing action is described. The swing action consists of a preliminary swing and a swing-back action. The objective of the swing action is to inject energy into the robot before the locomotion action. This action is described as follows.

At the initial state denoted by Sa, the arms are symmetric with length l_0, which is equal to the length of the stretched arm, and the torso is hanging down vertically. The robot swings up the torso (counterclockwise) while bending its elbow. The rear contact point denoted by O_- can be considered as the pivot point of the pendulum during the swing-up motion. Sb denotes the state when the pendulum angular velocity $\dot{\theta}_-$ is zero. After that, the robot swings back the torso (clockwise) with elbow-extending action synchronized with the pendulum motion. The arms are symmetric with length l_0 at the state denoted by Sc, during this swing-back action.

At this moment, the pivot point of the pendulum switches from the rear contact point O_- to the front contact point O_+. Then, the pendulum angular velocity $\dot{\theta}_+$ is zero at Sd. Immediately after Sd, the robot releases the rear bar and starts the locomotion action. The overview of the swing action is shown in Fig. 4.15. Note that the robot just repeats the swing-back action from the state Sb to Sd after the locomotion action.

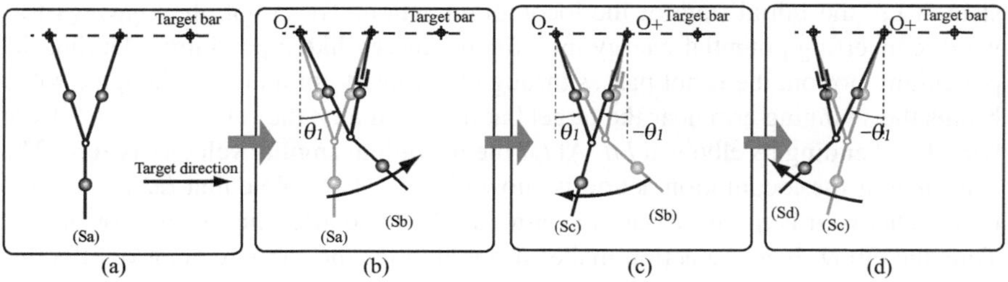

Fig. 4.15 Preliminary swing and swing-back action

The elbow-bending during the swing action is geometrically designed to realize a double pendulum motion as well as keep grasping the bar so that the model can be looked upon as a double pendulum consisted of link 1 and link 3. Figure 4.14(Left) shows the parameters of the swing action. The desired trajectories l_2^d and θ_2^d are given as functions of θ_1. θ_3^d is expressed as a sine wave function to inject energy into the robot. These parameters are given as follows:

$$x_p = l_0 \sin\theta_1,\ y_p = -l_0\cos\theta_1 \tag{4.21}$$

$$x_q = d_{bar},\ y_q = 0 \tag{4.22}$$

$$\begin{aligned} l_2^d &= \sqrt{(x_q - x_p)^2 + (y_q - y_p)^2} \\ &= \sqrt{(d_{bar} - l_0\sin\theta_1)^2 + (l_0\cos\theta_1)^2} \end{aligned} \tag{4.23}$$

$$\begin{aligned} \theta_2^d &= \cos^{-1}\left(\frac{x_q - x_p}{l_2}\right) - \theta_1 + \frac{\pi}{2} \\ &= \cos^{-1}\left(\frac{d_{bar} - l_0\sin\theta_1}{l_2}\right) - \theta_1 + \frac{\pi}{2} \end{aligned} \tag{4.24}$$

$$\theta_3^d = A_3\sin(\omega_3 t_3) \tag{4.25}$$

where A_3 is the amplitude, $\omega_3 = 2\pi/T_{swing}$ is the frequency, t_3 is zero at the grasping moment Sb.

A_3 and ω_3 are experimentally determined in order to pump enough energy to enable the robot to reach the target bar in the successive locomotion phase.

4.3.2.3 Locomotion Phase

This subsection describes the controller of the locomotion action. In locomotion phase, the objective is that the robot moves toward the target direction after releasing the rear bar, and grasp the front bar at the end of the action. This action is described as follows.

After releasing the rear bar at *Sd*, the robot changes posture to be in a suitable state at *La*, the initial state of the locomotion action. The robot, then, swings forward, converting potential energy into kinetic energy like a pendulum. During the pendulum motion, the robot passes through the lowest point at *Lb*. Then, the robot brings the swinging arm near the target bar at *Lc*. Finally, the robot grasps the front bar while bending its elbow at *Ld*. At *Ld* the pendulum angular velocity is zero. The overview of the locomotion action is shown in Fig. 4.16. Note that *Ld* is identical to *Sb*. That is, it is not necessary to swing up the torso after the locomotion action. Thus the elbow-bending action makes it possible for the robot to achieve effective and smooth brachiation.

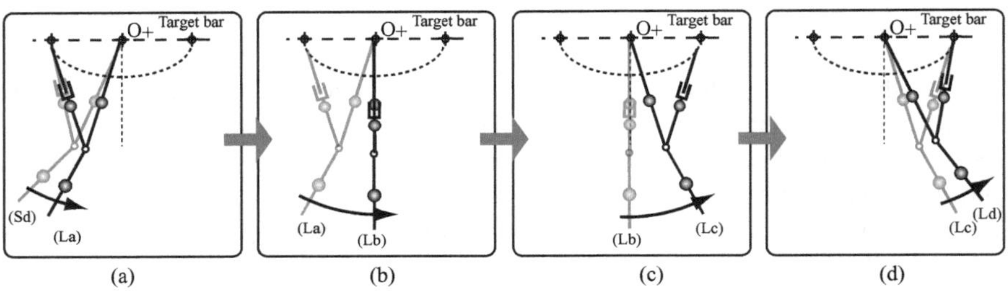

Fig. 4.16 Locomotion action

The elbow-bending action in the locomotion phase is designed to satisfy a desired hand trajectory of the swinging arm. Figure 4.14(Right) shows the parameters of the locomotion action. For simplicity, it is assumed that the trajectory can be described by an ellipse equation as follows:

$$\left(\frac{x_q}{d_{bar}}\right)^2 + \left(\frac{y_q}{d_c}\right)^2 = 1 \tag{4.26}$$

where d_c is the distance between the contact bar denoted by O_+ and the hand of the swinging arm at L_c.

Note that the locomotion action is designed so as to have symmetry along the y-axis. According to the symmetric property, the action is expected to behave like a stable pendulum. θ_2^d, l_2^d and θ_3^d are given as follows:

$$x_p = l_0 \sin\theta_1,\ y_p = -l_0 \cos\theta_1 \tag{4.27}$$

$$x_q = d_{bar}\frac{\theta_1}{|\theta_{1a}|},\ y_q = -d_c\sqrt{1-\left(\frac{\theta_1}{\theta_{1a}}\right)^2} \tag{4.28}$$

$$\begin{aligned} l_2^d &= \sqrt{(x_q-x_p)^2+(y_q-y_p)^2} \\ &= \sqrt{\left(d_{bar}\frac{\theta_1}{|\theta_{1a}|}-l_0\sin\theta_1\right)^2+\left(d_c\sqrt{1-\left(\frac{\theta_1}{\theta_{1a}}\right)^2}+l_0\cos\theta_1\right)^2} \end{aligned} \tag{4.29}$$

$$\begin{aligned} \theta_2^d &= \cos^{-1}\left(\frac{x_q-x_p}{l_2}\right)-\theta_1+\frac{\pi}{2} \\ &= \cos^{-1}\left(\frac{d_{bar}\frac{\theta_1}{|\theta_{1a}|}-l_0\sin\theta_1}{l_2}\right)-\theta_1+\frac{\pi}{2} \end{aligned} \tag{4.30}$$

$$\theta_3^d = \frac{\theta_{3a}}{\theta_{1a}}\theta_1 \tag{4.31}$$

where θ_{1a} and θ_{3a} are the angles at L_a respectively.

4.3.3 Experiment

In this section, we experiment two cases of continuous brachiation with a multilink robot to verify the effect of our control strategies described previously. In case 1, the robot brachiates continuously with straightening the elbow joint of the both arms. That is, the robot keeps the length of the swinging arm constant, $l_2^d = l_0$, during both the swing and locomotion phases. In this case, uncontrollable vibrations occur at the grasping moment due to the gripping problem. Therefore it is necessary to fix all joint angles in order to decrease the vibration caused by the collision and swing up the torso again in the swing phase. On the other hand, in case 2, the robot brachiates continuously due to the elbow-bending strategy described above.

4.3.3.1 Experimental Setup

In this section, the experiment is conducted using the Gorilla Robot III introduced in Chapter 3. The distance between bars d_{bar} remains constant at 0.4 [m] and all bars are set at the same height at 2.7 [m]. Figure 4.17 shows the sensors attached to each hand and the experimental setting.

The mass of arms m_1 or m_2 is 7.0 [kg], that of the torso m_3 is 15.0 [kg], the length of the stretched arm l_0 is 0.64 [m], and that of the torso l_3 is 0.85 [m]. Considering the limitations of the robot, we experientially set the parameters as follows: A_3 is 30.0 deg., ω_3 is $5.24(=2\pi/T_{swing}=2\pi/1.20)$ 1/s, and d_c is 0.07 [m]

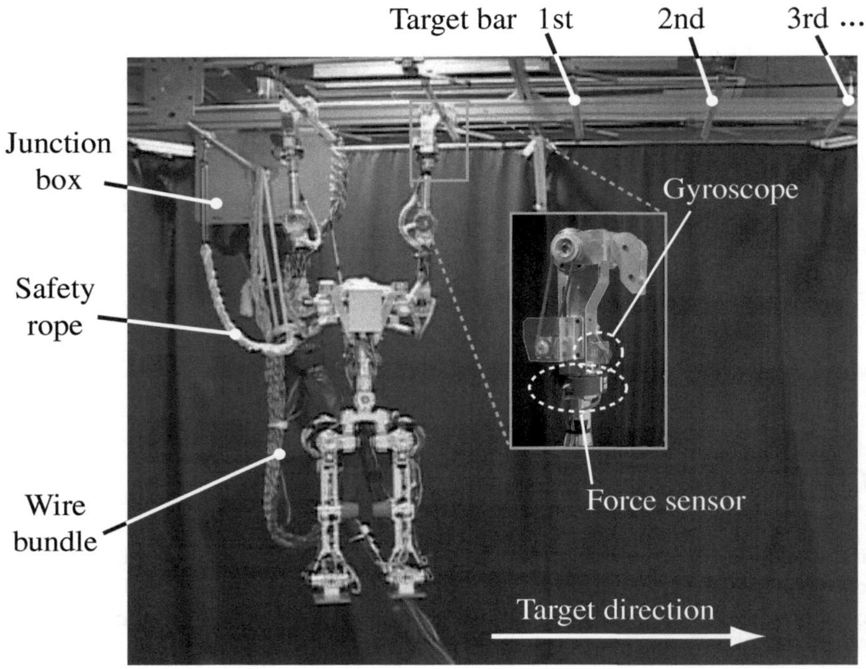

Fig. 4.17 Experimental setting

4.3.3.2 Results

Reaction forces and the pendulum angle, θ_p, in case 1 are respectively shown in Fig. 4.18 and 4.19. The pendulum angles at S_d and L_d are shown in Table 4.5. In this case, the graph of the pendulum angle indicates vibration in each swing phase as shown in Fig. 4.19. This motion was not smooth brachiation. Because the pendulum angle was decreasing due to the vibration in the swing phase, the robot failed to grasp the 3rd target bar.

In case 2, effective and smooth brachiation was achieved. Reaction forces and the pendulum angle are shown in Fig. 4.20 and 4.21 respectively. The pendulum angles at S_d and L_d are shown in Table 4.6. In this case, the graph of the pendulum angle shown in Fig. 4.21 is similar to the graph of a sine function. The pendulum angles at S_d and L_d are almost the same. This indicates the following two points:

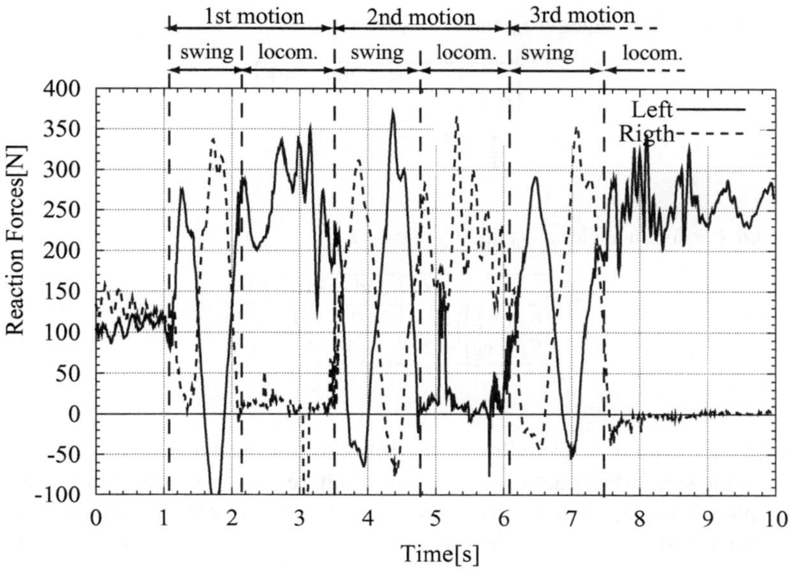

Fig. 4.18 Reaction forces in case 1

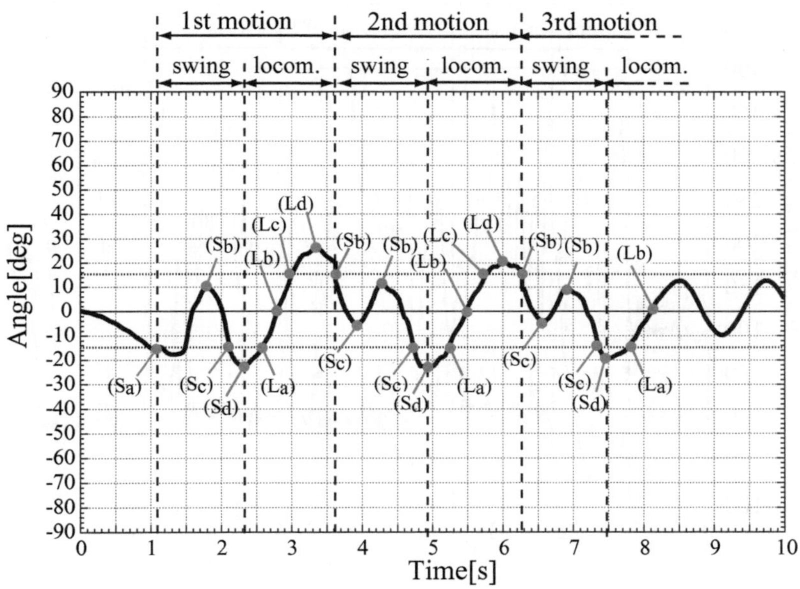

Fig. 4.19 Pendulum angle in case 1

Table 4.5 Pendulum angles at S_d and L_d in case 1

	1st	2nd	3rd
L_d [deg]	26.0	20.0	–
S_d [deg]	-24.0	-22.0	-20.0

Table 4.6 Pendulum angles at S_d and L_d in case 2

	1st	2nd	3rd	4th
L_d [deg]	22.0	21.0	20.0	20.0
S_d [deg]	-32.5	-34.0	-33.5	-34.0

1)the elbow-bending action allowed the robot to use the excess mechanical energy and start the swing-back action after L_d and 2)there was no waste of time during the swing phase: S_b - S_d. The controller described previously succeeded in improving its stability and achieved stable periodic motion. The robot achieved continuous brachiation effectively and smoothly as shown in Fig. 4.22.

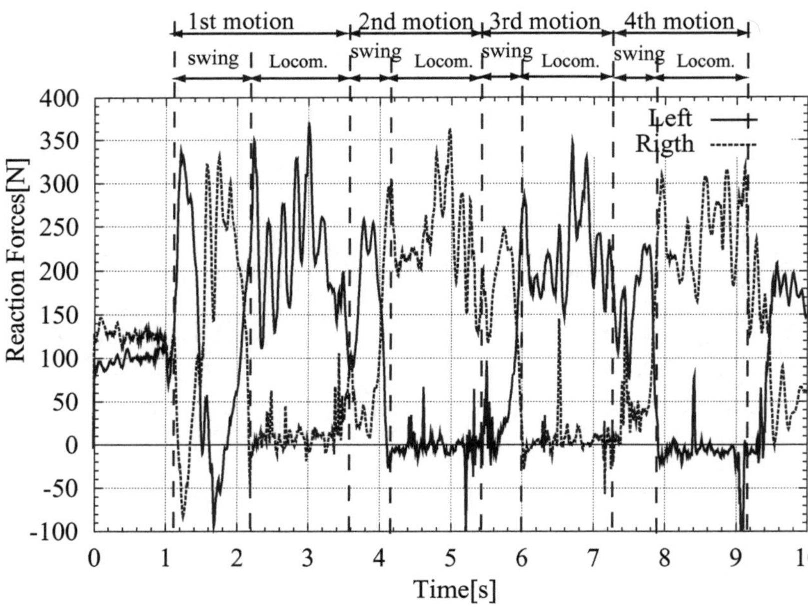

Fig. 4.20 Reaction forces in case 2

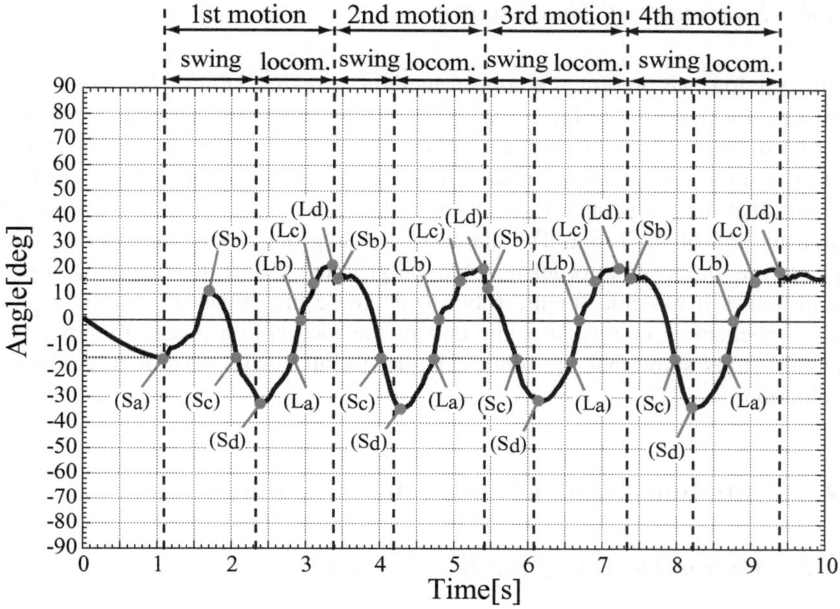

Fig. 4.21 Pendulum angle in case 2

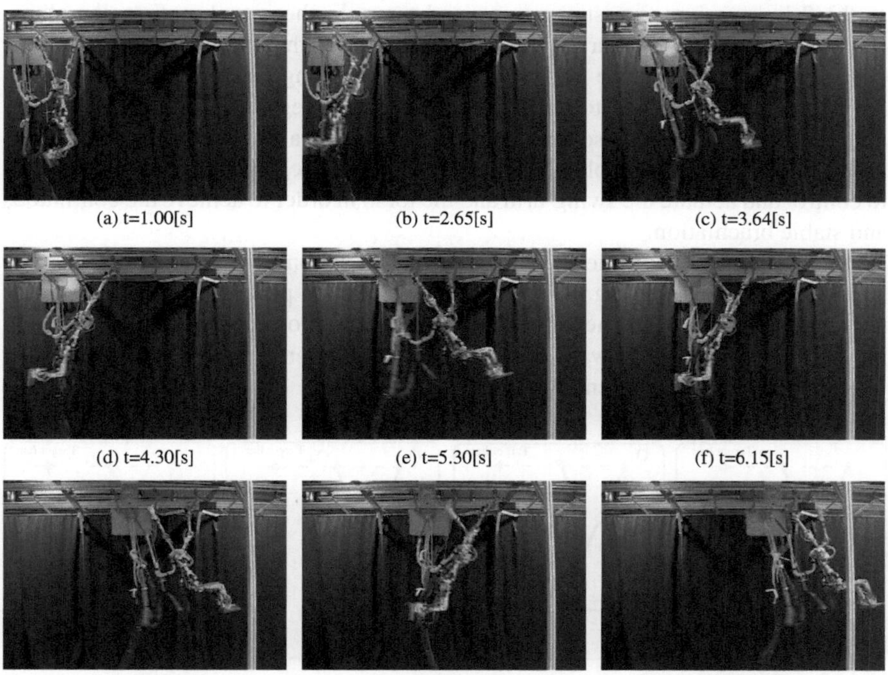

(a) t=1.00[s] (b) t=2.65[s] (c) t=3.64[s]

(d) t=4.30[s] (e) t=5.30[s] (f) t=6.15[s]

(g) t=7.48[s] (h) t=8.32[s] (i) t=9.35[s]

Fig. 4.22 Snapshots of continuous brachiation in case 2

4.3.4 *Summary of This Section*

In this section, we presented a control method to realize smooth brachiation. It is necessary for the robot to inject excess mechanical energy during the swing phase in order to improve its stability since disturbances reduce its energy through locomotion and the loss of energy results in failing to grasp the target bar. A gripping problem that the collision, which occurs at the gripping moment, affects the pendulum motion of the robot. The action of bending the elbow joint of the swinging arm was proposed in order to solve the gripping problem. The elbow-bending action makes it possible for the robot to decrease the impact forces and use the excess mechanical energy after the end of the locomotion phase. Finally experimental results show that the robot can successfully achieve smooth, continuous brachiation.

4.4 Continuous Brachiation on the Irregular Ladder

4.4.1 *Motion Design of the Brachiation*

In this chapter, in order to simplify the controller design, the whole robot is modeled as the 3-link: swing-arm, support-arm and torso. In the actual motion, the elbow of swing-arm is bended during locomotion; however, this chapter assumes that its effect on dynamics can be neglected since the bending angle is relatively small. Support-arm is controlled to be kept stretched. Both legs are actuated to be fixed and not moved so that torso and legs can be seemed a link. Note that the torque around a bar cannot be applied. Thus, the main issue regarding brachiation is how to control and actuate the swing-arm and the torso in order to achieve the continuous and stable brachiation.

The brachiation controller is composed of two actions: swing action (Fig. 4.23) and locomotion action (Fig. 4.24). In the swing action phase, the robot injects the energy by swinging back the torso prior to the locomotion action. In the locomotion phase, the robot moves toward the target direction after releasing the rear bar and grasps the front bar at the end of the action.

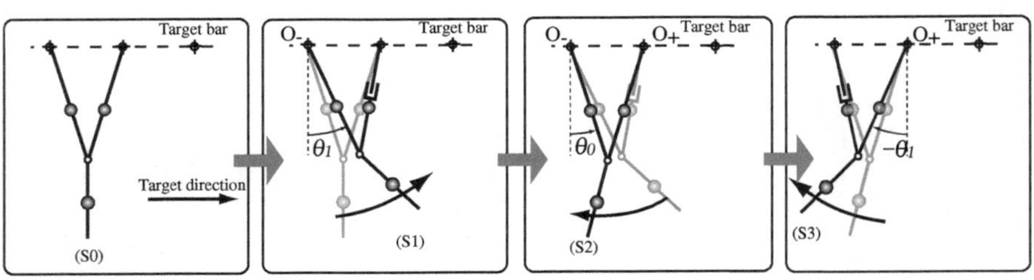

Fig. 4.23 Swing action

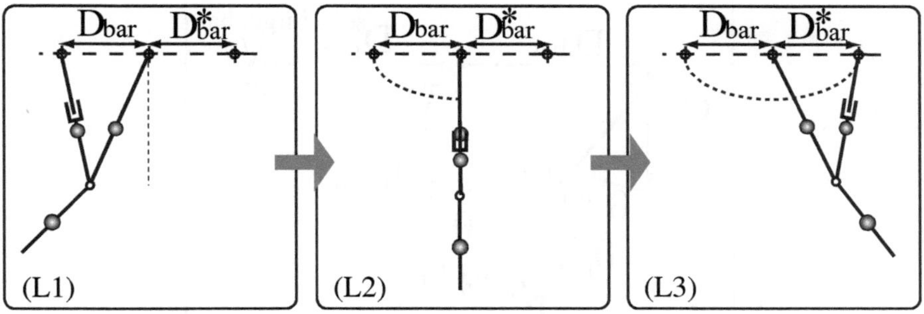

Fig. 4.24 Locomotion action

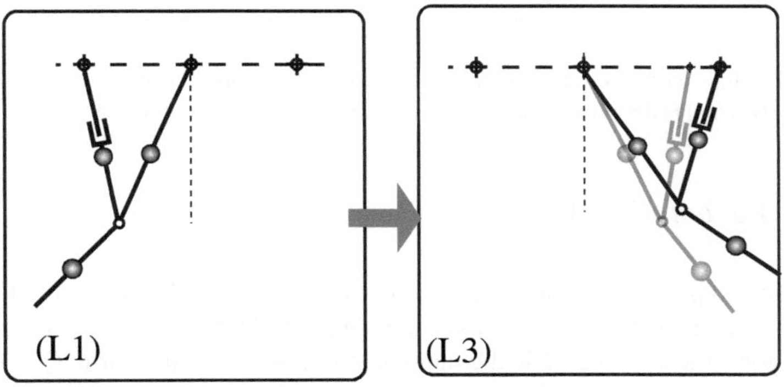

Fig. 4.25 Symmetry breaking

As for the uniform ladder problem, it is possible to design the locomotion action symmetrically along the graviational direction, that is, the first-half motion (from (L1) to (L2)) and the last-half motion (from (L2) to (L3)) are symmetrical. Under the condition of symmetric motion along the gravity force, the effects of the first-half motion and the last-half motion on the dynamics are antithetical, hence the mechanical energy in (L1) and (L3) are identical. We proposed previously the control method of the uniform ladder brachiation based on this symmetric property.

However, in the irregular ladder environment, symmetric motion cannot be designed as shown in Fig. 4.25. Thus, it is impossible to apply the same strategy as the uniform ladder brachiation. In order to achieve the irregular ladder brachiation, it is necessary to estimate the energy difference between (L1) and (L3) caused by symmetry breaking. In order to evaluate the energy difference, we apply PDAC (explained in Chapter 7) to the locomotion action. Swing motion is adjusted so as to satisfy the desired energy calculated from energy difference. The motion scheme is explained in detail below.

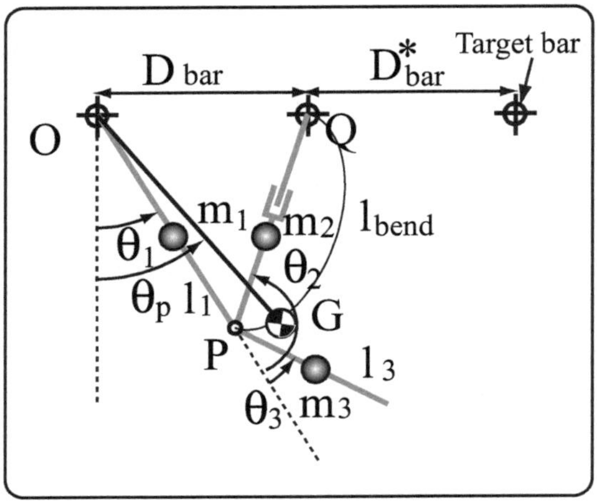

Fig. 4.26 3-link model. θ_1, θ_2 and θ_3 are the angles of support-arm, swing-arm and torso respectively. D_{bar} is the present bar-distance and D^*_{bar} is the target bar-distance.

4.4.2 Locomotion Action

As mentioned previously, this chapter employs 3-link model as shown in Fig. 4.26. Link 1 is stretched arm, link 2 is a swinging arm, and link 3 is a torso and legs. P is the position of a shoulder, O is the hand position of the support-arm, Q is the hand position of the swing-arm, and G is the position of the center of gravity. m_i, l_i, and θ_i are the mass, the length, and angle of the link i ($i = 1,2,3$). D_{bar} is the distnace between bars. θ_p is the pendulum angle. Note that the angle around the contact point cannot be actuated.

The first dynamic equation around the contact point, O, is derived as follows (Note that the angle around the contact point cannot be actuated i.e. $\tau_1 = 0$):

$$\frac{d}{dt}\left(M_1\dot{\theta}_1 + M_2\dot{\theta}_2 + M_3\dot{\theta}_3\right) - (h_1 + h_2 + h_3) = 0, \tag{4.32}$$

where

$$\begin{aligned} M_1 &= J_1 + J_2 + J_3 + m_1 a_1^2 + m_2 l_1^2 - 2m_2 l_1 a_2 \cos\theta_2 \\ &\quad + m_3 l_1^2 + m_3 a_3^2 + 2m_3 l_1 a_3 \cos\theta_3, && (4.33)\\ M_2 &= J_2 + m_2 a_2^2 - l_1 a_2 \cos\theta_2, && (4.34)\\ M_3 &= J_3 + m_3 a_3^2 + l_1 a_3 \cos\theta_3, && (4.35)\\ h_1 &= -(m_1 a_1 + m_2 l_1 + M_3 l_1) g \sin\theta_1, && (4.36)\\ h_2 &= -m_2 g a_2 \sin(\theta_2 - \theta_1), && (4.37)\\ h_3 &= -m_3 g a_3 \sin(\theta_1 + \theta_3). && (4.38) \end{aligned}$$

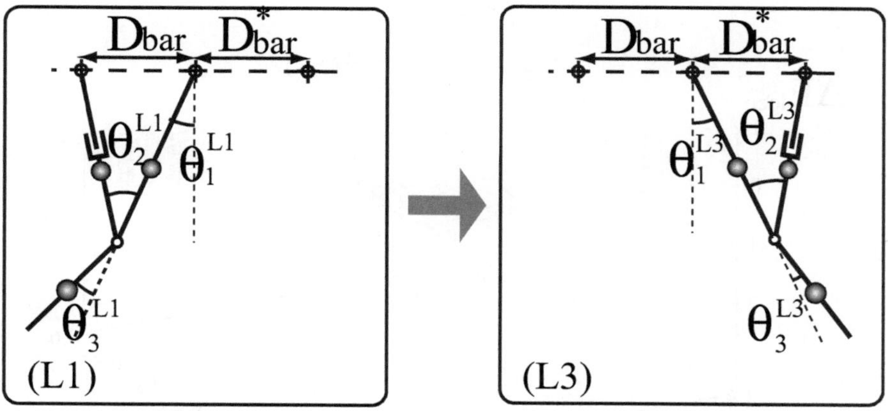

Fig. 4.27 The beginning and the ending of locomotion action. Superscripts denotes the state (L1) or (L3).

In accordance with the process of PDAC, the holonomic constraints between joints are designed. In order for the locomotion action and the swing action to switch smoothly, the constraint is necessary to satisfy the following conditions:

$$\theta_1 = \theta_1^{L1} \rightarrow \dot{\theta}_2 = 0, \dot{\theta}_3 = 0, \quad \theta_2 = \theta_2^{L1}, \theta_3 = \theta_3^{L1}. \tag{4.39}$$

$$\theta_1 = \theta_1^{L3} \rightarrow \dot{\theta}_1 = 0, \dot{\theta}_2 = 0, \dot{\theta}_3 = 0, \quad \theta_2 = \theta_2^{L3}, \theta_3 = \theta_3^{L3}. \tag{4.40}$$

In this chapter, we determine the constraint so as to satisfy Eqs. (4.39) and (4.40) as follows:

$$\theta_2 = A\left(\theta_1^2 - \theta_1^{L1}\right) + B = f(\theta_1), \tag{4.41}$$

$$\theta_3 = A'\left(\theta_1^2 - \theta_1^{L1}\right) + B' = g(\theta_1), \tag{4.42}$$

$$A = -\frac{2(\theta_2^{L1} - \theta_2^{L3})}{(\theta_1^{L1} - \theta_1^{L3})^2},$$

$$A' = -\frac{2(\theta_3^{L1} - \theta_3^{L3})}{(\theta_1^{L1} - \theta_1^{L3})^2}, \tag{4.43}$$

$$B = \frac{\left((\theta_1^{L3})^2 + (\theta_2^{L1})^2 - 2\theta_1^{L1}\theta_1^{L3}\theta_2^{L1} + \theta_1^{L1}\theta_2^{L1}\right)}{(\theta_1^{L1} - \theta_1^{L3})^2}, \tag{4.44}$$

$$B = \frac{\left((\theta_1^{L3})^2 + (\theta_3^{L1})^2 - 2\theta_1^{L1}\theta_1^{L3}\theta_3^{L1} + \theta_1^{L1}\theta_3^{L1}\right)}{(\theta_1^{L1} - \theta_1^{L3})^2}, \tag{4.45}$$

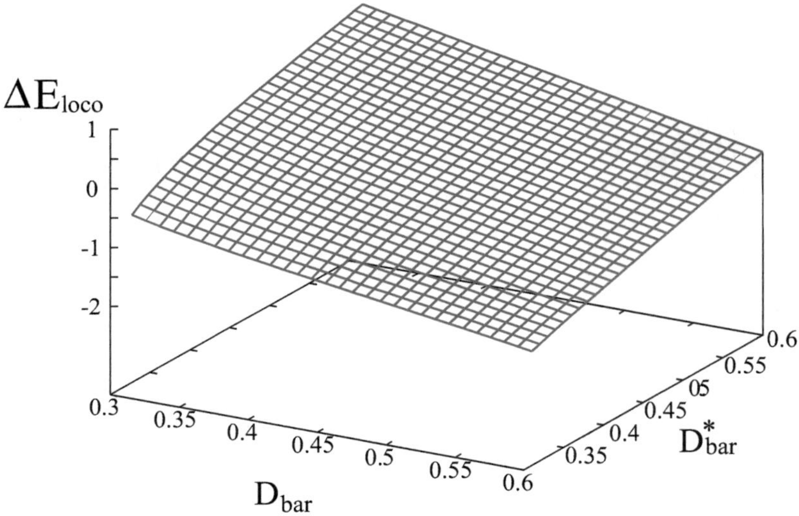

Fig. 4.28 Relationship among D_{bar}, D^*_{bar}, and ΔE_{loco}

$$\dot{\theta}_2 = \frac{df}{d\theta_1}\frac{d\theta_1}{dt} = f'(\theta_1)\dot{\theta}_1, \tag{4.46}$$

$$\dot{\theta}_3 = \frac{dg}{d\theta_1}\frac{d\theta_1}{dt} = g'(\theta_1)\dot{\theta}_1. \tag{4.47}$$

Substituting Eqs. (4.41)-(4.47) into Eq. (4.32) yields the following equation,

$$\frac{d}{dt}\left(M(\theta_1)\dot{\theta}_1\right) = h(\theta_1), \tag{4.48}$$

$$\tag{4.49}$$

where

$$M(\theta_1) = \hat{M}_1 + \hat{M}_2 f'(\theta_1) + \hat{M}_3 g'(\theta_1), \tag{4.50}$$

$$h(\theta_1) = \hat{h}_1 + \hat{h}_2 f'(\theta_1) + \hat{h}_3 g'(\theta_1), \tag{4.51}$$

$$\hat{M}_i = M_i \,|_{(\theta_2,\theta_3)=(f(\theta_1),g(\theta_1))}, \tag{4.52}$$

$$\hat{h}_i = h_i \,|_{(\theta_2,\theta_3)=(f(\theta_1),g(\theta_1))}, \tag{4.53}$$

$$(i = 1,2,3).$$

Finally, the phase around contact point is obtained as below,

$$\dot{\theta}_1 = \frac{1}{M(\theta_1)}\sqrt{\int 2M(\theta_1)h(\theta_1)d\theta_1} \tag{4.54}$$

$$:= \frac{1}{M(\theta_1)}\sqrt{\left(D(\theta_1)+C\right)} \tag{4.55}$$

$$(C = -D(\theta_1^{L3}))$$

$$:= F(\theta_1). \tag{4.56}$$

The $\dot{\theta}_1^{L1}$ can be calculated by substituting θ_1^{L1} into Eq. (2.27). By use of $\dot{\theta}_1^{L1}$, it is possible to estimate the energy difference between (L1) and (L3) as below,

$$\Delta E_{loco} = \frac{1}{2}(m_1 a_1^2 + J_1)(\dot{\theta}_1^{L1})^2. \tag{4.57}$$

Figure 4.28 depicts the relationship among the present bar-distance D_{bar}, the target bar-distance D_{bar}^*, and the energy difference ΔE_{loco}.

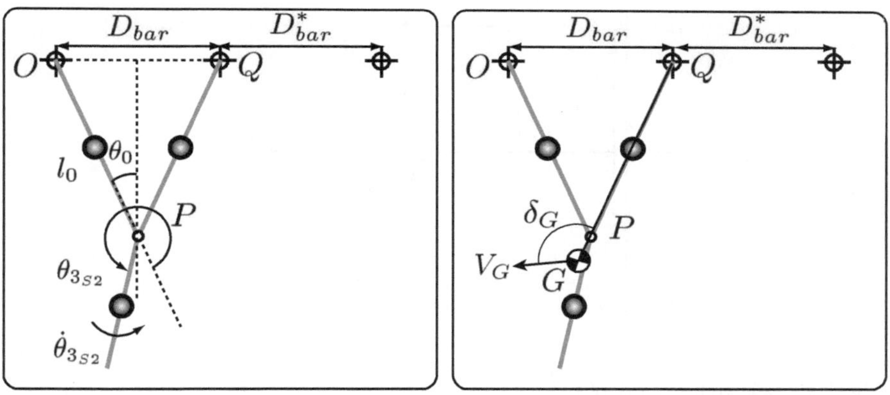

Fig. 4.29 Parameter setting at $S2$.

4.4.3 Swing Action

In this subsection, the controller of the swing action is described. At the initial state denoted by $S0$, the arms are symmetric with length l_0, which is equal to the length of the stretched arm, and the torso is hanging down vertically. The robot swings up the torso (counterclockwise) while bending its elbow. The rear contact point denoted by O can be considered as the pivot point of the pendulum during the swing-up motion. $S1$ denotes the state when the pendulum angular velocity is zero. After that, the robot swings back the torso (clockwise) with elbow-extending action synchronized with the pendulum motion. The arms are symmetric with length l_0 at the state de-

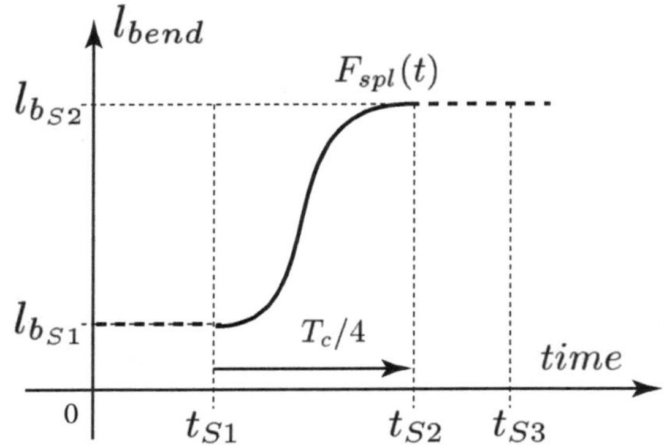

Fig. 4.30 Desired trajectory of l_{bend}.

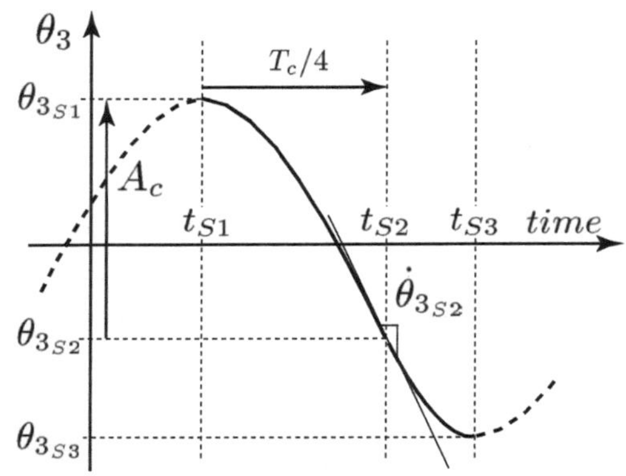

Fig. 4.31 Desired trajectory of θ_3.

noted by $S2$, during this swing-back action. At this moment, the pivot point of the pendulum switches from the rear contact point O to the front contact point Q. Then, the pendulum angular velocity is zero at $S3$. Immediately after $S3$, the robot releases the rear bar and starts the locomotion action. The overview of the swing action is shown in Fig. 4.23. Note that the robot just repeats the swing-back action from the state $S1$ to $S3$ after the locomotion action.

The elbow-bending during the swing action is geometrically designed to realize a double pendulum motion as well as keep grasping the bar so that the model can be looked upon as a double pendulum consisted of link 1 and link 3. The desired trajectories l_b^d is given as a cubic spline function $F_{spl}(t)$ shown in Fig. 4.30. The boundary conditions at t_{s1} and t_{s2} are specified as follow: $l_b^d(t_{s1}) = l_{2s1}, l_b^d(t_{s2}) = l_o, \dot{l_b}^d(t_{s1}) = \dot{l}(t_{2s}) = 0$. θ_2^d is given as function of θ_1 and l_b^d. θ_3^d is expressed by using a cosine wave function($t_{s1} \leq t_{s2}$) and a cubic spline $F'_{spl}(t)(t_{s2} \leq t)$ as shown in Fig. 4.31. These parameters are described as follows:

$$x_p = l_0 \sin\theta_1,\ y_p = -l_0 \cos\theta_1, \tag{4.58}$$

$$x_q = d_{bar},\ y_q = 0, \tag{4.59}$$

$$l_b^d = F_{spl}(l_{2s1}, l_{2s2}, T_2, t), \tag{4.60}$$

$$\theta_2^d = \cos^{-1}\left(\frac{x_q - x_p}{l_2}\right) - \theta_1 + \frac{\pi}{2}$$

$$= \cos^{-1}\left(\frac{d_{bar} - l_0 \sin\theta_1}{l_2}\right) - \theta_1 + \frac{\pi}{2}, \tag{4.61}$$

$$\theta_3^d = A_s \cdot \cos\left(\frac{2\pi}{T_s}(t - t_{s1})\right), \tag{4.62}$$

$$A_s := \theta_{3s1} - \theta_{3s2}, \tag{4.63}$$

$$\dot{\theta}_{3s2} = -A_s \cdot \frac{\pi}{T_s}, \tag{4.64}$$

where D_{bar} is the distance between bars, l_o is the length of a straight arm, A_s is the amplitude and T_s is the period of the oscillation.

The input parameters A_s and T_s are determined by the energy based control method. The angular velocity, $\dot{\theta}_{3S2}$, is controlled so that the robot obtains the enough energy to reach the target bar in the successive locomotion phase. The posture at S2 affects its motions since the center of rotation of the pendulum switches from the rear contact point O to the front contact point Q.

In this section, θ_{3S2} is given to satisfy the following condition,

$$QG \cdot V_G = 0, \tag{4.65}$$

where QG is the position vector of the center of gravity, V_g is the velocity vector of the center of gravity. Figure 4.29 shows these vectors. This equation means the velocity vector V_G is orthogonal on the position vector QG so that the loss of the angular momentum at the moment $S2$ is zero.

The target energy at S2 is estimated by the following equation,

$$E_{S2}^d = E^* + \Delta E_{loco}, \tag{4.66}$$

where E^* is the potential energy at L3.

The kinetic energy(K_{S2}) and potential energy(P_{S2}) at S2 are given as follows:

$$E_{S2} = P_{S2} + K_{S2}, \tag{4.67}$$

$$P_{S2} = P_1 + P_2 + P_3, \tag{4.68}$$

$$P_1 = m_1 g \frac{l_0}{2} \cos\theta_0, \tag{4.69}$$

$$P_2 = P_1, \tag{4.70}$$

$$P_3 = m_3 g \left(l_0 \cos\theta_0 + \frac{l_3}{2} \cos(\theta_0 + \theta_{3_{S2}}) \right), \tag{4.71}$$

$$K_{S2} = \frac{1}{2} m_3 \left(\frac{l_3}{2} \dot{\theta}_{3_{S2}} \right)^2, \tag{4.72}$$

where

$$\theta_0 = \sin^{-1}\left(\frac{D_{bar}}{2l_0} \right). \tag{4.73}$$

The required energy at S2 is obtained from Eqs. (4.66) and (4.67),

$$K_{S2}^d = E_{S2}^d - P_{S2}^d \tag{4.74}$$

$$= E^* + \Delta E_{loco} - P_{S2}^d. \tag{4.75}$$

By substituting Eq. (4.75) into Eq. (4.72), the desired angular velocity is acquired as follows:

$$\dot{\theta}_3^d = -\frac{2}{l_3} \sqrt{\frac{2}{m_3} \left(E^* + \Delta E_{loco} - P_{S2}^d \right)}. \tag{4.76}$$

From above equations, the desired period, T_s, is decided as the following equation,

$$T_s = l_3(\theta_{3_{S1}} - \theta_{3_{S2}}) \pi \sqrt{\frac{m_3}{2\left(E^* + \Delta E_{loco} - P_{S2}^d\right)}}. \tag{4.77}$$

4.4.4 Experiment

In this section, we validate the proposed method for the irregular ladder problem by the experiment using the Gorilla Robot III. In order to confirm the validity of the proposed method, we performed the experiment of the irregular ladder brachiation. In this chapter, as the experimental environment, the bar-distance is set up to be 4.0[m] and 3.0[m] alternately. Note that the robot is given the information of the bar distance and that how to acquire that information (for example, with a camera) is future work.

As a result, the stable continuous brachiation on the irregular ladder was achieved. Table I shows the value of T_s employed in the experiment and Table II shows the

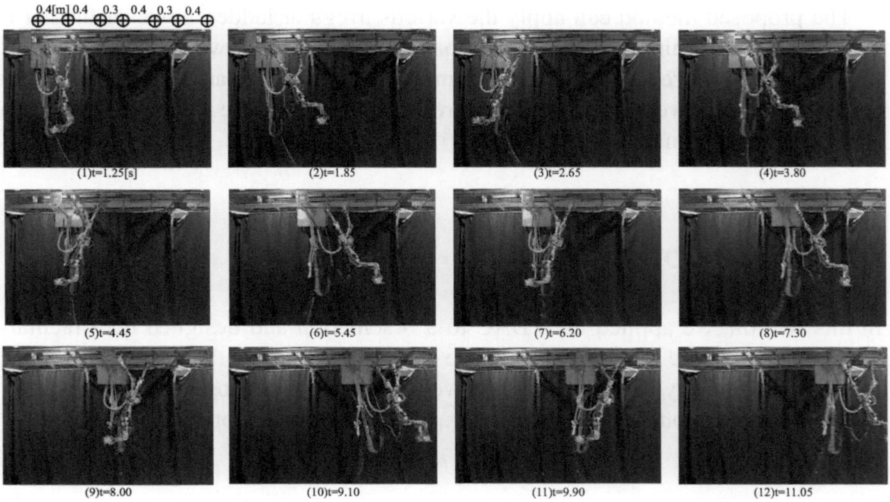

Fig. 4.32 Snapshots of the irregular ladder brachiation

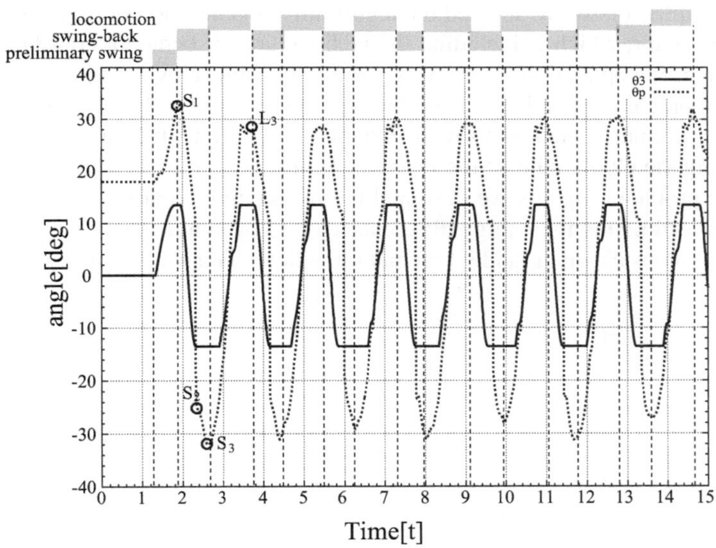

Fig. 4.33 Experimental results

pendulum angle, θ_p at S3 and L3 of the ith action. As shown in this table, θ_p^{L3} increases compared to θ_p^{S3} under the codition of $D^*_{bar} = 0.4$, on the other hand, θ_p^{L3} decreases compared to θ_p^{S3} under the codition of $D^*_{bar} = 0.3$. Figure 4.32 shows the snapshots of the experiment and Fig. 4.33 depicts the experimental results of the pendulum angle, θ_p, and the torso angle, θ_3.

The proposed method can apply the various irregular ladder, however the hardware capability limits the robot performance. For example, we found that it was impossible to realize the brachiation from 3.0[m] to 5.0[m] bar-distance. It is conceivable that improvement of the hardware or proposal of more efficient method can enhance the performance.

4.4.5 Summary of This Section

In this section, we applied the PDAC to a brachiation and designed the brachiating controller on the irregular ladder. The controller estimates the energy difference caused by symmetry breaking. Finally, we validated the proposed controller by realizing the continuous irregular ladder brachiation with the Gorilla Robot III.

4.5 Summary

In this chapter, firstly we showed the algorithm for generation of dynamically dexterous behavior, which is including the architecture for behavior coordination and adjustment. We applied the learning algorithm to achieve two types of brachiation: overhand brachiation and side-hand brachiation by Gorilla Robot II. Secondly, we presented a control method to realize smooth continuous brachiation. The action of bending the elbow joint of the swinging arm was proposed in order to solve the gripping problem. Finally, we designed the brachiating controller on the irregular ladder by estimating the energy difference caused by symmetry breaking and realized continuous irregular ladder brachiation with the Gorilla Robot III.

Chapter 5
Quadruped Walking

5.1 Realization of a Crawl Gait

5.1.1 Motion Design of a Crawl Gait

Keeping static stability is necessary condition in order to realize a crawl gait. Thus, a torso is moved so that center of gravity (COG) exists in the support polygon formed by support legs. Y-coordinate of the landing position of each leg is set as constant in the Cartesian coordinate system fixed on a ground as shown in Fig. 5.1. The Y axis of the COG motion is moved as a function of sin, and the X axis of the COG motion is move so that the velocity of COG is 0 when COG cross the segment of the support polygon. There are two times when the velocity of the X axis is 0 in one walking cycle. The height of COG is constant, that is, the robot walks keeping the constant height from a ground. This desired COG trajectory is written as follows:

$$X_G(t) = c_3 t^3 + c_2 t^2 + c_1 t^1 + c_0, \tag{5.1}$$

$$y_G(t) = A_G \sin\left(\frac{2\pi}{T} t\right) + Y_{G0}, \tag{5.2}$$

$$Z_G = h_G, \tag{5.3}$$

where A_G is amplitude of lateral oscillating movement, T is walking cycle, h_G is COG height, and S is stride length. c_0, c_1, c_2, c_3 are given following conditions:

$$\begin{gathered} X_G(0) = X_{G0}, \quad X_G(\frac{T}{2}) = \frac{S}{2} + X_{G0} \\ \left.\frac{dX_G}{dt}\right|_{t=0} = 0, \quad \left.\frac{dX_G}{dt}\right|_{t=\frac{T}{2}} = 0 \end{gathered} \tag{5.4}$$

where X_{G0}, Y_{G0} are initial position of X- and Y-coordinate COG respectively. The position at the base of each leg decided by COG trajectory $(X_r^{leg}, Y_r^{leg}, Z_r^{leg})$ are expressed as follows:

T. Fukuda et al.: Multi-Locomotion Robotic Systems, STAR 81, pp. 117–137.
springerlink.com © Springer-Verlag Berlin Heidelberg 2012

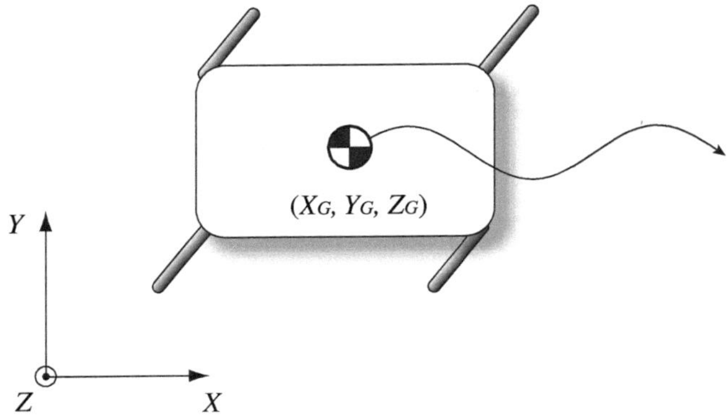

Fig. 5.1 Trajectry of COG(Center Of Gravity)

$$X_r^{leg}(t) = X_G(t) + a^{leg}, \tag{5.5}$$
$$Y_r^{leg}(t) = Y_G(t) + b^{leg}, \tag{5.6}$$
$$Z_r^{leg}(t) = Z_G + c^{leg}, \tag{5.7}$$

where $a^{leg}, b^{leg}, c^{leg}$ are the position of the base of each leg from COG at X, Y, Z direction. FL, FR, RL, and RR express front left leg, front right leg, rear left leg, and rear right leg respectively. leg comes in FL, FR, RL, RR.

Next, the tip position of each leg $(X_e^{leg}, Y_e^{leg}, Z_e^{leg})$ are derived. The tip position move stride length in quarter walking cycle, because the duty ratio of the crawl gait is 0.75. Then, the tip positions of each leg are expressed as following equations.

First walking cycle

$$X_e^{leg_{sw}}(t) = g(\frac{S}{2} + X_r^{leg_{sw}}(t) - X_e^{leg_{sw}}{}_0, t) + X_e^{leg_{sw}}{}_0, \tag{5.8}$$
$$Y_e^{leg_{sw}}(t) = Y_e^{leg_{sw}}{}_0, \tag{5.9}$$
$$Z_e^{leg_{sw}}(t) = h_{up} \sin(\frac{4\pi}{T} t), \tag{5.10}$$

$$X_e^{leg_{sp0}}(t) = X_e^{leg_{sp0}}{}_0, \tag{5.11}$$
$$Y_e^{leg_{sp0}}(t) = Y_e^{leg_{sp0}}{}_0, \tag{5.12}$$
$$Z_e^{leg_{sp0}}(t) = 0, \tag{5.13}$$

$$X_e^{leg_{sp1}}(t) = \frac{S}{2} + X_r^{leg_{sp1}}(\frac{T}{4} m), \tag{5.14}$$
$$Y_e^{leg_{sp1}}(t) = Y_e^{leg_{sp1}}{}_0, \tag{5.15}$$
$$Z_e^{leg_{sp1}}(t) = 0, \tag{5.16}$$

$$\begin{aligned}
&(0 \le t < \frac{T}{4}) \\
&\quad leg_{sw} : RR \quad leg_{sp0} : FR, RL, FL \quad m = 0, \\
&(\frac{T}{4} \le t < \frac{T}{2}) \\
&\quad leg_{sw} : FR \quad leg_{sp0} : RL, FL \quad leg_{sp1} : RR \quad m = 1, \\
&(\frac{T}{2} \le t < \frac{3}{4}T) \\
&\quad leg_{sw} : RL \quad leg_{sp0} : FL \quad leg_{sp1} : RR, FR \quad m = 2, \\
&(\frac{3}{4}T \le t < T) \\
&\quad leg_{sw} : FL \quad leg_{sp1} : RR, FR, RL \quad m = 3,
\end{aligned}$$

n-th walking cycle

$$X_e^{leg_{sw}}(t) = g(S, t-(n-1)T) + \frac{S}{2} + X_r^{leg_{sp1}}(\frac{T}{4}m), \tag{5.17}$$

$$Y_e^{leg_{sw}}(t) = Y_e^{leg_{sw}}{}_0, \tag{5.18}$$

$$Z_e^{leg_{sw}}(t) = h_{up}\sin(\frac{4\pi}{T}(t-(n-1)T)), \tag{5.19}$$

$$X_e^{leg_{sp0}}(t) = (n-2)S + \frac{S}{2} + X_r^{leg_{sp1}}(\frac{T}{4}m), \tag{5.20}$$

$$Y_e^{leg_{sp0}}(t) = Y_e^{leg_{sp0}}{}_0, \tag{5.21}$$

$$Z_e^{leg_{sp0}}(t) = 0, \tag{5.22}$$

$$X_e^{leg_{sp1}}(t) = (n-1)S + \frac{S}{2} + X_r^{leg_{sp1}}(\frac{T}{4}m), \tag{5.23}$$

$$Y_e^{leg_{sp1}}(t) = Y_e^{leg_{sp1}}{}_0, \tag{5.24}$$

$$Z_e^{leg_{sp1}}(t) = 0, \tag{5.25}$$

$$
\begin{aligned}
&((n-1)T \le t < (n-1)T + \frac{T}{4})\\
&\qquad leg_{sw} : RR \quad leg_{sp0} : FR, RL, FL \quad m = 0,\\
&((n-1)T + \frac{T}{4} \le t < (n-1)T + \frac{T}{2})\\
&\qquad leg_{sw} : FR \quad leg_{sp0} : RL, FL \quad leg_{sp1} : RR \quad m = 1,\\
&((n-1)T + \frac{T}{2} \le t < (n-1)T + \frac{3}{4}T)\\
&\qquad leg_{sw} : RL \quad leg_{sp0} : FL \quad leg_{sp1} : RR, FR \quad m = 2,\\
&((n-1)T + \frac{3}{4}T \le t < nT)\\
&\qquad leg_{sw} : FL \quad leg_{sp1} : RR, FR, RL \quad m = 3,
\end{aligned}
$$

where $X_e^{leg}{}_0, Y_e^{leg}{}_0$ are initial positions of each leg, h_{up} is the tip height from the ground, leg_{sp0} is the next swing leg, leg_{sp1} is the previous swing leg, leg_{sw} is the swing leg. $g(\lambda, \tau)$ is the function that the value changes from 0 to λ while time τ changes from 0 to $\frac{T}{4}$, the beginning and end of time derivative is 0. m is the integer number changing 0 to 3 according to cycle time.

5.1.2 Joint Trajectory of the Leg

The original point of the Cartesian coordinate is set at the base of the leg in order to calculate the time dependent joint trajectory of each leg. The new coordinate is expressed as follows:

$$x_e^{leg} = X_e^{leg} - X_r^{leg}, \tag{5.26}$$

$$y_e^{leg} = Y_e^{leg} - Y_r^{leg}, \tag{5.27}$$

$$z_e^{leg} = Z_e^{leg} - Z_r^{leg}. \tag{5.28}$$

Frames are defined for the inverse kinematics as shown in Fig. 5.2. Transformation matrix of the frames are as follows:

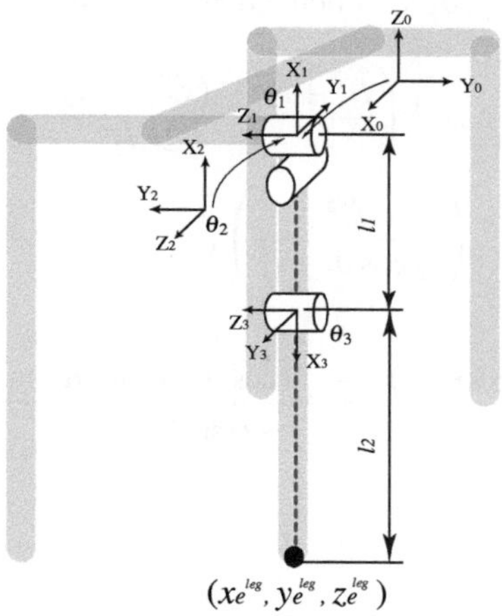

Fig. 5.2 Definition of coordinate

$$T_1^0 = \begin{bmatrix} \cos(\frac{\pi}{2}+\theta_1) & -\sin(\frac{\pi}{2}+\theta_1) & 0 & 0 \\ 0 & 0 & -1 & 0 \\ \sin(\frac{\pi}{2}+\theta_1) & \cos(\frac{\pi}{2}+\theta_1) & 0 & 0 \\ 0 & 0 & 0 & 1 \end{bmatrix}, \tag{5.29}$$

$$T_2^1 = \begin{bmatrix} \cos\theta_2 & -\sin\theta_2 & 0 & 0 \\ 0 & 0 & -1 & 0 \\ \sin\theta_2 & \cos\theta_2 & 0 & 0 \\ 0 & 0 & 0 & 1 \end{bmatrix}, \tag{5.30}$$

$$T_3^2 = \begin{bmatrix} \cos\theta_3 & \sin\theta_3 & 0 & l_1 \\ 0 & 0 & 0 & 0 \\ -\sin\theta_3 & \cos\theta_3 & -1 & 0 \\ 0 & 0 & 0 & 1 \end{bmatrix}. \tag{5.31}$$

$x_e^{leg}, y_e^{leg}, z_e^{leg}$ are expressed as follows:

$$x_e^{leg} = \sin\theta_1 \cos\theta_2 (l_2 \cos\theta_3 + l_1) - l_2 \cos\theta_1 \sin\theta_3, \tag{5.32}$$

$$y_e^{leg} = \sin\theta_2 (l_2 \cos\theta_3 + l_1), \tag{5.33}$$

$$z_e^{leg} = -\cos\theta_1 \cos\theta_2 (l_2 \cos\theta_3 + l_1) - l_2 \sin\theta_1 \sin\theta_3. \tag{5.34}$$

From Eq. 5.32-5.34, $\theta_1, \theta_2, \theta_3$ are obtained as follows:

$$\theta_3 = \pi - \cos^{-1}\left(\frac{l_1{}^2 + l_2{}^2 - (x_e^{leg2} + y_e^{leg2} + z_e^{leg2})}{2l_1 l_2}\right), \tag{5.35}$$

$$\theta_2 = \sin^{-1}\left(\frac{y_e^{leg}}{l_2\cos\theta_3 + l_1}\right), \tag{5.36}$$

$$\theta_1 = 2\tan^{-1}\left(\frac{\cos\theta_2(l_2\cos\theta_3 + l_1) - \sqrt{(}\cos^2\theta_2(l_2\cos\theta_3 + l_1)^2 + l_2{}^2\sin^2\theta_3 - x_e^{leg2}}{-l_2\sin\theta_3 + x_e^{leg}}\right). \tag{5.37}$$

5.1.3 *Estimation of Walking Energy*

The walking energy of the crawl gait is obtained as the sum of the energy to move torso and to move the swing legs. The swing is modeled as a pendulum.

5.1.3.1 Energy to Move the Torso

By using the robot mass m, the force of X-axis F_X, and the force of Y-axis F_Y, the force to move the torso (COG) is expressed as follows:

$$F_X = m\ddot{X_G}(t), \tag{5.38}$$

$$F_Y = m\ddot{Y_G}(t). \tag{5.39}$$

The force of Z-axis is not considered, because there is no motion in Z-axis. The energy to move the torso is obtained as follows:

$$E_G = \int \left(F_X\dot{X_G}(t) + F_Y\dot{Y_G}(t)\right) dt. \tag{5.40}$$

5.1.3.2 Energy to Move Swing Leg

The swing leg motion is same as stance leg. The time dependent trajectory is given from Eq. (9.252),

$$\theta = A\exp^{\sqrt{\frac{g}{l}}t} + B\exp^{-\sqrt{\frac{g}{l}}t} - \frac{\tau_k}{mgl}. \tag{5.41}$$

By differentiating in time,

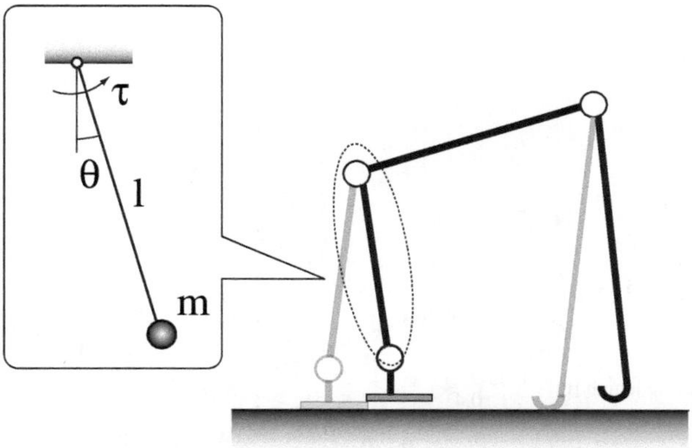

Fig. 5.3 Model of swing leg

$$\dot{\theta} = A\sqrt{\frac{g}{l}}\exp^{\sqrt{\frac{g}{l}}t} - B\sqrt{\frac{g}{l}}\exp^{-\sqrt{\frac{g}{l}}t}, \tag{5.42}$$

$$\ddot{\theta} = A\frac{g}{l}\exp^{\sqrt{\frac{g}{l}}t} + B\frac{g}{l}\exp^{-\sqrt{\frac{g}{l}}t}. \tag{5.43}$$

The torque is derived from dynamics of the pendulum,

$$\tau = ml^2\ddot{\theta} + mgl\sin\theta. \tag{5.44}$$

The energy to move swing leg is expressed as follows:

$$\begin{aligned} E^{sw}_{sagi} &= \int \tau\dot{\theta}dt \\ &= \int (ml^2\ddot{\theta} + mgl\sin\theta)\dot{\theta}dt \\ &= \int \left(ml^2\left(A\frac{g}{l}\exp^{\sqrt{\frac{g}{l}}t} + B\frac{g}{l}\exp^{-\sqrt{\frac{g}{l}}t}\right) + mgl\sin\theta\right) \\ &\quad \cdot \left(A\sqrt{\frac{g}{l}}\exp^{\sqrt{\frac{g}{l}}t} - B\sqrt{\frac{g}{l}}\exp^{-\sqrt{\frac{g}{l}}t}\right)dt. \end{aligned} \tag{5.45}$$

5.1.3.3 Energy Map of the Crawl Gait

The walking energy of the crawl gait E_{crawl} is the sum of energy to move the torso and to move the swing leg,

$$E_{crawl} = E_G + E^{sw}_{sagi}. \tag{5.46}$$

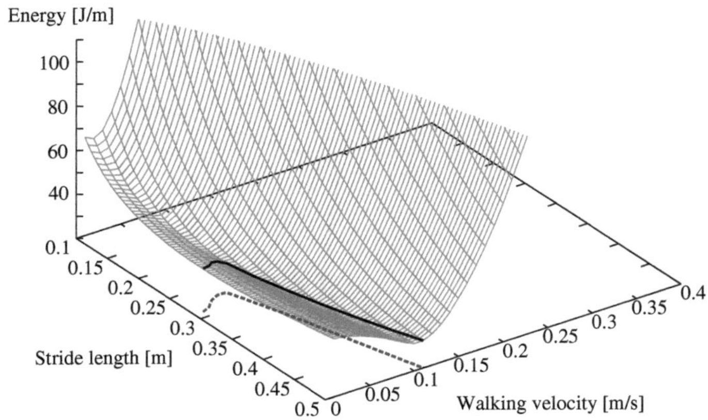

Fig. 5.4 Crawl gait

Figure 5.4 shows the 3-D map of the energy according to stride length and walking velocity calculated from above equation. In this map, the continuous line shows the minimizing points according to walking velocity; also dotted line shows the projected line on the plane composed of stride length axis and walking velocity axis. There is the stride to minimize the walking energy according to the walking velocity as shown in Fig. 5.4. When we chose optimum stride length, the waking cycle is also decided. Thus robot can realize efficient walk by use of these parameters.

5.1.4 Experiment

In this subsection, the quadruped walking experiment based on crawl gate is conducted using the Gorilla Robot III introduced in Chapter 3.

The desired walking velocity is set as 0.10[m/s]. The optimum stride length given form the energy map is more than 0.30[m], that is limitation stride length of the Gorilla Robot III. Thus, the stride length is set as 0.30[m], the walking cycle of this condition is 3.0[s].

Figure 5.5 shows the joint angles of this experiment. The walking energy calculated from experimental data is 224.1[J/m]. The snapshots of the experiment is shown in Fig. 5.6.

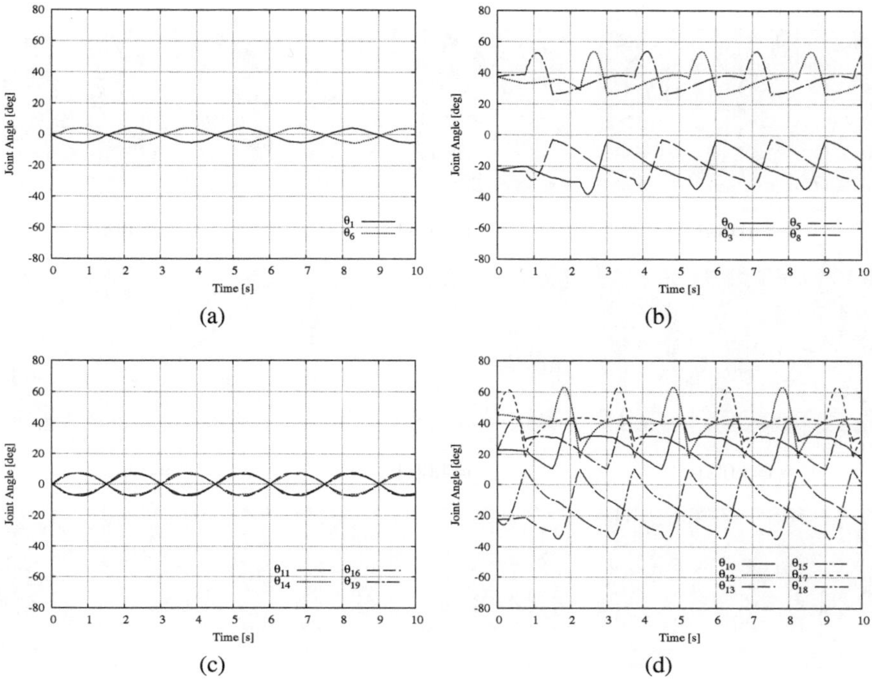

Fig. 5.5 Experimental results of quadruped walking in crawl gait (desired walking velocity: 0.10[m/s]). (a) time vs. upper joint angle in the lateral plane, (b) time vs. upper joint angle in the sagittal plane, (c) time vs. lower joint angle in the lateral plane, (d) time vs. lower joint angle in the sagittal plane

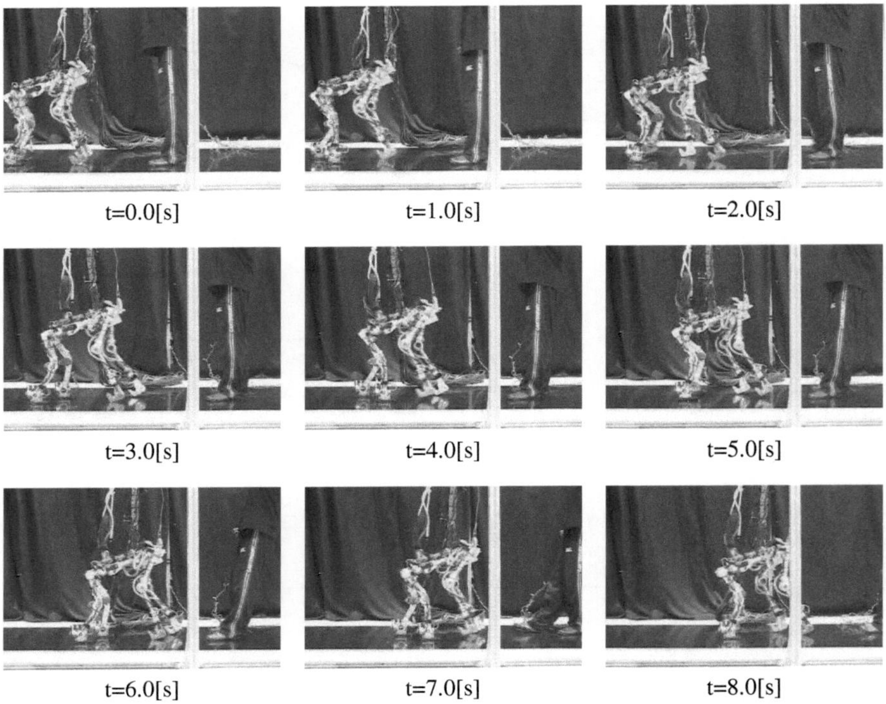

Fig. 5.6 Snapshots of quadruped walking in crawl gait (desired walking velocity: 0.10[m/s])

5.2 Joint Torque Evaluation of the Gorilla Robot on Slopes as Quadruped Hardware

5.2.1 *Structure of Gorilla Robot III*

A characteristic of the Gorilla Robot III link structure is that its arm length is longer than its leg length. In other words, its front leg length is longer than its rear leg length when it is postured for quadruped locomotion (Fig. 5.7(a)). The parameters of the Gorilla Robot III in Fig. 5.7 are shown in Table 5.1.

5.2.2 *Basic Gait Pattern*

The intermittent crawl gait is adopted as the basic gait pattern in this work, because the intermittent crawl gait has been successfully used by robots walking on steep slopes and it is widely used control method [283]. The intermittent crawl gait pattern is described below.

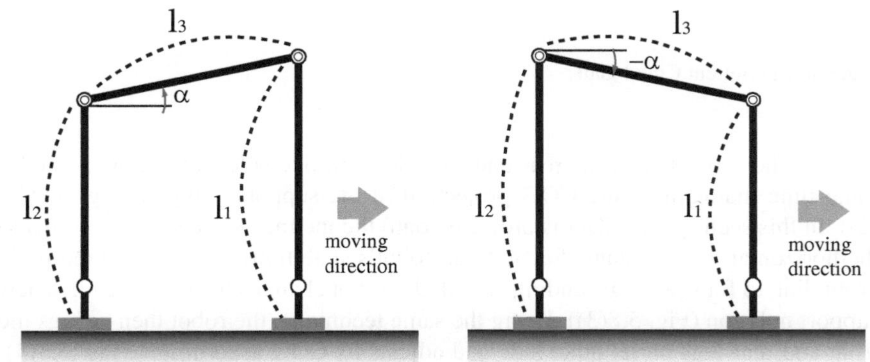

(a) case1: the length of the rear leg is shorter than the length of front leg ($l_1 > l_2$)

(b) case2: the length of the rear leg is longer than the length of fore leg ($l_1 < l_2$)

Fig. 5.7 Definition of Link Parameter

Table 5.1 Parameters of Gorilla Robot III

Parameter	Units	Value
l_1	m	0.68
l_2	m	0.52
l_3	m	0.45
α	deg	22.8

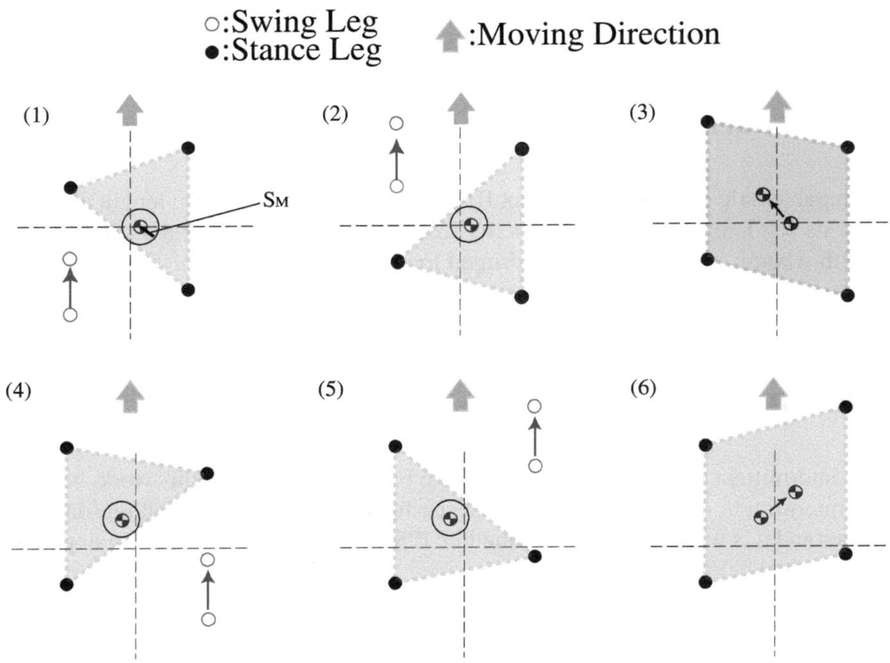

Fig. 5.8 Intermttent Crawl Gait

First, the robot swings its rear and front legs on one side forward while at the same time maintaining the COG projection in the support polygon (Fig. 5.8(1)-(2)). In this section, the COG is projected onto the inclined surface via a normal to the horizontal ground plane. Second, the robot's state transitions from three-point grounding to four-point grounding. Third, the robot changes the COG inside a new support polygon (Fig. 5.8(3)). Using the same technique, the robot then swings the front and rear legs on its other side and adjusts its COG accordingly (Fig. 5.8(4)-(6)). This shows that the Intermittent Crawl Gait is capable of performing a forward motion that follows a zigzag COG trajectory.

Next, we will describe the design method of a zigzag COG trajectory. The COG trajectory is determined by ensuring the projecting point of the COG is located among the static stable domains provided by the support polygon. In this section, the position of the projective point of the COG is decided by setting up the stability margin. The stability margin degree is the distance from the projecting point of the COG to the neighborhood of the nearest support polygon (S_M in Fig. 5.8), and is the evaluation index of the simplest static stability [234]. Thus, by establishing the stability margin, the projecting point of the COG is determined, and the COG trajectory is projected.

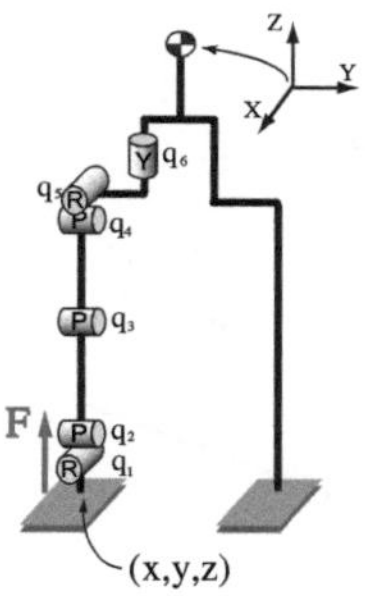

Fig. 5.9 Link Model

5.2.3 Evaluation of Joint Torque in Quadruped Walk on a Slope

In this section, we explain the derivation method of joint torques and the cost function used to evaluate simulation results.

Derivation Method of Joint Torque

The intermittent crawl gait is composed of two phases; three point ground contact and four point ground contact. In the four point ground contact phase, joint torque has to be calculated using the dynamics, because the motion of the COG has to be considered. We calculated joint torque using "open dynamics engine (ODE)" [263]. ODE is a library used for simulating articulated rigid body dynamics without the calculation of complex equations. In the three point ground contact phase, the joint torque can be calculated by the statics because the COG motion can be ignored. Calculations of static torque are expressed as follows: In general, in order to control a position and a posture, six or more degrees of freedom are needed. Thus, link structure consists of six degrees of freedom in each leg as shown in Fig. 5.9. The origin of the coordinate axes is placed at the COG position. Additionally, we express a position vector on the top of the leg by, $X = (x, y, z)^T$ and a joint angle vector by, $q = (q_1, q_2, q_3, q_4, q_5, q_6)^T$. Then, forward kinematics is expressed as follows:

$$X = f(q). \tag{5.47}$$

Small displacements of the joint angles and the position of the end effector are expressed as follows:

$$\delta X = J(q)\,\delta q, \tag{5.48}$$

where $J(q)$ is a Jacobean matrix, δX is a small displacement vector of position, δq is a small displacement vector of angle. The principle of virtual work provides the relation between the force from an end effector and the joint torque as follows:

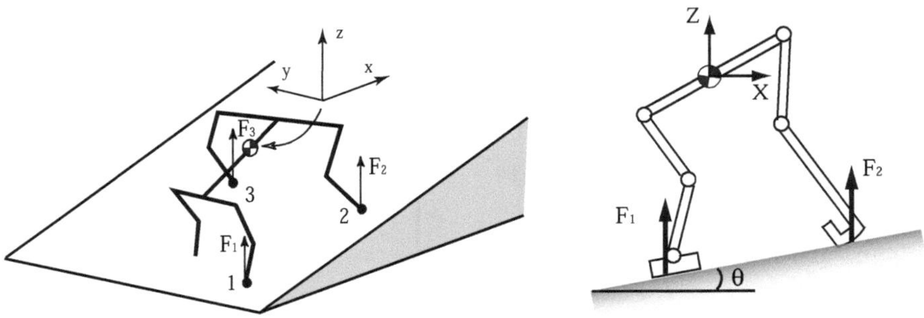

(a) Three-dimensional figure (b) Two-dimensional figure

Fig. 5.10 Definition of Coordinates and Grounding Points

$$\tau = J(q)^T F, \tag{5.49}$$

where $\tau = (\tau_1, \tau_2, \tau_3, \tau_4, \tau_5, \tau_6)^T$ is a vector of joint torques, $F = (f_x, f_y, f_z)^T$ is a vector of ground-reaction-force. Eq. (5.49) gives a joint torque vector from a vector of ground-reaction-force.

Next, we will explain the calculation method of the ground-reaction-force vector in order to derive the torque vector. As shown in Fig. 5.10, the right-handed standard coordinates is set at the position of the center of gravity and the X axis parallels horizontal surface. We define the three ground contact points as the 1, 2 and 3, the position vectors from the COG to each ground contact point as $r_1 = (x_1, y_1, z_1)^T$, $r_2 = (x_2, y_2, z_2)^T$, $r_3 = (x_3, y_3, z_3)^T$, ground-reaction-forces of each grounding point as F_1, F_2, F_3.

The ground-reaction-forces vector parallels the gravity vector due to the condition that the COG doesn't move and the legs are vertically in contact with the ground in transverse plane; hence components of F_1, F_2, F_3 are expressed as follows:

$$F_1 = (\|F_1\| \sin\theta, 0, \|F_1\| \cos\theta)^T \tag{5.50}$$
$$F_2 = (\|F_2\| \sin\theta, 0, \|F_2\| \cos\theta)^T \tag{5.51}$$
$$F_3 = (\|F_3\| \sin\theta, 0, \|F_3\| \cos\theta)^T, \tag{5.52}$$

where θ is a slope angle, $\|F_1\|, \|F_2\|, \|F_3\|$ are normed vector of F_1, F_2, F_3. From equilibrium of force and momentum, Eqs. (5.53) and (5.54) are derived as follows:

$$\sum_{i=1}^{3} F_i = Mg \tag{5.53}$$
$$\sum_{i=1}^{3} F_i \times r_i = 0, \tag{5.54}$$

where $g = (0, 0, -g)$ is a gravitational acceleration vector.

Simultaneous equations with respect to $||F_1||$, $||F_2||$, $||F_3||$ are given by substituting the components of the respective vector to Eqs. (5.53) and (5.54), as follows:

$$||F_1|| + ||F_2|| + ||F_3|| = Mg \tag{5.55}$$

$$y_1||F_1|| + y_2||F_2|| + y_3||F_3|| = 0 \tag{5.56}$$

$$\sum_{i=1}^{3}(x_i\cos\theta - z_i\sin\theta))||F_i|| = 0. \tag{5.57}$$

From Eqs. (5.55)-(5.57), $||F_1||, ||F_2||$ and $||F_3||$ are obtained as follows:

$$||F_1|| = \frac{Mg\left[y_3\left(x_2\cos\theta + z_g\sin\theta\right) - y_2\left(x_3\cos\theta + z_g\sin\theta\right)\right]}{A} \tag{5.58}$$

$$||F_2|| = \frac{Mg\left[-y_3\left(x_1\cos\theta + z_g\sin\theta\right) - y_1\left(x_3\cos\theta + z_g\sin\theta\right)\right]}{B} \tag{5.59}$$

$$||F_3|| = \frac{1}{y_3 - y_2}\left(y_2 - ||F_1||\right), \tag{5.60}$$

where

$$\begin{aligned} A = &\left(y_2 - y_3\right)\left(x_1\cos\theta + \left(y_3 - y_1\right)z_g\sin\theta\right) \\ &+ \left(y_1 - y_2\right)\left(x_2\cos\theta + z_g\sin\theta\right) + \left(x_3\cos\theta + z_g\sin\theta\right) \end{aligned} \tag{5.61}$$

$$\begin{aligned} B = &\left(y_2 - y_3\right)\left(x_1\cos\theta + \left(y_3 - y_1\right)z_g\sin\theta\right) \\ &+ \left(y_1 - y_2\right)\left(x_2\cos\theta + z_g\sin\theta\right) + \left(x_3\cos\theta + z_g\sin\theta\right). \end{aligned} \tag{5.62}$$

Ground reaction force vectors can be calculated by substituting $||F_1||, ||F_2||$ and $||F_3||$ in Eqs. (5.58)-(5.60) into Eqs. (5.50), (5.51) and (5.52). In addition, the toque vector of each leg can be calculated by substituting F_1, F_2 and F_3 into Eq. (5.49).

Evaluated Value

We define a torque cost function as squared torque divided by the stride length S as follows:

$$C_s = \frac{1}{S}\int_0^{T_c} \tau^T\, \tau dt, \tag{5.63}$$

where T_c is the cycle time of walking, $\tau = (\tau_i)^T\,(i = 1, \cdots, 24)$ is the 24 degrees of joint torque vector that is the sum of all joint torque in the robot. Squared torques are used by Channnon *et al.* as a cost function to derive the optimal biped gait that enables a smooth walking and low energy consumption [36]. In addition, Kiguchi *et al.* make use of squared torque as an evaluated function of walking energy under the assumption that most of the energy is consumed in acceleration and deceleration motions [147]. In this section, we also use C_s as a cost function for the criteria of energy reduction and smooth walking motion.

5.2.4 Simulation Analysis

In this section, the simulation analysis is conducted with regard to relations among the limb length of the robot, the slope angle and the torque cost function C_s.

Simulation Setting

To that end, we first defined several parameters related to the walking motion. When walking, we assume that the robot will take largest stride S possible as long as its physical constraints allow it to do so. In order to provide the uniform simulation conditions for a respective parameter set, the moving velocity is set to V [m/s], constantly. Thus, the time of a cycle T_c [sec] is expressed as $T_c = \frac{S}{V}$. In other words, the walking cycle time will decrease as the walking stride length increases. Moreover, we set the stability margin to S_M=0.05[m] in order to design the COG trajectory. In what follows, the employed physical parameters of the simulations are adopted from the structure of "Gorilla Robot III. (Fig. 3.9 and 3.10)". The limb lengths l_1 and l_2 are variable values under the constraint that $l_1 + l_2 = L(const.)$ (Fig. 5.7). Let L be 1.2[m]; l_3 be 0.45 [m]; and mass of robot be 24[kg]. The angle of torso α is determined from limb leg length as follows:

$$\alpha_1 = \arcsin\left(\frac{l_1 - l_2}{l_3}\right). \tag{5.64}$$

Here, let the range of motion of the front leg S_f, the range of motion of the rear leg S_r. When angle of torso is α_1, the relation of range is $S_f > S_r$ or $S_f < S_r$ except the condition of $l_1 = l_2$. Therefore, stride length is fitted to the shorter range of motion. Alternatively, if we set the angle of torso α_2 as the front and rear range are equal such as $S_f = S_r$, stride length is longest. In the basic gait pattern, joint torque is influenced by static torque supporting the COG in the three point contact phase while dynamic torque moves the COG during the four point contact phases. In this research, since the moving velocity is constant, a walking cycle becomes longer as stride length increases. Since the effects of static torque increase as a walking cycle lengthens, the required torque that the angle of torso is α_1 is smaller than the torque that the angle of torso is α_2. The angle of torso is set by α_1, because the purpose of this research is the reduction of joint torque. Table 5.2 shows these parameters.

In the simulation, Cs is calculated based on various conditions where the "Ratio of Rear leg Length (RRL)" and slope angle are changed. Here, RRL is defined as follows:

$$RRL = \frac{l_2}{l_1}, \tag{5.65}$$

where, l_1 is the length of front leg length, and l_2 is the length of rear leg length (See Fig. 5.7).

Table 5.2 Parameter Setup

Parameter	Units	Value
Total Link Length, L	m	1.2
Velocity, V	m/s	0.05
Stability Margin, Sm	m	0.05
Mass of Robot, M	kg	24
Time of Cycle, T	sec	Variable
Stride Length, S	m	Variable
Angle of Torso, α	deg	Variable

The trajectory of a joint angle is calculated by inverse kinematics obtained from the COG position and a torso posture. We assume that the COG of a robot is fixed in the robot body. The joint angle is controlled by the PID controller. Simulation results are evaluated in the following two cases.

$$Case1 : RRL < 1,$$
$$Case2 : RRL > 1.$$

In other words, *Case1* is the situations where the length of the rear leg is longer than that of the front leg (Fig. 5.7(a)), *Case2* is the situations where the length of the rear leg is shorter (Fig. 5.7(b)).

Simulation Result

Figure 5.11 shows the cost function *Cs* of each RRL and the slope angle based on the simulation. If we focus on a two-dimensional plane with the constant RRL, it can be confirmed that the slope angle minimizing *Cs* will be small as RRL is larger. The simulation result is qualitatively evaluated in two cases in the section 5.2.4. Traveling upslope, the C_s of *Case1* is larger than *Case2*. Alternatively, the C_s of *Case2* is larger than *Case1* when traveling downslope. From this result, if the walking motion is slow like the intermittent crawl gait, quadruped animals which have shorter rear leg lengths can walk easily on an upslope.

Here, we consider the simulation result. As shown in a Fig. 5.12(a), both legs are set to τ torque of the sum total output by each leg. Since the force used for kicking the ground can be expressed as $F_1 = (\tau/l_1^{'})$, $F_2 = (\tau/l_2^{'})$, the short leg can take out the large kicking force with the same torque. On the slope, since the ratio of the projecting point of the COG is $R_1 : R_2$ as shown in Fig. 5.12(b), the ratio of the load concerning fore and hind legs is set to $R_2 : R_1$. Therefore, a robot with short hind legs can take out impellent force with the same output. On upslope, since the load on hind legs increases as the slope angle increases, a robot with a small ratio of hind leg has advantages in upslope conditions.

Furthermore, in this section, since the simulation is analyzed under constant movement velocity conditions, the length of the walking cycle and the stride length

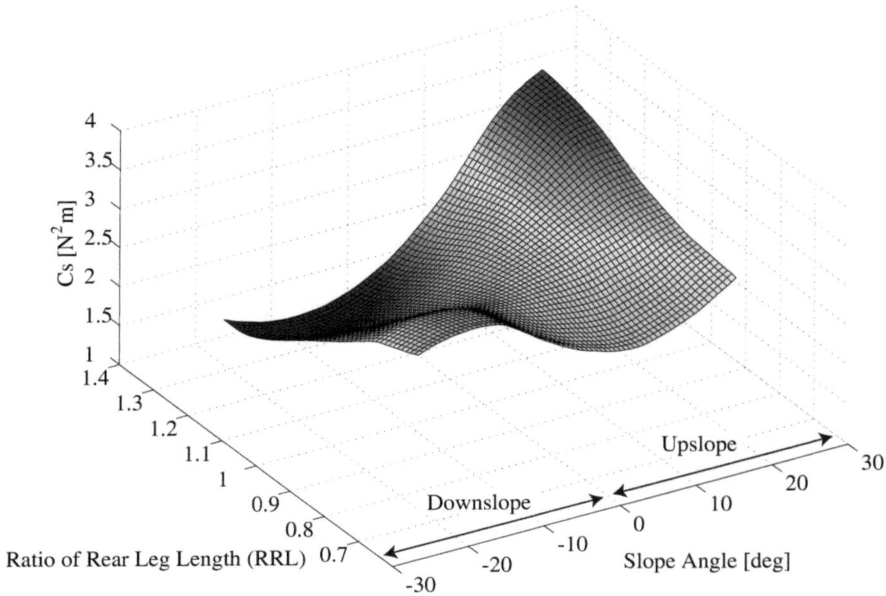

Fig. 5.11 Simulation Result

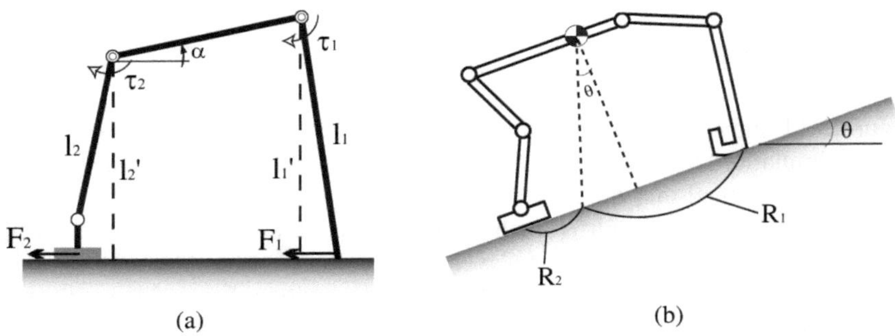

Fig. 5.12 Relationship between Torque and Length of Legs

increase in tandem. Thus, an increase of stride length leads to the increasing time in the three point contact phase. Due to the absence of COG motion during the three point contact phase, the increasing time of this phase implies further influence of static torque for C_s. Contrariwise, a decrease of stride length can speed up walking motion, which extends the influence of dynamic torque for C_s. The RRL can change the influence of static and dynamic torque, because stride length geometry is decided by RRL. In other words, optimal RRL can be expressed as the physical structure that minimizes the sum of static and dynamic torque.

5.2.5 *Experiment*

In this section, we compare cost function C_s obtained by simulation with the cost function C_s^{ex} obtained by the experimental data of an actual robot.

Evaluated Value of Experiment

In our experimental environment, an experimental joint torque τ_i^{ex} is proportional to the order voltage v_i for the motor driver; also, the proportionality constant is $\tau_r/3$, where τ_r is the rated torque. Then, the τ_i^{ex} is given as follows:

$$\tau_i^{ex} = \frac{\tau_r}{3} v_i m, \tag{5.66}$$

where m is the gear ratio.

τ_i^{ex} is calculated each sampling time during one cycle of quadruped walking from Eq. (5.66). Experimental torque vector τ^{ex} can be calculated from τ_i^{ex}, and experimental cost C_s^{ex} is calculated by Eq. (5.63). The experiments were carried out based on the understanding that all parameters would be equal to those used in the simulation. However, the link length is fixed at $l_1 = 0.52, l_2 = 0.68$ based on the Gorilla Robot III, because link length of this robot cannot be modified, and the C_s^{ex} is measured experimentally when the slope angle of a slope is set at every 2.5[deg] in -15[deg] to 15[deg].

Experimental Result

Experimental results are shown in Table 5.3. A snapshot of one experiment is shown in Fig. 9.23. Figure 5.14 shows the experimental results and simulation results that the RRL is 0.76 (Fig. 5.11), the same as gorilla robot III. Experimental results confirmed that C_s^{ex} is close to C_s when the slope angle is -15[deg] to 15[deg]. In particular, the experimental result that C_s in 5 [deg] slope is less than 0 [deg] slope agrees with the simulation result to be most characteristic. These experimental results confirm that the simulation result agrees with the results obtained from an actual robot on a small slope angle and that the RRL can be indexed to reduce C_s of each slope angle.

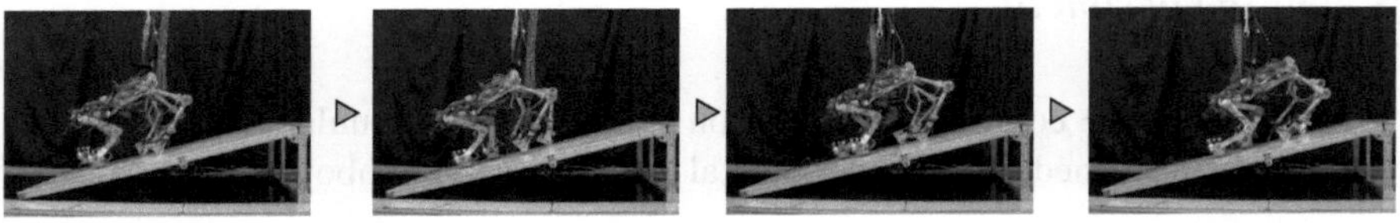

(a) t = 0.0 [sec] (b) t = 3.7 [sec] (c) t = 7.5 [sec] (d) t = 11.2 [sec]

Fig. 5.13 Snapshots of Experiment (slope Angle: 15[deg])

Table 5.3 Experimental Results

Slope Angle [deg]	C_s^{ex} [$\times 10^4$N^{2}m]
-15	4.00
-12.5	3.73
-10	3.50
-7.5	3.34
-5	3.28
-2.5	3.11
0	3.04
2.5	2.88
5	2.77
7.5	2.87
10	3.20
12.5	3.34
15	3.90

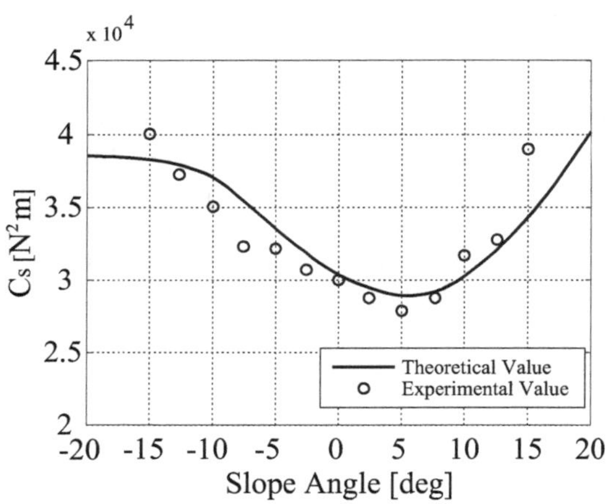

Fig. 5.14 Comparison between simulation and experiment

5.3 Summary

In this chapter, at first, the motion design of crawl gait was conducted and crawl gait using Gorilla Robot was realized. The motion is designed based on COG trajectory maintaining a static stability. The walking energy is analyzed and we gave the optimum walking cycle and stride length according to the desired walking velocity. Then, the optimum crawl gait was realized by use of the energy map.

Next, the optimal ratio of limb length of a quadruped robot designed to reduce joint torque on a slope was derived by analysis. Numerical simulations analyzed the torque cost function C_s in each limb length and each slope angle for a robot walking on a slope and the optimal Ratio of Rear leg Length (RRL) was derived. The results indicate that the C_s increases as the slope angle increases if the length of the rear leg is longer than that of the front leg. Alternatively, the C_s decreases as the slope angle increases, if the length of the rear leg is shorter. In other words, the robot that has rear leg lengths that are shorter than its front leg lengths, such as Gorilla Robot III, requires reduced C_s as going forward upslope and going backwards downslope. Additionally, experimental results conducted with an actual robot showed that results of numerical simulation were sufficiently precise to be of future use.

Chapter 6
Ladder Climbing Motion

6.1 Model of Ladder Climbing

6.1.1 Basic Motion Model

The pitch and roll joints which is used for this ladder climbing locomotion are shown in Fig. 6.1(a). The idlling arm and leg motion are designed in sagittal plane for avoiding the collision with any rungs.

$$(y_a - y_{pos})^2 + (z_a - z_{pos})^2 = s^2, x_a = 0 \tag{6.1}$$

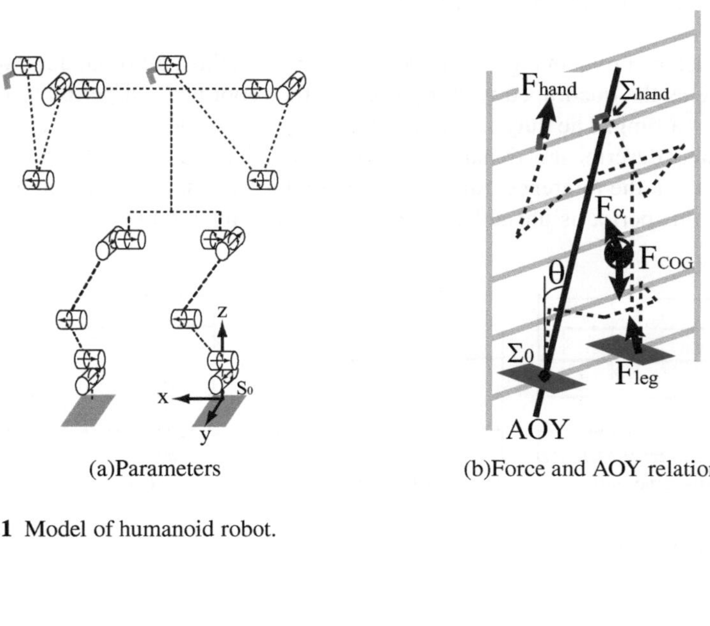

(a)Parameters (b)Force and AOY relation.

Fig. 6.1 Model of humanoid robot.

T. Fukuda et al.: Multi-Locomotion Robotic Systems, STAR 81, pp. 139–151.
springerlink.com © Springer-Verlag Berlin Heidelberg 2012

x_a, y_a, z_a are the reference trajectory of the hand and the foot, s is the trajectory parameter which is the half distance of climbing and y_{pos}, z_{pos} is the center of trajectory circle. The robot keeps the climbing motion with a constant distance between the torso and the ladder. The foot on the ladder has the gradient which the reaction force directs to COG and avoids sliding into the back of a ladder.

6.1.2 Ladder Climbing Gait

The MLR which climbs vertical ladder uses all fours as quadruped does. As the gait supporting the robot more than two points in quadruped walking, there are three supporting legs walking like transverse gait and two supporting legs walking like pace and trot gait. We propose to realize these gaits in vertical ladder climbing in order to extend the mobitily in various environment. The order of the step in each gait is shown in Fig. (6.2).

6.1.2.1 Transverse Gait

When the robot climbs the ladder each one step respectively about all fours, it's static gait because the body is always sustained by three points. At this time, the duty factor is 0.75.

6.1.2.2 Pace Gait with Constant Velocity

The duty factor of dynamic gait is under 0.5. However, this gait of ladder climbing is not same as the quadruped walking because the robot needs a period to release a hand from a rung. The duty factor of the pace gait in ladder climbing is over 0.5 and we cannot describe it is dynamic gait. Even though this gait is not dynamic, it is difficult to track the reference trajectory because the robot supports the body at two points and this period is unstable. The climbing velocity need to be slowed not to

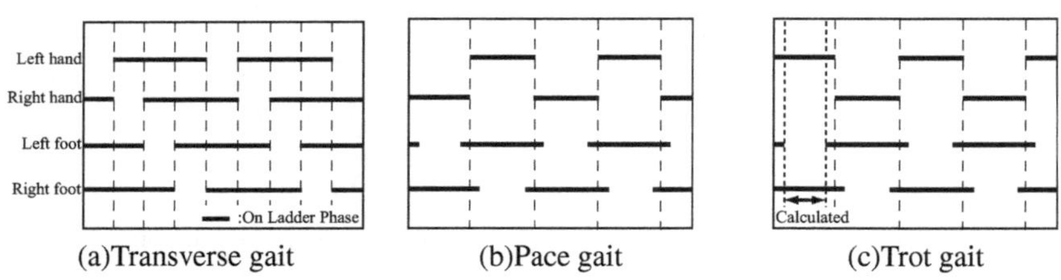

Fig. 6.2 Gait pattern.

cause any disturbance. On this gait, the operator needs to decide the climbing initial velocity and a one step climbing time heuristically and has to look for the relevant conditions that the robot does not collide with a ladder.

6.1.2.3 Trot Gait with Acceleration

The duty factor of this gait is also over 0.5. When behavior of the gravity momentum by inclination of AOY influences ladder climbing motion, it is difficult for the robot to track the reference trajectory only by the posture control because a large yawing momentum is induced. We suppose that the acceleration force which is added at COG also influences for the determination of vertical ladder climbing motion to reduce gravity momentum. We describe it in detail at next section.

6.1.3 Body Yawing Momentum

The dynamics is completely different between a vertical ladder and a step ladder. When the robot climbs a step slope or tilt ladder, the robot locomotes like biped and quadruped walking. In this case, the Zero-Moment Point(ZMP) to confirm the stability must be considered.However, ZMP is not related to the stability of vertical ladder climbing because the COG is always out from the supporting polygon of soles. We remark the momentum around the AOY to design the stable climbing motion and foot posture.

The momentum which is a rotational factor around the axis is calculated as follows on the basis of Σ_0.

$$\mathbf{M} = R_\theta \cdot \Sigma \left(\mathbf{r}_a \times \mathbf{F}_a\right) \tag{6.2}$$

$\mathbf{M}$: The sum of all momentum around AOY
R_θ : Rotational matrix around Y-axis
θ : The AOY inclination
$\mathbf{r}_a$: The coordinate at Σ_a
$\mathbf{F}_a$: The force vector at Σ_a

The yawing momentum around AOY calculated from Eq.(6.2) is as follows:

$$M_{yaw} = a_1 \sin\theta + a_2 \cos\theta \tag{6.3}$$

$$\theta = \arctan\left(\frac{x_{hand}}{z_{hand}}\right) \tag{6.4}$$

$$a_1 = -z_{hand}F_{y_{hand}} + y_{COG}(F_{z\alpha} - Mg) \tag{6.5}$$

$$a_2 = x_{hand}F_{y_{hand}} + x_{leg}F_{y_{leg}} - y_{COG}F_{x\alpha} \tag{6.6}$$

M is the total weight of the MLR and r_α is the position coordinate and F_{r_a} is r-direction element of force. When the robot keeps the motion of $M_{yaw} = 0$, it represents no momentum around the AOY, we suppose that the MLR can realize each gait stably. We decide the foot posture of each gait for the MLR with considering the momentum.

6.1.3.1 Three Supporting Climbing(Foots on Ladder)

In this static climbing($F_{hand} = 0$, $F_\alpha = 0$), M_{yaw} is given by

$$M_{yaw} = -y_{COG} Mg \sin\theta + x_{leg} F_{y_{leg}} \cos\theta \tag{6.7}$$

When the reaction force $F_{y_{leg}}$ is applied well, the body is maintained at three points and the robot can climb the ladder stably. $M_{yaw} = 0$, that is $F_{y_{leg}} = \frac{y_{COG} Mg}{x_{leg}} \tan\theta$.

The vector of $F_{y_{leg}}$ is to COG and the value is minus, and y_{COG} is also minus from the coordinates. From this equation,the value of the inclination of AOY θ must be plus to maintain the body. It means the robot posture must be maintained $x_{hand} > 0$. The shape of this foot posture is like trapezoid.

6.1.3.2 Three Supporting Climbing(Hands on Ladder)

In this static climbing($F_{leg} = 0$, $F_\alpha = 0$), M_{yaw} is given by

$$\begin{aligned} M_{yaw} = (-z_{hand} F_{y_{hand}} - y_{COG} Mg) \sin\theta \\ + x_{hand} F_{y_{hand}} \cos\theta \end{aligned} \tag{6.8}$$

$M_{yaw} = 0$ and $\theta > 0$ is maintained in this gait, that is $F_{y_{hand}} = \frac{-y_{COG} Mg \sin\theta}{x_{hand} \cos\theta + z_{hand} \sin\theta} > 0$. This momentum is generated from the force of supporting hands. However, this motion easily becomes unstable because the stiffness of robot with a lot of DOF is weak.

6.1.3.3 Two Supporting Climbing(Pace Gait with Constant Velocity)

In this static two supporting climbing without any acceleration($F_{hand} = 0$, $F_{leg} = 0$, $F_\alpha = 0$), M_{yaw} is

$$M_{yaw} = -y_{COG} Mg \sin\theta \tag{6.9}$$

This equation means the momentum of this gait does not take when the robot maintains the posture of the inclination of AOY θ = 0. The intervals of hands and foots are kept same during the robot climbing in this gait. The shape of this foot posture is like rectangle.

6.1.3.4 Two Supporting Climbing(with Acceleration)

In this type of gait($F_{hand} = 0$, $F_{leg} = 0$), the force of acceleration F_{α} and M_{yaw} are given by

$$F_{z_{\alpha}} = M\alpha \cos\phi, \; F_{x_{\alpha}} = M\alpha \sin\phi \tag{6.10}$$

$$M_{yaw} = y_{COG}(-Mg\sin\theta + M\alpha \sin(\theta - \phi)) \tag{6.11}$$

The acceleration value at COG without gravity and direction are shown α and ϕ.

As the condition of constraint of this gait, the acceleration is constant. The operator chooses the initial parameters including the initial velocity. Then, the value of acceleration, direction for the COG trajectory and one step time for vertical climbing are calculated uniquely from the following equations. The equation of x_{COG} and z_{COG} trajectory are as follows:

Table 6.1 Parameter of Climbing and Swinging Directions

	Parameters		
DIRECTION	Initial Velocity	Destiny COG Position	Gripping position
Climbing(z)	v_z	z_{des}	z_{hand}
Swinging(x)	v_x	x_{des}	x_{hand}

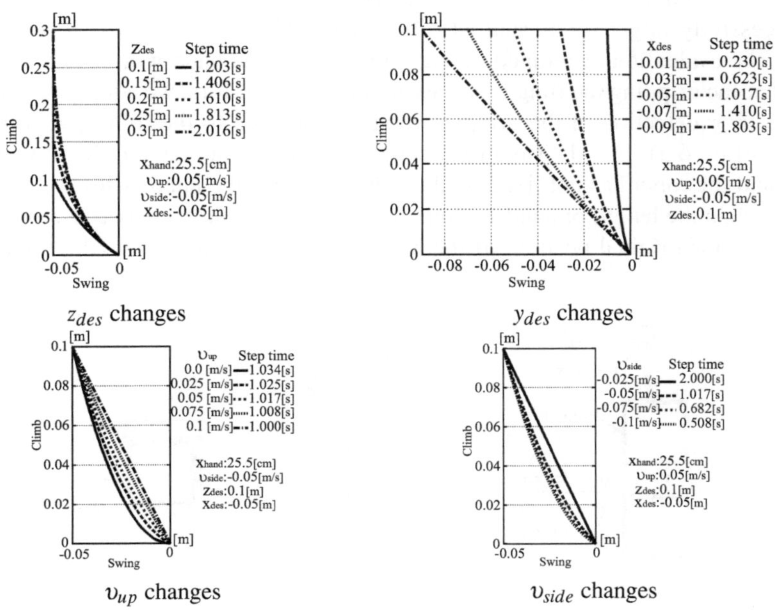

Fig. 6.3 Trajectory of COG Changing Each Parameter

$$x_{COG} = \int_0^t \left(\int (\alpha \cos\phi - g)\, dt + v_{up} \right) dt \tag{6.12}$$

$$z_{COG} = \int_0^t \left(\int (\alpha \sin\phi)\, dt + v_{side} \right) dt \tag{6.13}$$

The parameters to calculate the reference COG trajectory from Eq. 6.4,6.11-6.13 are shown in TABLE 6.1. The calculated results are in Fig. 6.3. The operator chooses the suitable parameters heuristically for the reference COG trajectory, acceleration and one step time which the robot can realize stably in this climbing motion.

6.1.4 Error Recognition and Escape Motion

In the ladder climbing motion, following failure cases of gripping the rung may occur:

- The robot hand fails to grip the rung.
- The hand grips the rung insufficiently.

The robot can recognize whether the hand grips a rung appropriately without any external sensors by considering the output voltage from the hand. In the case of that the robot successfully grips the rung, some voltage output from the hand is returned. However, if the less voltage output is returned, it implies that the gripping rung failed. On the other hand, if the tip of the finger reached the rung but failed to catch it successfully, larger voltage output will be retuned because the hand is overloaded. Thus, the evaluation of retuned output voltage from the hand will enable to judge whether the gripping the rung is successful or not. Therefore, we adopt this scheme to the error detection of the gripping rung and proceed to the following recovery process(Fig. 6.4). In 1-1, it is in the state where the hand has interfered with the rung and the robot has not fixed on the ladder. Even if the hand raises the distance shorter than the length of hand 1st link, the tip is still on the rung because the robot returns to the original position. If the hand raises more than the length of hand 1st

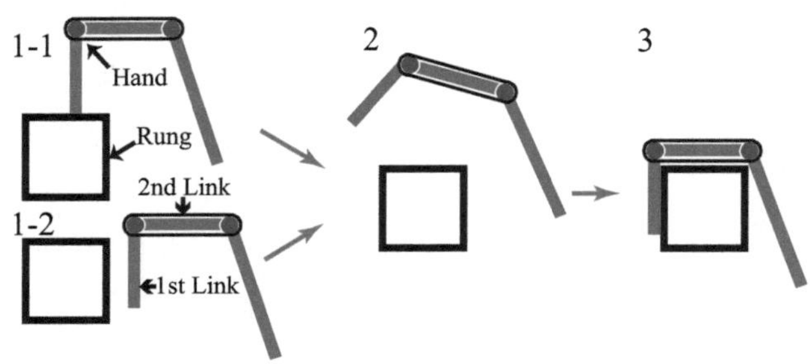

Fig. 6.4 Recovery Motion Model for Gripping Rung

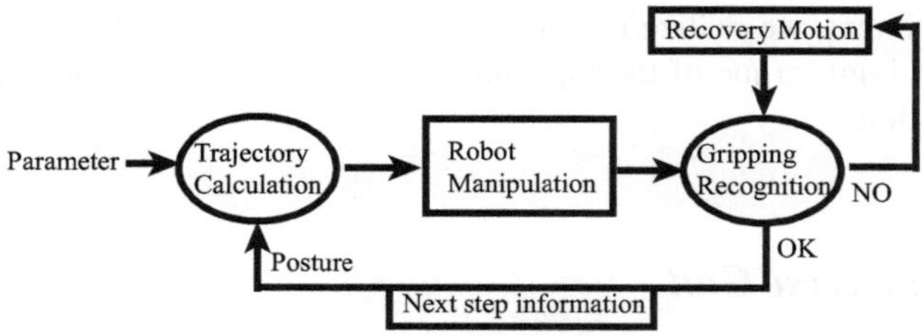

Fig. 6.5 The Control Flow in Continuous Ladder Climbing

link, the hand is not interfered with the rung to re-grip in the situation 1-1 and 1-2. Then the hand moves the length of a hand 2nd link with opening to the ladder back. At last, it returns to the original position with closing. The hand detects whether it is well grip the rung from the output voltage and repeat a series of this operation until it succeeds. In this case, we need to consider the robot posture in the same condition that the three supporting climbing(foot on ladder), that is the $x_{hand} \geq 0$, because the hand is free but the foot is on the rung in recovery motion. Considering this case, we decided the robot posture which the foot interval is wider than hand interval.

The control flow of ladder climbing is as follows(Fig. 6.5).

1. The operator gives the suitable parameters for basic motion.
2. The computers calculate the AOY inclination, the reference COG trajectory and one step period from the parameters.
3. The robot tracks the reference COG trajectory.
4. It judges the gripping situation from the hand output voltage.
5. It operates a recovery motion or not, then computer calculates the gripping position as a parameter from the robot posture.
6. Back to 2).

We confirmed that this control flow is useful to operate the continuous vertical ladder climbing experimentally.

6.2 Experiment

We conducted experiments with Gorilla III introduced in Chapter 3 to validate the proposed control method. The experimental condition is as follows:

- The ladder is set up vertically, and the interval of rungs of the ladder is constant 0.2[m] with each other.
- Cross section of rung is square of $2 \times 2[\text{ cm}^2]$ and is covered with a rubber plate for slip prevention.

In this experiment, we realize the three types of ladder climbing gait and observe the output of joint torque of the hips and the shoulders to estimate the stability of climbing motion.

6.2.1 Transverse Gait

In the experiment of ladder climbing of transverse gait, x_{hand} is 0.045[m] and the MLR keeps the foot posture like trapezoid not to cause the body yawing. The climbing speed is 1.67[cm/s] and one step cycle is 12.0[s]. Figure 6.6 shows the experimental snapshot of the ladder climbing with transverse gait. From the figure, it is shown that the MLR stably climb the ladder with no yawing moment. Since the robot supports the body at three points in this gait, the shoulder and the hip joint torques were well dispersed and the trapezoid foot posture for the motionw is considered suitable to the stable climbing of the ladder(Fig. 6.7).

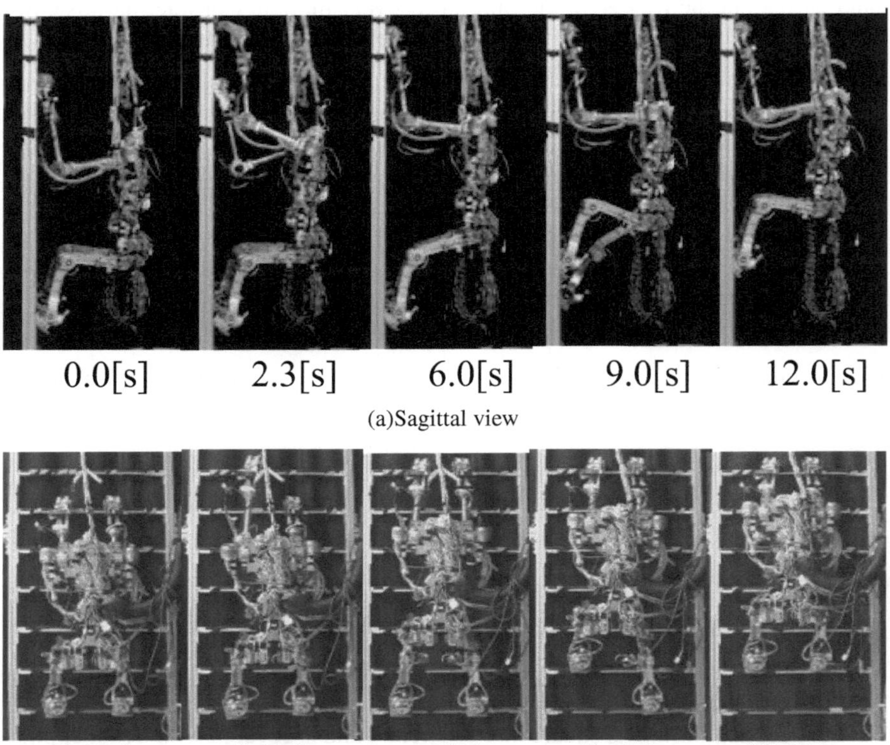

0.0[s] 2.3[s] 6.0[s] 9.0[s] 12.0[s]

(a)Sagittal view

0.0[s] 2.3[s] 6.0[s] 9.0[s] 12.0[s]

(b)Lateral view

Fig. 6.6 Snapshots of Transverse gait.

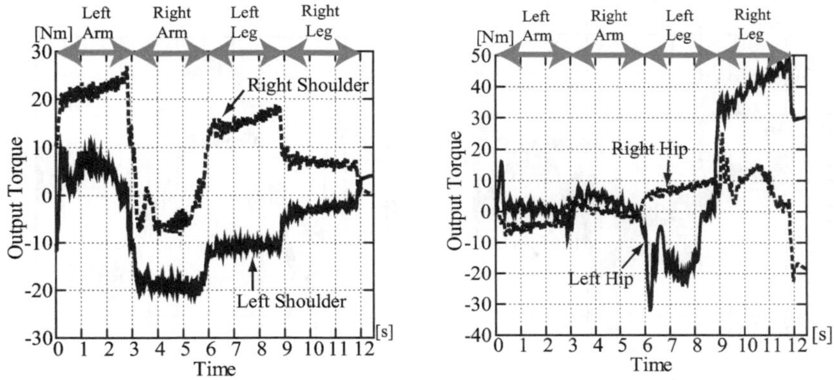

(a)Shoulder (b)Hip

Fig. 6.7 Output Torque of Transverse Gait

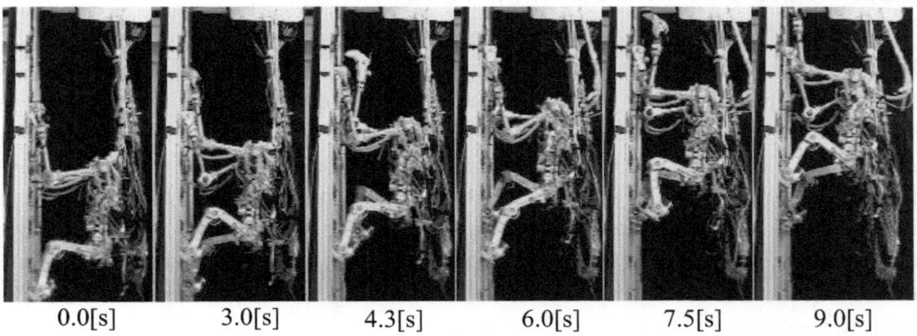

(a)Sagittal view

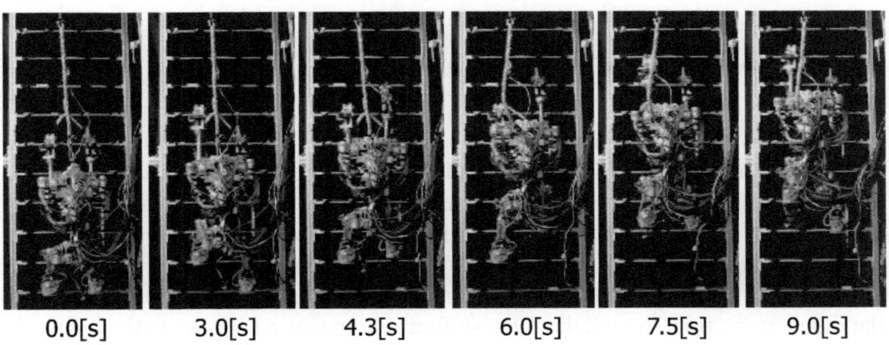

(b)Lateral view

Fig. 6.8 Snapshots of Pace Gait with Constant Velocity

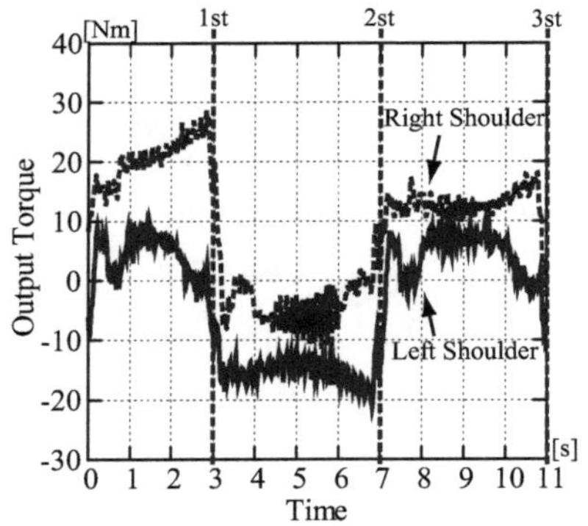

(a)Shoulder

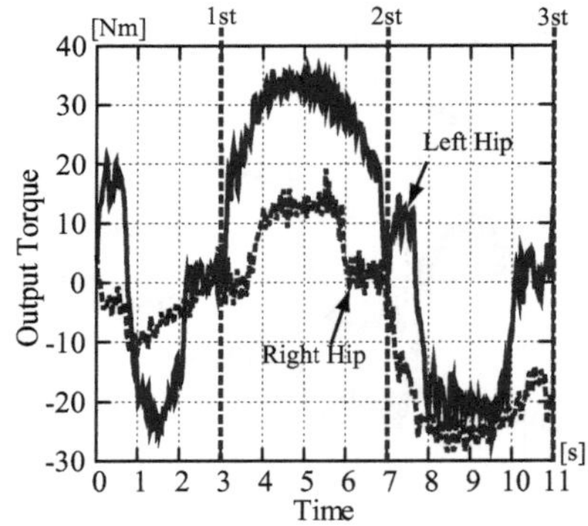

(b)Hip

Fig. 6.9 Torque result of Pace gait with constant velocity

6.2.2 Pace Gait with Constant Velocity

In the case of pace gait, the MLR keeps the foot posture like rectangle($x_{hand} = 0.0$) which the interval of the hands and the foots equal. The climbing speed is set to 5[cm/s] which was determined heuristically taking account not to cause any disturbance. The pace gait cycle is almost one third of static gait. Figure 6.8 shows the experimental snapshot of the ladder climbing with pace gait. From the figure, it is shown that any disturbance was not caused, and the MLR stably climb the ladder continuously. The result of the periodic output torque in Fig. 6.9 also shows that any disturbance is not caused whiles the robot climbing in this gait and the rectangle posture is considered suitable for this pace gait.

6.2.3 Trot Gait with Acceleration

This gait has an AOY with inclination. The parameters to determine the posture and motion are shown in TABLE 6.2 and the calculated values for each step are on TABLE 6.3. This ladder climbing motion with trot gait is realized with the flow of motion planning and control(Fig. 6.5). Figure 6.10 shows the experimental snapshot of the ladder climbing with trot gait. From the figure, it is shown that the MLR climbed each one step faster than other gaits with acceleration and more stable with the recovery motion, though the foots shifted side to side a little when the hand grip a rung. The result of the periodic output torque in Fig. 6.11 also shows the validity of the flow of this motion and the problem of the foot stablilzing. We checked the result of the error recognition to continue the climbing motion from the output voltage of the hand. The MLR recognized the successful hand gripping at the time where the step number and the enclosure is drawn in Fig. 6.12. From these results, the flow of this gait motion with reference COG trajectory, error recognition and

Table 6.2 Parameter of Up and Side Directions

		Parameters		
Step	Direction	Initial Velocity[m/s]	Destiny COG Position[m]	Gripping position[m]
1st	Cilmb	0.05	0.1	1.0
	Swing	-0.05	-0.05	0.245
2st	Cilmb	0.05	0.2	1.0
	Swing	0.05	0.05	-
3st	Cilmb	0.05	0.2	1.0
	Swing	-0.05	-0.05	-

Table 6.3 The Calculated Parameters for Each Step in Trot Gait

Step number	Gripping position[m]	Step period[s]
1st	0.245	2.244
2st	0.237	2.616
3st	0.213	2.568

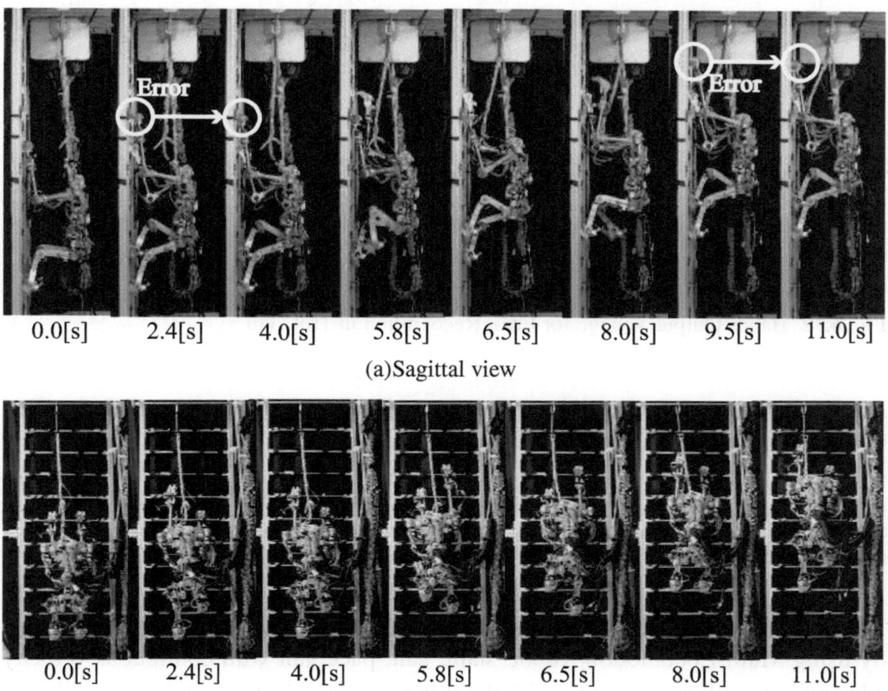

Fig. 6.10 Snapshots of Trot gait with accerelation velocity

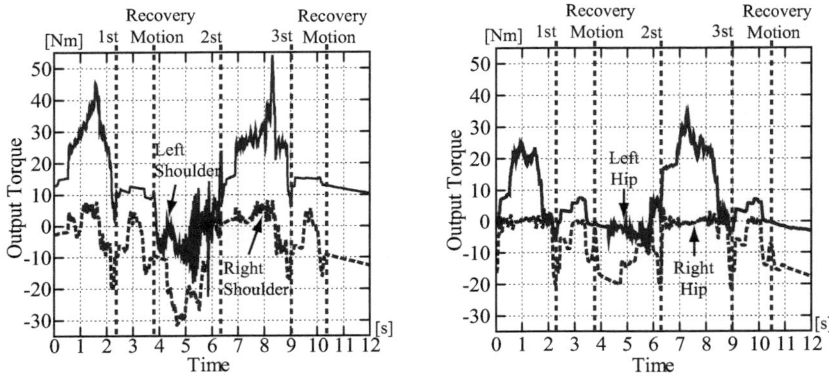

(a)Shoulder (b)Hip

Fig. 6.11 Output Torque of Trot Gait with Acceleration

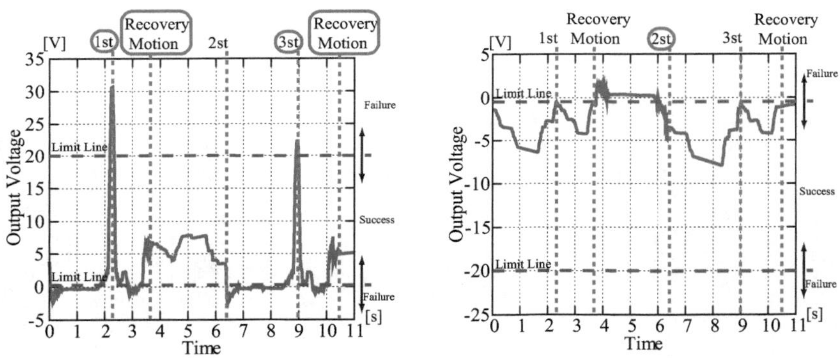

(a)Left hand (b)Right hand

Fig. 6.12 Hands Output Voltage for Error Recognition in Trot Gait

recovery motion sufficiently functioned to maintain the body and stable climbing on the ladder for the MLR.

6.3 Summary

In this chapter, we realize the various types of vertical ladder climbing locomotion by using Multi-Locomotion Robot: static gait, pace gait with continuous velocity and trot gait with acceleration. The stability of vertical ladder climbing of MLR in static gait and pace gait is maintained with the posture control considering the momentum around axis of yawing. Even if the axis has inclination, the reference COG trajectory and acceleration is calculated to determine the motion which can maintain the stability of posture on the ladder. The control flow with the error recognition from output voltage and recovery motion were well operated and the MLR realized the continuous ladder climbing in the trot gait.

Additional use of some other sensors would improve the stability of climbing furthermore. Also, we are attempting to realize a transition from the biped walking posture to the ladder climbing mode using an external sensor.

Chapter 7
Transition Motion from Ladder Climbing to Brachiation

7.1 Motion Design

7.1.1 Environment Statement

The environmental information is given beforehand and includes the intervals and shape parameters of the rungs on the ladder. The experimental environment, shown in Fig. 7.1, is as follows:

1. A rigid ladder is set up vertically.
2. The cross section of each rung on the vertical ladder is a 2×2 cm^2 square, and the rungs are covered with rubber plates for slip prevention.
3. Another ladder that connects with the top of the vertical ladder is set up horizontally to simulate tree limbs for primates' brachiation.
4. Each rung on the horizontal ladder is a cylinder (ϕ2.5 cm) connected to the beams with two rubber suspension units, as shown in Fig. 7.1. The mechanical specifications are shown in table 1, where α is the deformation angle of the suspension unit, as shown in Fig. 7.1.
5. The intervals between rungs on the vertical and horizontal ladders are 20 cm and 40 cm, respectively.

Table 7.1 Rubber suspension unit.

Type	Torque M in N.m at $\angle\alpha$					
	5°	10°	15°	20°	25°	30°
DR-A 18×50	3.2	7.5	12.5	18.3	25.0	34.4

T. Fukuda et al.: Multi-Locomotion Robotic Systems, STAR 81, pp. 153–171.
springerlink.com © Springer-Verlag Berlin Heidelberg 2012

7.1.2 Motion Planning

The motion flow is as follows:

Step I: Move the right hand to grasp the rung's center with two legs stepping on the vertical ladder,
as shown in Fig. 7.2 a.
Step II: Adjust the contact situation and relax the left hand's grip preparing for turning waist, as
shown in Fig. 7.2 a.
Step III: Turn a body to approach a horizontal ladder on the opposite side of the vertical ladder, and
grasp the second horizontal rung with left hand, as shown in Fig. 7.2 a-c.
Step IV: Adjust the orientation of the joint on the right shoulder preparing for the motion of next
step, as shown in Fig. 7.2 c.
Step V: Grasp the first horizontal rung with the right hand, as shown in Fig. 7.2 c-d.
Step VI: Move feet away from the vertical ladder, and adjust the posture to initialize brachiation,
as shown in Fig. 7.2 d-f.

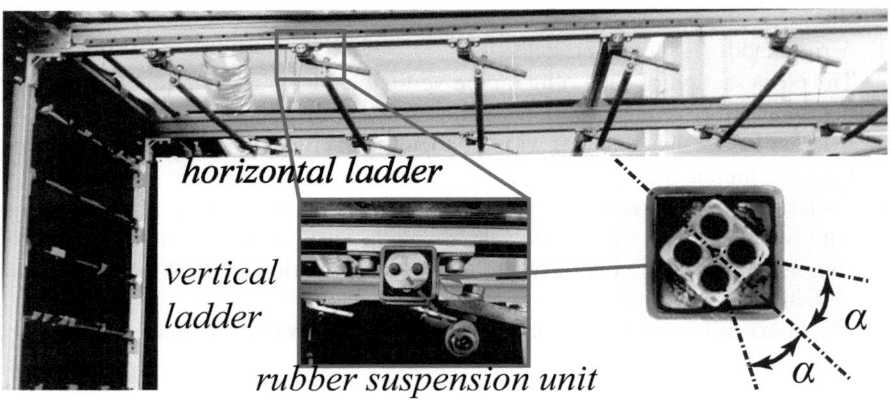

Fig. 7.1 Ladders and rubber suspension unit.

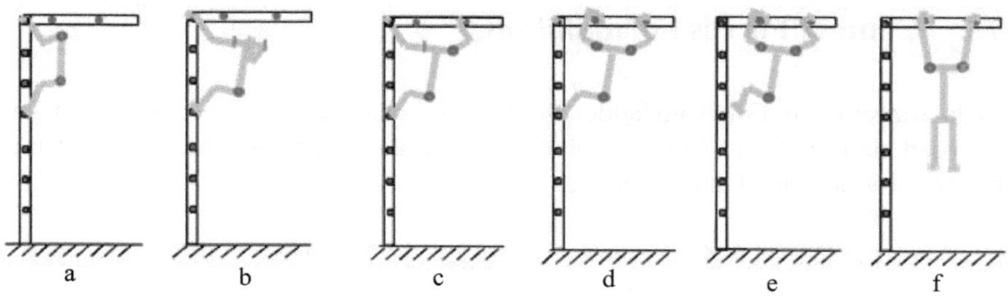

Fig. 7.2 Transition motion from ladder climbing to brachiation.

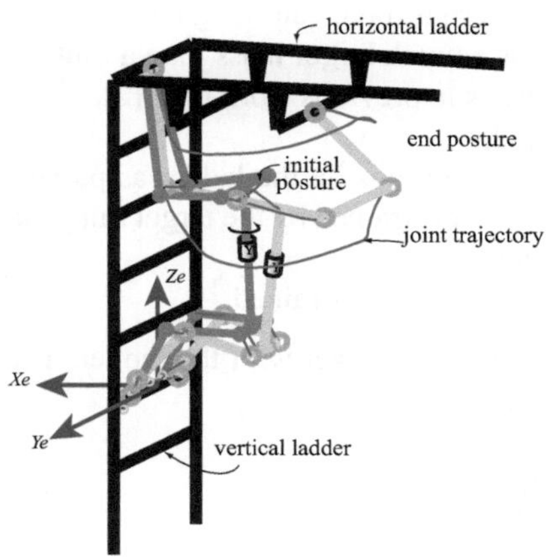

Fig. 7.3 Motion trajectory in motion step III.

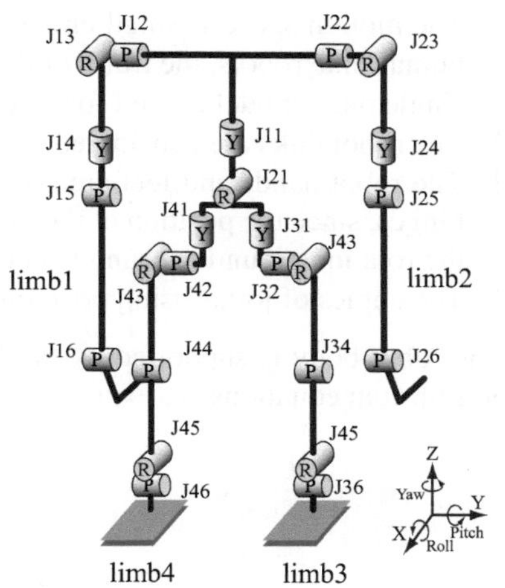

Fig. 7.4 Joint numbers of Gorilla Robot III.

7.1.3 Transition Motion

The motion of the end effectors in free space is easy to realize (e.g., motion steps I, IV, V, and VI) using a position control based on the inverse kinematics. The robot's stability in step III is maintained with three end effectors supporting its body, while one hand transfers from a rung on the vertical ladder to a new horizontal rung behind the robot, as shown in Fig. 7.3. The link structure of the MLR is shown in Fig. 7.4. In order to balance the supporting forces of the two feet, the positions of the supporting points are designed to be symmetric with respect to the center plane of the ladders (X-Z plane), where the right hand grasps the center of a rung and the two feet are placed symmetrically on a lower rung. The center of the robot body is also designed to move in the center plane. Then, the turning around motion is determined. The designed trajectory of each joint is shown in Fig. 7.3.

7.2 Contact Forces Formulation

In the transition motion from ladder climbing to brachiation designed in section 7.1, a closed chain is formed by the robot links and the ladder. A force model of this transition is described in this section.

7.2.1 *Assumptions and Equilibrium Equations*

There are some assumptions in this transition.

1. The motion speed is small enough that the inertial force can be ignored. As for human-like robots, the trunk of the MLR has a much larger mass than a limb. A single mass model is used for force analysis, as in previous work [11, 12].
2. The robot links are considered infinitely rigid.
3. The robot hands and feet are constrained by the rungs on the ladder as passive hinges, since the position of the end effector is constrained by the target rung and the rotation around the rung is not constrained.
4. The angles of joints using position control are strictly restrained.

The robot body is supported by the limbs that are in contact with the ladder. The equilibrium equations are expressed as follows:

$$G+\sum_{i=1}^{r} f_i = 0, \tag{7.1}$$

$$p_{cog} \times G+\sum_{i=1}^{r} p_i \times f_i = 0, \tag{7.2}$$

where f_i is a 3×1 supporting force vector acting at end effector i, i=1, 2, 3, and 4 respectively indicate the right hand, left hand, left foot, and right foot, G is a 3×1 gravity vector, p_i is a 3×1 position vector denoting the position of end effector i, and p_{cog} is a 3×1 position vector of the robot's center of gravity (COG). The supporting force at each end effector is balanced by the torques on the limb in a static situation. The torques of the joints on limb i are expressed as follows:

$$\tau_i = J_i^T(\Theta) f_i, \tag{7.3}$$

where τ_i is an $n_i \times 1$ torque vector generated by the joint motors of limb i, $\tau_i = [\tau_{i1}, \tau_{i2}, ..., \tau_{in_i}]$, n_i is the number of joints on limb i, $J_i^T(\Theta)$ is the $n_i \times 3$ transposed Jacobian matrix of limb i, Θ is the angle vector denoting the angles of the robot joints, and n_i is the number of DOF of limb i.

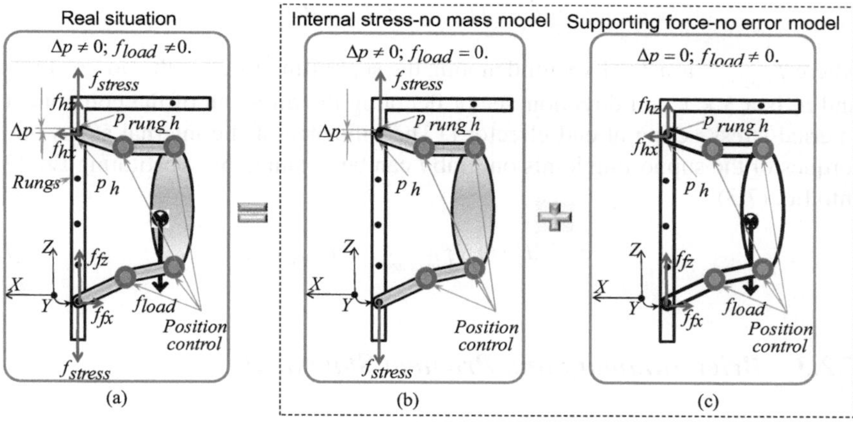

Fig. 7.5 Decomposition of force, internal stress, and equilibrium force system.

7.2.2 *Supporting Forces Decomposition*

The supporting forces of the robot end effectors are not only determined by the robot posture and robot load, but are also influenced by the internal stress. The real supporting forces can be represented as the combination of internal stress and the ideal supporting force, ignoring the position error, as shown in Fig. 7.5.

A no-mass model is used for stress analysis (Fig. 7.5 (b)). In the no-mass model, we only consider the forces caused by the deformation of the mechanical structure, and the influence of the robot mass is ignored. In this model $f_{load} = 0$ and $\Delta p \neq 0$, where f_{load} is the total load of the robot body generated by the robot's mass and Δp is the distance error between the robot's end effectors and the target supporting points in the environment. Since the robot's end effectors are constrained by the target rungs, the elastic deformation is equal to Δp. The internal stress between the robot's hand and feet is proportional to the distance error:

$$f_{stress} = -k\Delta p, \tag{7.4}$$

where f_{stress} is the magnitude of the internal stress between the supporting hand and supporting feet. The direction is along the vertical ladder (in the line of the Z axis), and k is the stiffness of the environment between the target supporting points of the hand and two supporting feet.

A no-error model is used for the supporting force calculation (Fig. 7.5 (c)). In the no-error model, the distance between the robot's supporting hand and feet is just equal to the distance between the target rungs. In this model, $\Delta p = 0$. The supporting forces can be calculated based on the robot load, position of COG, and the positions of the supporting end effectors. In a real situation, the supporting forces are a combination of the no-mass and no-error models (see Fig. 7.5 (a)):

$$f_i = f_{i_{no\_error}} + f_{stress} u_i, \tag{7.5}$$

where $f_{i_{no\_error}}$ is a 3×1 vector denoting the supporting force of the no-error model and u_i is a 3×1 unit direction vector denoting the direction of the component of internal stress acting at end effector i. The influence of the internal stress on the torques of the supporting joints on limb i can be obtained by substituting Eq. (7.5) into Eq. (7.3):

$$\tau_i = J_i^T(\Theta)(f_{i_{no\_error}} + f_{stress} u_i). \tag{7.6}$$

7.2.3 Brief Summary and Problem Statement

In brief, the position error Δp influences the internal stress f_{stress} (Eq. (7.4)), and then the internal stress f_{stress} affects the supporting force (Eq. (7.5)) and the torque of the supporting joint (Eq. (7.6)). Thus, there are two problems in traditional position control caused by the position error: the end effector cannot grasp or step on the target supporting point, and the torques of some joints go over their limits. From Eq. (7.4), we can see that the internal stress f_{stress} is not only determined by the position error Δp but also influenced by the stiffness of the contact system k. The internal stress problem is not significant when the robot moves in the horizontal ladder environment, because passive compliance schemes are used to reduce the stiffness of the contact system, as in previous works [15-17]. However, the stiffness of the vertical ladder is very high in our experiment. In this case, even a small position error will cause a large change in internal stress. Thus, a load-allocation control is proposed to adjust the internal stress in the turning around motion (step II and step III) when the robot is constrained by the vertical ladder, as shown in Fig. 7.2 a-c.

7.3 Load-Allocation Control

7.3.1 Concept of Load-Allocation Control

For a multilimb robot, if more than one limb is constrained by a rigid environment, the robot allocates the motors' loads by adjusting the internal stress inside the closed chain formed by the robot links and the environment. We call this the load-allocation control. Load-allocation control is useful for a motion with multiple contacts, especially in case where the positions of the end effectors are constrained strictly by a rigid environment. The type of strict position constraint can be an actual joint, such as those in a parallel link robot, a contact with adhesion forces, such as a suction cup or magnet, or a contact with a high friction coefficient, such as those used for walking. In this section, we take the transition motion from ladder climbing to brachiation as an example to explain how to realize an optimal load-allocation control in multi-contact robot motion.

7.3.2 *Objective Function and Constraints*

To solve an optimal load-allocation problem, an objective function has to be minimized while respecting the essential constraints. Several issues can be taken into account. Under some conditions energy efficiency can be an issue, while for others safety and stability may be more critical. In this section, the safety of the robot's motion is more important, because it is dangerous if the robot falls down from the ladder.

A load factor is used to evaluate the working conditions of a joint motor. To determine the load factor, we first measure the output torque and then divide it by the maximum torque from the motor manual:

$$r_{ij} = \frac{|\tau_{ij}|}{\tau_{ij_{max}}}, \tag{7.7}$$

where r_{ij} is the load factor of joint ij on limb i, τ_{ij} is the order torque of joint ij, i =1, 3, and 4, j =1, 2, ..., 6, and $\tau_{ij_{max}}$ is the maximum torque of joint ij. Note that in our experiment, the actual output torque corresponds well to the order torque (validated by the experimental result, as shown in Fig. 7.15), thus the load factor is calculated using the order torque divided by the maximum torque.

If the load factor is more than 1.0, the motor is overloaded. A motor is expected to burn out if it is overloaded continuously for a certain time. Even though an overload protector is used in our experiment, a motor's power will be cut off if it is overloaded for more than 1.5 s, and then the motor will be out of control. Thus, it is important to minimize the maximum load factor of the joint motors for safety. The objective function is given as a *min-max* function as follows:

$$\text{minimize} \quad r_{fail} = \max(r_{11}, r_{12}, \ldots, r_{ij}, \ldots, r_{46}), \tag{7.8}$$

subject to

$$\begin{aligned}
&\text{Equilibrium equations:} && \text{Eq.(7.1) and Eq.(7.2)}, \\
&\text{Robot kinematics:} && p_i = Kin_i(\Theta), && (7.9)\\
&\text{Torque limitation:} && -\tau_{ij_{max}} \le \tau_{ij} \le \tau_{ij_{max}}, && (7.10)\\
&\text{Contact constraint:} && f_{fz} > 0, && (7.11)
\end{aligned}$$

where r_{fail} , in Eq. (7.8) is the chance of failure defined as the maximum load factor of all the motors, $Kin_i(\Theta)$ is the kinematic function of limb i, Eq. (7.10) and Eq. (7.11) are respectively the torque limitation of each joint and the contact constraint of the robot's feet, and f_{fz} is the force acting at the robot's feet in the Z direction.

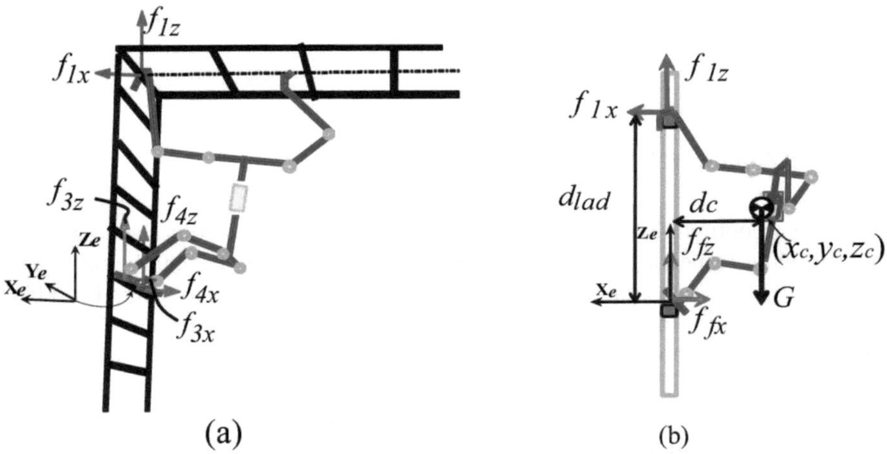

Fig. 7.6 Force model in turning around motion. (a) 3-D. (b) 2-D.

Since the robot's feet have no grippers like the hands do, only a supporting force directed toward the foot surface can be supplied.

As mentioned in Sec. 7.1.3, the positions of the supporting points are designed to be symmetric with respect to the center plane of the ladders, and the center of the robot body is also designed to move in the center plane, as shown in Fig. 7.3. Based on the equilibrium principle, the reaction forces of the two feet are expected to be equal ($f_3 = f_4$) under the condition that there is no position error between the robot foot and its target rung. The condition $f_3 = f_4$ can also be satisfied in an actual experiment if the position errors are compensated (validated by the experimental results in Sec. 7.4.3). Since the load-allocation control is able to compensate the position error between the end effectors and the environment, the 3-D force model shown in Fig. 7.6 (a) can be simplified to a 2-D model, as shown in Fig. 7.6 (b), in the process of load-allocation control. The force supplied by the robot's feet is expressed as:

$$f_f = f_3 + f_4, \tag{7.12}$$

where f_f is the total force supplied by the robot's feet and f_3 and f_4 are respectively the supporting force of the left foot and right foot. Note that the supporting force in the Y direction was controlled to be 0 to prevent the foot from sliding sideways on the rung. By solving the equilibrium equations (7.1) and (7.2) subject to $f_3 = f_4$, the supporting forces are calculated as

$$f_1 = [f_{1x}, f_{1y}, f_{1z}]^T = \left[-\frac{Gx_c}{d_{lad}}, 0, f_{1z}\right]^T$$
$$:= \mathbf{F}_1(x_c, f_{1z}), \text{and} \tag{7.13}$$
$$f_3 = f_4 = \frac{G - f_1}{2} = \left[\frac{Gx_c}{2d_{lad}}, 0, \frac{G - f_{1z}}{2}\right]^T$$
$$:= \mathbf{F}_{3,4}(x_c, f_{1z}), \tag{7.14}$$

where (x_c, y_c, z_c) is the coordinate of the COG relative to the environment coordinate system and d_{lad} is the distance between the two target rungs of the robot's hand and feet (see Fig. 7.6). From Eq. (7.13) and Eq. (7.14), we can see that the supporting forces in the X direction are determined by G, x_c, and d_{lad}. G and d_{lad} are constant values, and x_c is a time-based function that is calculated according to the robot's posture and the mass parameters of the robot structure. Thus, the forces in the X direction are determined for a designed robot motion. However, the forces in the Z direction are not unique. Even though the total force in the Z direction is equal to robot's gravitational force G, the feet's supporting force f_{fz} will decrease as the hand's supporting force f_{1z} increases.

7.3.3 Generation of Optimized Supporting Forces

The evaluated value (r_{fail}) in the objective function (Eq. (7.8)) can be changed by adjusting f_{1z}. The load factor r_{ij} which is a variable in the objective function, can be expressed as follows by substituting Eq. (7.13), Eq. (7.14), and Eq. (7.3) into Eq. (7.7):

$$r_{ij} = \frac{|J_{ij}^T(\Theta)\mathbf{F}_i(x_c, f_{1z})|}{\tau_{ijmax}}$$
$$:= \mathbf{F}_{r_{ij}}(f_{1z}, t), \tag{7.15}$$

Since the position variables, such as Θ and x_c, change according to time t, the load factor of each supporting joint can be expressed as a function of f_{1z} and t. Figure 7.7 shows the chance of failure, which is defined as the maximum load factor of all the joints in the objective function, Eq. (7.8). The optimized supporting forces are obtained when the chance of failure r_{fail} is minimized under the torque limitation and contact constraint, as shown in Fig. 7.7:

$$f_{1z_{opt}} = arcmin(r_{fail}), \tag{7.16}$$

where $f_{1z_{opt}}$ is the optimized supporting force of the right hand in the Z direction. We obtain $f_{1z_{opt}}$ by using a numerical search. The optimized supporting forces of the hand and feet are obtained by substituting Eq. (7.16) into Eq. (7.13) and Eq. (7.14):

$$f_{1_{opt}} = \left[-\frac{Gx_c}{d_{lad}}, 0, f_{1z_{opt}} \right]^T, \tag{7.17}$$

$$f_{3,4_{opt}} = \left[\frac{Gx_c}{2d_{lad}}, 0, \frac{G - f_{1z_{opt}}}{2} \right]^T. \tag{7.18}$$

7.3.4 Load-Allocation Algorithm

7.3.4.1 Determining the Degrees of Freedom

Pennestri et al. listed some definitions of the DOF. "*In mechanics, DOF are the set of independent displacements and/or rotations that specify completely the displaced or deformed position and orientation of the body or system*" [223]. This is a fundamental definition relating to systems of moving bodies in mechanical engineering, aeronautical engineering, robotics, structural engineering, etc. The number of DOF is also defined as "*the number of inputs which need to be provided in order to create a predictable output*" [210]. A multiple DOF mechanism, such as a robot, will have a constrained motion for all its links as long as the necessary number of inputs is supplied to control all its DOF [210]. This means that in order to create a desired motion, the number of inputs should be equal to the number of DOF for kinematic

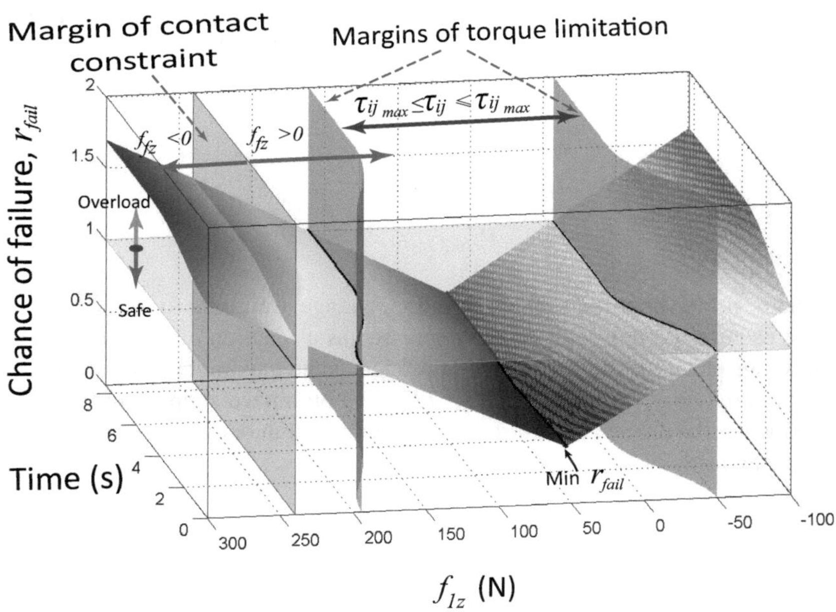

Fig. 7.7 Chance of failure and constraint margins in the turning around motion.

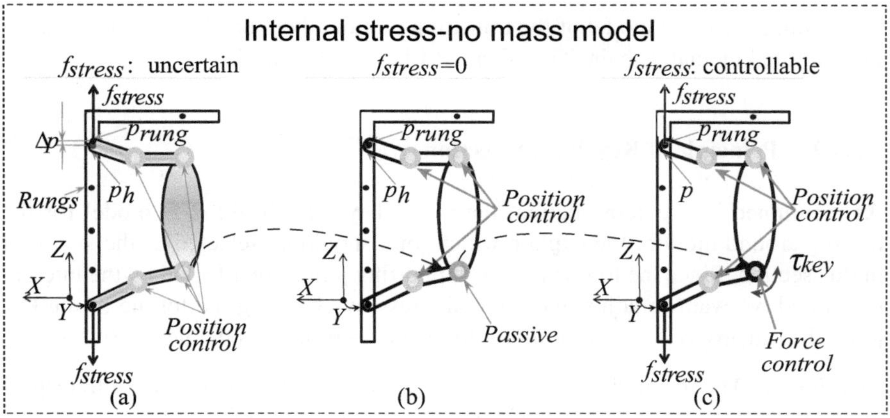

Fig. 7.8 Load-allocation measure by adjusting the internal stress using a key joint.

chains or mechanisms. It is convenient to control a robot's motion by providing a designed angle for all the joints using position control for a serial-link robot. However, sometimes the number of robot joints is not equal to the number of DOF of the kinematic chain. The number of DOF can be calculated based on Gruebler's equation as follows:

$$M = 3N_L - 2N_J - 3N_G, \tag{7.19}$$

where M is the number of DOF or mobility, N_L is the number of links, N_J is the number of joints, and N_G is the number of grounded links.

7.3.4.2 Determination of Control Type for Robot Joints

Since position errors are unavoidable, internal stress appears if the number of position-controlled active joints ($N_{\theta_{input}}$) is greater than the number of DOF (M) of the kinematic chain. That is, if

$$N_{\theta_{input}} - M > 0. \tag{7.20}$$

The internal stress can be reduced to be 0 by allowing the redundant robot joints to be passive (Fig. 7.8 (b)). Furthermore, the internal stress can be adjusted by changing the redundant joints from position control to force control (Fig. 7.8 (c)). Since these joints are very important for adjusting the internal stress, we call them key joints. The number of key joints is equal to the number of redundant joints:

$$N_{key} = N_{redun} = N_{\theta_{input}} - M. \tag{7.21}$$

In the simplified 2-D model of the transition motion, there are 4 robot joints, and 3 DOF of the kinematic chain. Thus there is 1 key joint (Eq. (7.21)).

7.3.4.3 Principle of Key Joint Selection

As mentioned in previous section, there is 1 key joint in the 2-D model for the turning around motion. We explain which joint should be selected as the key joint in this section. Since the torques of the supporting joints are affected by the internal stress and we want to adjust the internal stress by changing the torque of the key joint, the sensitivity of the joint motor to the internal stress is defined as follows:

Definition 1: The sensitivity s_{ij}, is the partial derivative of a function τ_{ij} with respect to the internal stress f_{stress} divided by the maximum torque of joint ij.

$$s_{ij} = \left| \frac{1}{\tau_{ijmax}} \frac{\partial \tau_{ij}}{\partial f_{stress}} \right|, \tag{7.22}$$

where s_{ij} is the sensitivity of joint ij denoting the degree to which the supporting joint (τ_{ij}) is affected by the internal stress (f_{stress}), i =1, 3, and 4, j =1, 2, ..., 6. From Eq. (7.22) we know that the joint that has the maximum value of s_{ij} is the most affected by the internal stress. That joint is easy to overload and is also good at adjusting the position of the end effector. Considering the importance of safety and efficiency of error compensation, we select the joint most affected by internal stress as the key joint.

Definition 2: Joint ij^* is referred to as a key joint if

$$s_{ij^*} = \max(s_{11}, s_{12}, ..., s_{46}). \tag{7.23}$$

If there are more than one s_{ij} that have the maximum value, then we compare their derivatives $\dot{s}_{ij}$, and chose the joint that has a smaller derive priority. The sensitivity s_{ij} is related to the robot posture. The sensitivities of the supporting joints in the designed turning around motion are shown in Fig. 7.9. In this motion, the hip joint, which has the maximum sensitivity, is selected as a key joint for the 2-D model. The corresponding key joints in the real experiment are J32 and J42, which are the pitch joints of the left and right hips.

7.3.4.4 Control Chart

The control chart is shown in Fig. 7.10. First, the joint angles are calculated using inverse kinematics. The waist angle θ_{11} changes monotonically in the turning around motion. Hence along this motion this angle θ_{11} can be used for an alternative representation instead of time t. $\theta_k(\theta_{11})$ is the designed angle of each joint. Then, the torques of the joints are adjusted by applying the desired torque to the key joints.

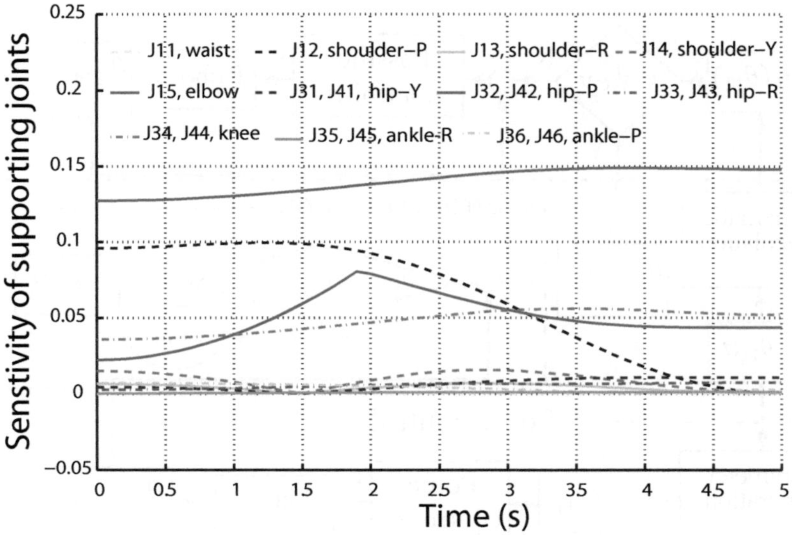

Fig. 7.9 Sensitivity of supporting joints in the turning around motion.

The position errors are evaluated using the angle error of the key joint θ_{key_e}. The desired torque of the key joint is obtained by substituting Eq. (7.18) into Eq. (7.3):

$$\tau_{key_{opt}} = J_{32}^{T}(\Theta) f_{3_{opt}}, \tag{7.24}$$

where $J_{32}^{T}(\Theta)$ is the transpose of the 3×1 geometric Jacobian vector of the hip joint (J32) and $f_{3_{opt}}$ is the 3×1 optimized supporting force vector of the robot's foot.

7.4 Eexperiment Results and Discussion

This transition motion starts after the MLR climbs up to the top of the vertical ladder. In this experiment, the robot climbs up using a pace gait [170]. The climbing speed of the pace gait is about 0.1 [m/s]. Figure 7.11 shows snapshots of the transition motion. Consider the robot moving from its initial posture to a goal posture in a certain time. Using inverse kinematics, the set of joint angles corresponding to the goal posture was calculated. The set of joint angles for the initial posture is also known. Then, a unique smooth trajectory of joint angles was generated using a cubic polynomial under the initial and final angle constraints and velocity constraints. The trajectories of the joints are shown in Fig. 7.12.

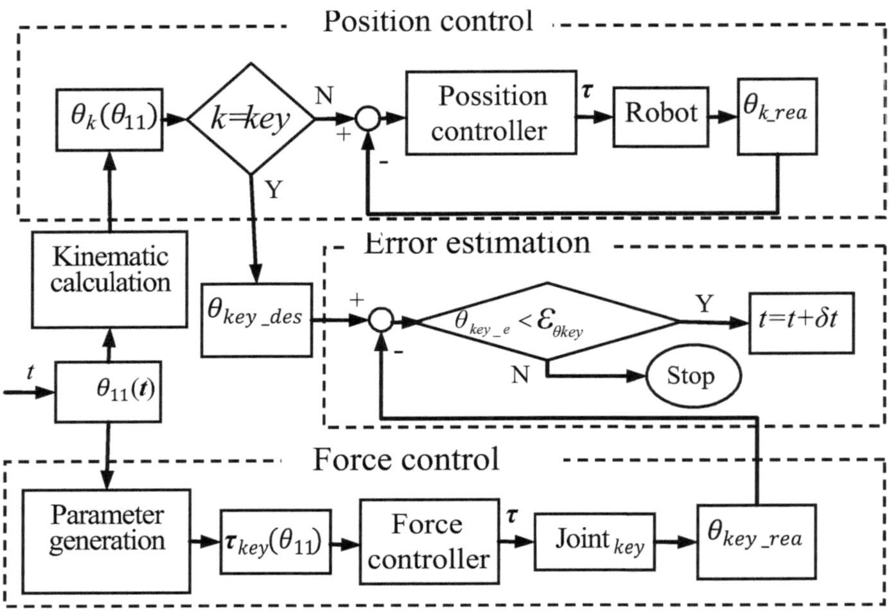

Fig. 7.10 Control chart for load-allocation. Here t is the time parameter, k is the joint number, $\theta_k(\theta_{11})$ are the angle functions of joint k about θ_{11}, θ_{k_des} and θ_{k_rea} are the designed angles and real angles of joint k, and e and τ are the angle error and torque of the joint.

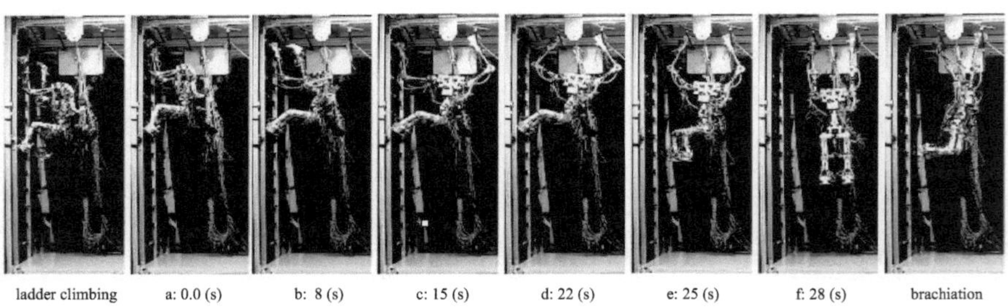

Fig. 7.11 Snapshots of transition motion from ladder climbing to brachiation.

7.4.1 Validating the Assumptions and Load-Allocation Ability

Assumption 1 in Sec. 7.2 is that the inertial force is ignored. The trajectory of the COG is calculated based on the masses and positions of the robot's parts (such as the links, reducers, and motors). Figure 7.13 shows the trajectory of the robot's COG in the designed motion. The acceleration of the COG, as shown in Fig. 7.14, is calculated based on the COG's time domain trajectory. The inertial force is

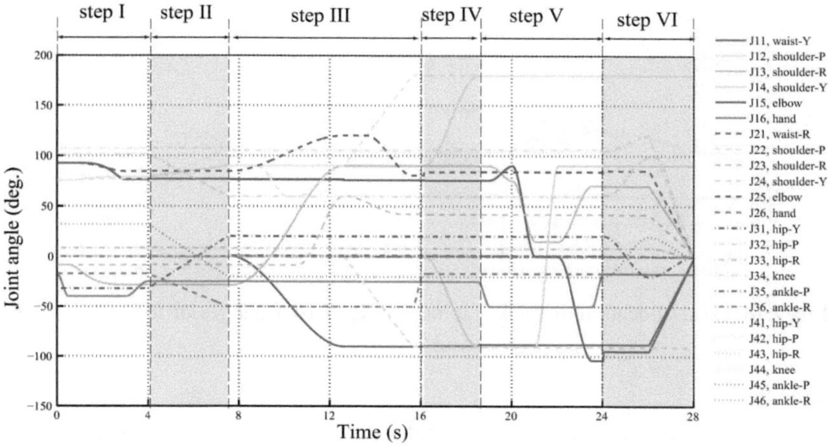

Fig. 7.12 Trajectories of robot joints in the transition motion. Note that steps I, II, ... VI are the steps of the motion flow defined in Sec. 7.1.2.

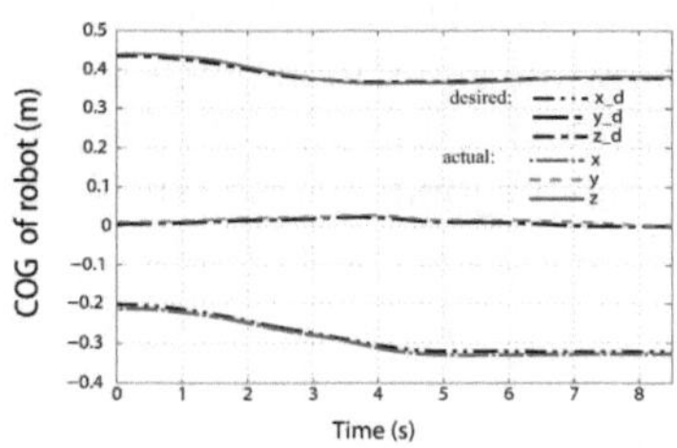

Fig. 7.13 Trajectory of COG in turning around motion.

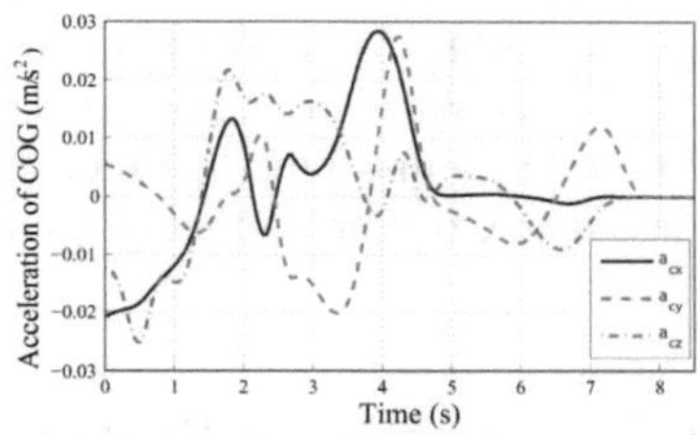

Fig. 7.14 Acceleration of COG in turning around motion.

$$f_{inetia} = -ma, \tag{7.25}$$

where a is a 3×1 vector denoting the acceleration of the COG. Figure 7.14 shows the acceleration of robot's COG. Since the absolute value of the inertial force is very weak relative to the gravity G ($\frac{|f_{inetia}|}{G} \leq 3.64\%$), the influence of the inertial force is small.

The allocation of the robot's load is adjusted by changing the torques of key joints for the rigid closed chain. Here, we take the starting posture of the turning around motion as an example to verify the load-allocation method. For safety, we change the torque of the key joints (J32 and J42) from 0% to 80% of τ_{max} ($\tau_{32_{max}} = \tau_{42_{max}} = 33.3$ Nm). Figure 7.15 shows the changes of the forces according to τ_{32} and τ_{42}. As τ_{32} and τ_{42} increase, f_{3z} and f_{4z} increase and f_{1z} decreases. This means that the robot feet supply more supporting force, while the load on the right hand decreases. Since the forces in the X direction, which is perpendicular to the internal stress, are obtained based on the equilibrium equations, f_{1x}, f_{3x}, and f_{4x} are determined by the

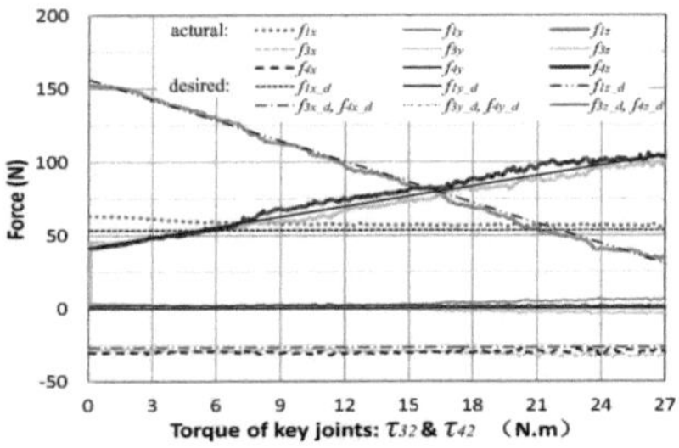

Fig. 7.15 Load-allocation by the key joints.

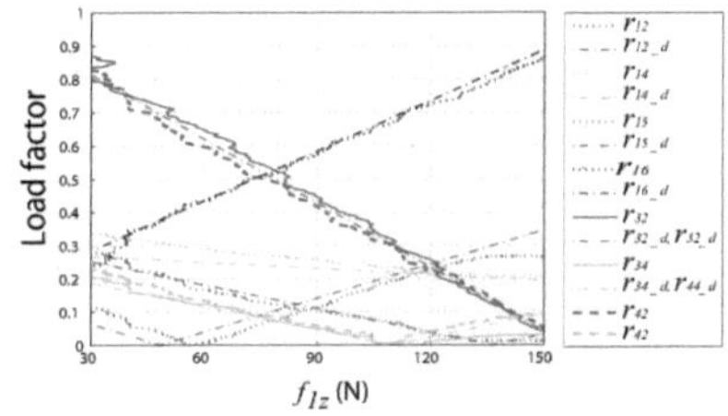

Fig. 7.16 Changes of load factors according to f_{1z}.

posture and are independent of internal stress (see the X components of Eq.(7.13) and Eq. (7.14)). Thus, the forces f_{1x}, f_{3x} and f_{4x} are constant for a certain posture. Since the equilibrium forces in the Y direction are not needed to balance the robot's load and moment (G is in the Z direction, and the COG moves in the X-Z plane) and the internal stress along the rung can be easily reduced as the robot moves its foot perpendicular to the rung, f_{1y}, f_{3y} and f_{4y} are close to 0. The changes of the load factors according to f_{1z} as shown in Fig. 7.16 follow the simulation result well, so the optimized parameters generated based on simulation results are valuable. Note that in order to make the figure clear, the load factors of joints whose sensitivities are less than 0.001 are not included in Fig. 7.16.

7.4.2 Discussion of Failures with Position Control

Failures of robot motion happen easily with position control. This is because in position control, position errors between the robot end effector and its supporting rung are unavoidable, and these errors directly influence the supporting forces of each end effector. Sometimes the motor will be overloaded due to the imbalance of supporting forces. Figure 7.17 shows the load factor when the left hip motor is overloaded in an experiment with position control. Figure 7.18 shows that most of the supporting forces are supplied by the left foot in this experiment. This illustrates that the left foot steps down too hard on the supporting rung, and the motor for the left hip in the pitch direction is overloaded in step III at 11.49 s. In order to prevent the motor from burning out, the motor's power was switched off automatically by the overload protector after the motor was overloaded for 1.5 s. This resulted in the motor becoming out of control, and the robot failed the turning around motion in this experiment.

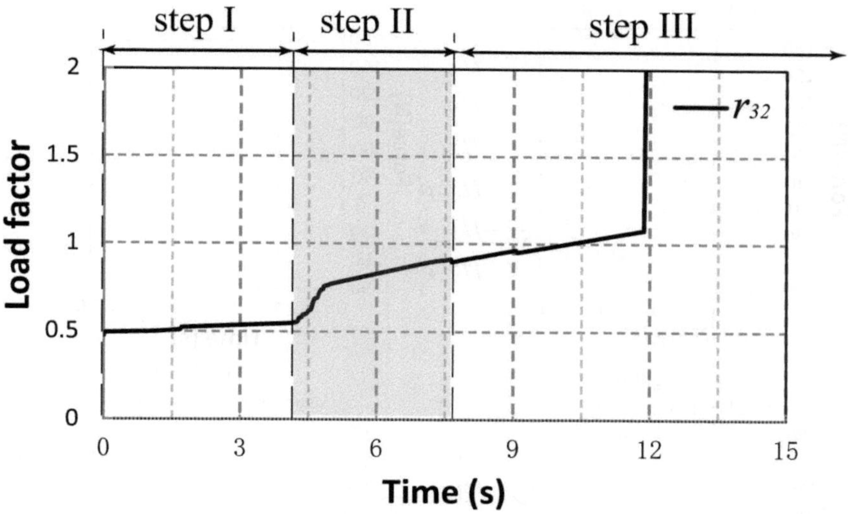

Fig. 7.17 Left hip motor overload.

7.4.3 Experiment Results with Load-Allocation Control

The internal stress can be released by making the key joints passive. An experimental result in which the initial position error was similar to that in the experiment in Sec. 7.4.2 was selected for comparison. The chance of failure with passive key joints is shown in Fig. 7.19 (a). The supporting forces of each end effector are shown in Fig. 7.20. The position errors between the robot feet and their supporting rung are compensated in step II by making the key joints passive. From these experimental results, we find that the left foot supplied more supporting force than the right foot in the beginning, but after step II the large supporting forces supplied by the left foot decreased. The actual supporting forces shown in Fig 7.20 illustrate that the supporting forces of the left and right feet are almost the same. This is because the uncertainty in the supporting forces caused by the position errors vanishes, and then the forces of the two feet meet the ideal relation $f_3 = f_4$ which is described in Sec. 7.3.2. However, the passive key joints have no ability to allocate the robot load. By observing the supporting forces in Fig. 7.20, we find that most of the robot load is supported by the right hand, and the motors on right arm were working at a high risk of being overloaded.

Since compliant rubber suspension units are used to fix the rungs on the horizontal ladder, the internal stress problem is not significant when a robot end effector is in contact with the horizontal rung. Thus, the robot can just use position control to realize its motions in this case. The optimal load-allocation control was applied when the robot end effectors were constrained by the rigid vertical ladder (step II and step III). From Fig. 7.21 we find that the desired supporting forces were accurately produced by providing the optimized torques to the key joints (J32 and

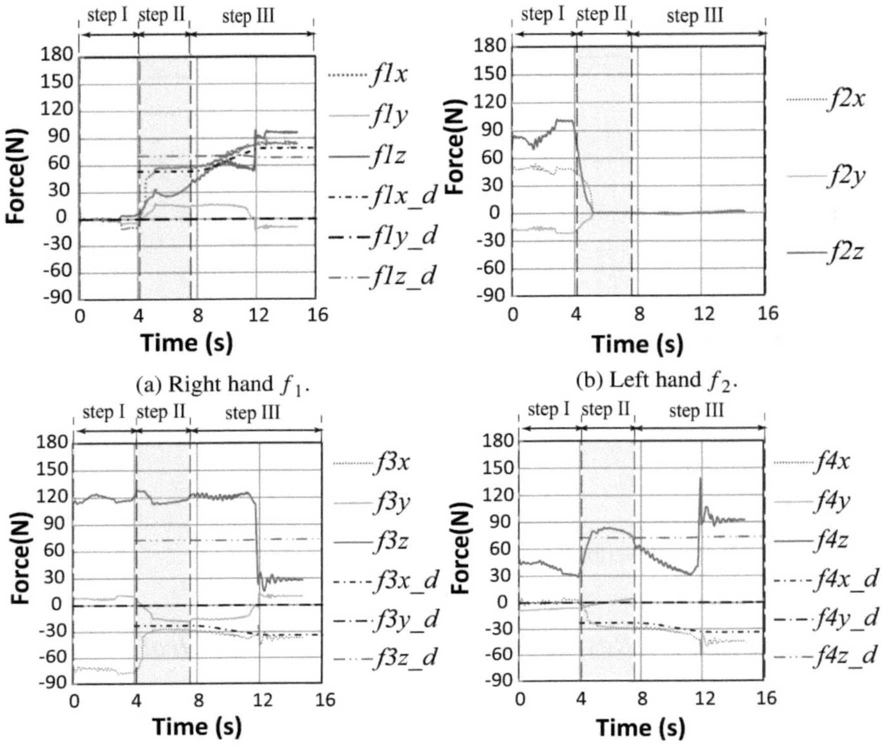

(a) Right hand f_1.

(b) Left hand f_2.

(c) Left foot f_3.

(d) Right foot f_4.

Fig. 7.18 Supporting forces in transition motion with position control.

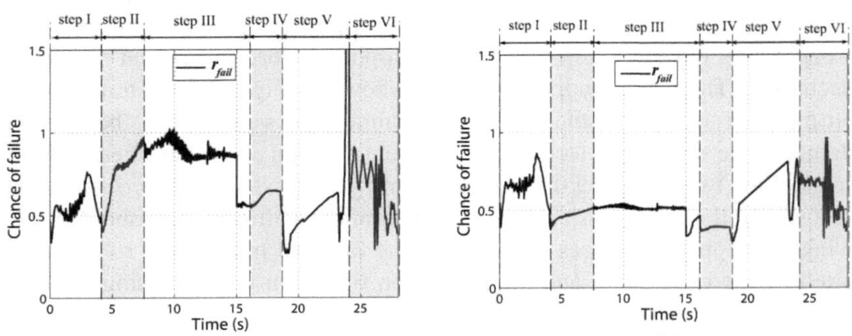

(a)

(b)

Fig. 7.19 Chance of failure with internal stress adjusted. (a) Control by making the key joints passive. (b) Optimal load-allocation control.

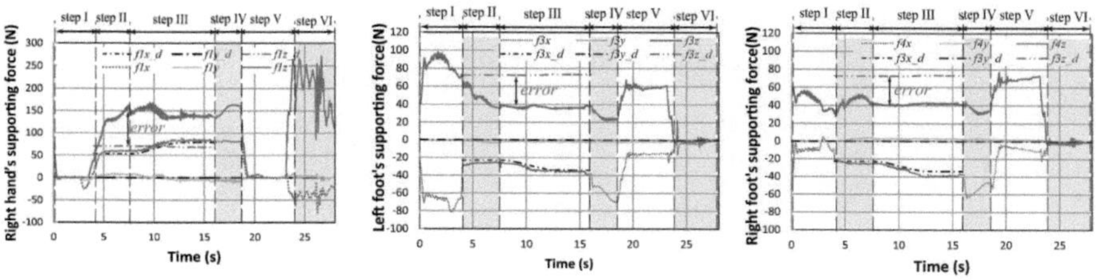

(a) Right hand f_1. (b) Left foot f_3. (c) Right foot f_4.

Fig. 7.20 Supporting forces when making the key joints passive ($\tau_{32} = \tau_{42} = 0$).

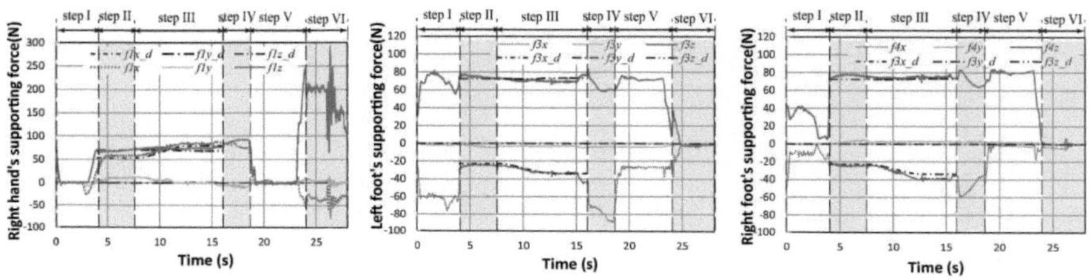

(a) Right hand f_1. (b) Left foot f_3. (c) Right foot f_4.

Fig. 7.21 Supporting forces when providing the desired torque to the key joints.

J42). The internal stress was adjusted by using load-allocation control. Then, each supporting limb was supplied a suitable supporting force. This procedure reduced the chance of failure. The chance of failure with optimal load-allocation control is shown in Fig. 7.19 (b).

7.5 Summary

In this chapter, a transition motion is designed in which the multi-locomotion robot begins the brachiation motion from the ladder climbing posture. A load-allocation algorithm was proposed to cope with the unbalanced load distribution among the supporting limbs. The robot is successfully prevented from falling down from the ladder by setting a suitable torque at the key joint. The position errors in the initial posture are compensated by the key joint. The difference between the angle of the key joint and the expected angle is used to evaluate the position error. We confirmed that the transition motions with the load-allocation algorithm were robust and stable through the experiments with the multi-locomotion robot.

The load-allocation control in this study is developed based on a simplified single closed chain 2-D model for the MLR. In the future, we would like to extend the control algorithm to a general multiple closed chain 3-D model and then modify the configuration of the control algorithm to make it useful for control of other robots.

Chapter 8
Locomotion Transition Based on Walking Stabilization Norm Using Bayesian Network

8.1 Introduction

In this chapter, we aim to realize robust locomotion in unknown test courses, so robots recognize a slope or a step and plan to locomotion. In the next place they need to know whether they realize the plan or not. In robotics system there is uncertainness. Since it influence on realization of performance, we have to deal with uncertainty. This uncertainty is classified into four categories. First one is the uncertainty caused by motion. For example, it's approximation of motion algorithm. Most robots have models to simplify calculating dynamics. So this gives robot systems uncertainty because there are difference between a reality robot shape and a robot model. Second uncertainty is about recognition is accuracy of sensors, effective ranges of sensor or abstraction of environment. Third uncertainty comes from controller (software). If the controller is not good for robot motion, the robot has a lot of error in moving. And fourth uncertainty is about hardware. For example, reliance on consumption of motor or breakdown of motor, and reliance on sensors with noise have uncertainty. In this chapter, we propose the way of estimation of uncertainty in robot system by use of the Bayesian Networks. Uncertainty in robot system limits robot locomotion modes. And robot can get adaptation to environmental or conditions.

8.2 Sensor System and Locomotion Mode

As a sensor for recognition of slope, a laser range finder is installed at the neck of the robot (see Fig. 8.1). Its angular resolution is 0.36 [deg], scan angular range is 240 [deg], scan time is 100 [ms], and maximum range of detection is 4.0 [m]. The rotation axes of motors are pitch and yaw axes. In addition a web camera is also installed next to the laser range finder.

T. Fukuda et al.: Multi-Locomotion Robotic Systems, STAR 81, pp. 173–185.
springerlink.com © Springer-Verlag Berlin Heidelberg 2012

In biped walk, we model the robot as a 3D inverted pendulum shown in Fig. 8.2, same as the work [135]. The supporting point of the pendulum is assumed to be point-contact. Then, only the heeling force f and the gravity act on Center of Gravity (COG). In this chapter, we use crawl gait as a quadruped walking described in Chapter 5. In this gait, the idling leg changes, left rear leg, left front leg, right rear leg, and right front leg, in that order (see Fig. 8.3). It is designed in order that three feet always contact the ground, COG moves within the triangle which is formed by the three supporting feet. The transition from biped to quadruped posture is made keeping static balance. Before transiting the posture between biped and quadruped stance, the robot stops walking.

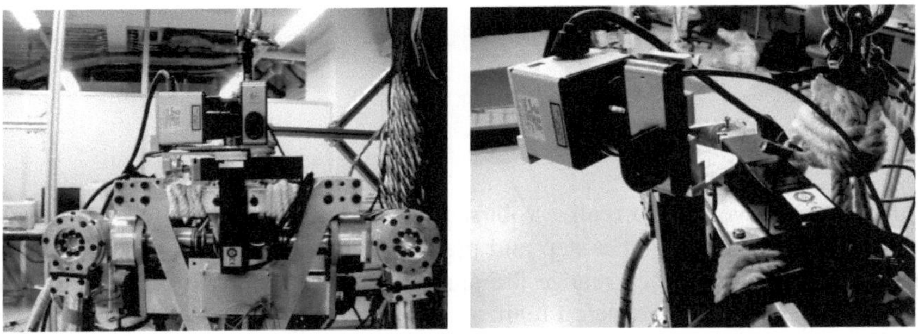

Fig. 8.1 Laser range finder

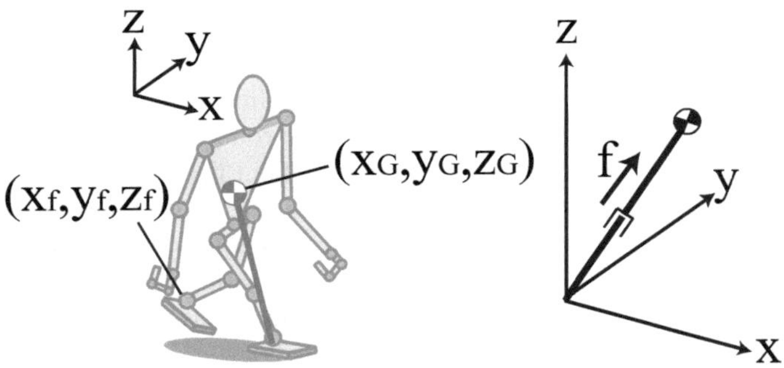

Fig. 8.2 Robot modeled as 3D penulum

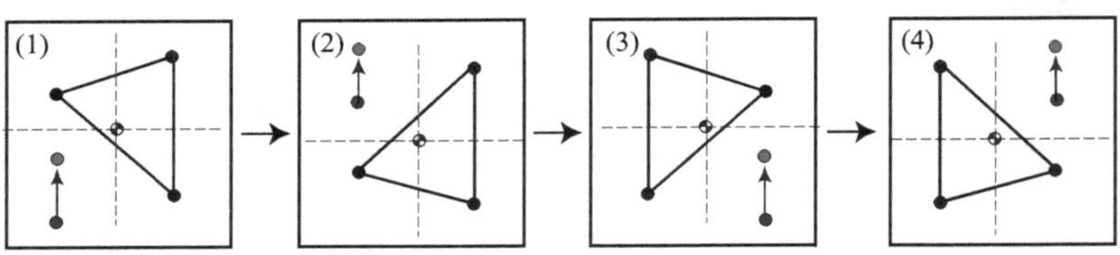

Fig. 8.3 Crawl gait

8.3 Locomotion Stabilization

In this chapter, locomotion stabilization is executed along algorithm shown in Fig. 8.4

As prospection for locomotion, robots determine parameters about locomotion mode, walking velocity, direction or numbers of paces. This robot has biped walking and quadruped walking as locomotion mode, and on flat it travels in biped walk because flat ground is easy to walk. If on slope or rough ground it is impossible to move in biped state, robots select quadruped walking to be more robust. In this research, we propose recognition and planning using a laser range finder. A laser range finder enable robot to recognize slopes or steps. So robot modifies the position of landing or COG position adapting to environmental.

As a feedback for locomotion, this stabilization scheme has evaluation of stability based on internal condition. Robot estimates a risk of falling down using parameters which have uncertainty. If the risk of falling down is high, the robot changes walking velocity or direction, or selects internal models. But if still high, it changes locomotion modes. In this chapter, we propose the method of estimating the risk of falling down using Bayesian Networks. In estimating it, we set "Robot Model Reliability (Reliability of Internal states)" and "Environmental Model Reliability (Reliability of External dynamics)". Reliability of a robot model shows how far difference between reality motion and locomotion algorithm is, or physical abilities of robot. For example, if the robot has motor trouble, this is low and the risk of falling down is high. Reliability of an environmental model shows how accurately a robot recognizes environment. If robots move in dark, it does not get information of environment, so this parameter is low and the risk of falling down is high. In biped and quadruped walking, the robot evaluates both reliabilities, estimate the risk of falling down and attain an optimum gait adapting to the environments or the conditions.

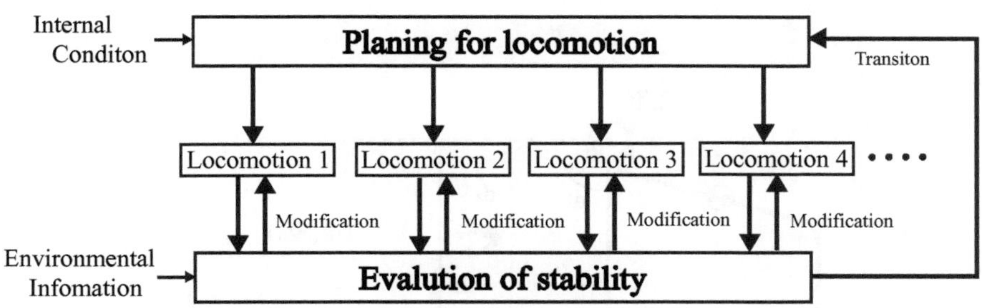

Fig. 8.4 Locomotion Stabilization Scheme

8.4 Stabilization Based on External Information

8.4.1 Recognition of Ground

We propose the way of calculating a gradient of ground which the robot directs to, and the boundary between two planes (between a flat and a slope, a flat and a wall) using the laser range finder. The gradient and the boundary determine how the robot transfers from a start to a goal. Fig. 8.5 and Fig. 8.6 are schematic showings that the robot measures an unknown ground. The laser runs in parallel with the sagittal plane. And the neck motor rotates around yaw axes, so the robot can acquires infomation about 3D surface. Therefore, we define the system of Cartesian coordinates so that their origin is located at the point directly under the laser range finder and design a method to recognize a landform in the x-y-z 3D space. The robot need to get (x_i, y_i, z_i) to estimate the landform. In Fig. 8.5, α_i is the angle of laser i to the downward direction, and di is the distance between the center of the laser range finder and the point on the ground surface. These data are obtained from the laser range finder. In addition, h is the height of the laser range finder. Then, a point on the ground surface can be described as (s_i, z_i). x_i and y_i are caluculated from s_i using the angle ofthe neck motor, β (Fig. 8.6). β varies from -20[deg] to 20[deg] arranged to 5[deg].

$$\begin{cases} x_i = s_i cos\beta \\ y_i = s_i sin\beta \end{cases}, \quad i = 1, 2, \cdots, m \quad . \tag{8.1}$$

The equation of the ground surface is extracted by least-square method as $z = ax + by + c$. a, b, and c are fixed numbers. Then, the robot modifies the position of landing and the COG trajectory with this equation [252].

$$\begin{cases} s_i = d_i sin\alpha_i \\ z_i = h - d_i cos\alpha_i \end{cases}, \quad i = 1, 2, \cdots, m \quad . \tag{8.2}$$

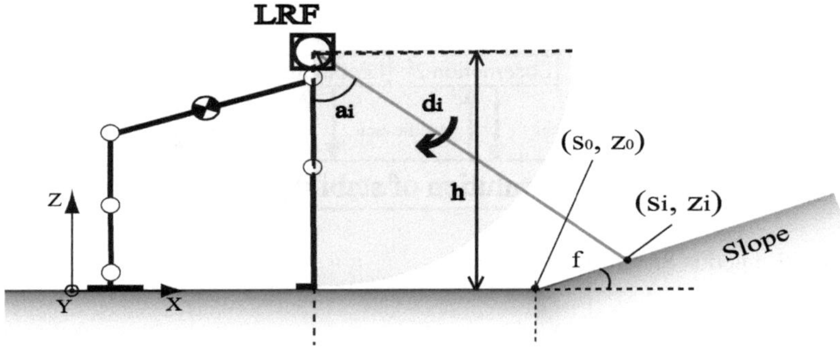

Fig. 8.5 Measuring slope 1

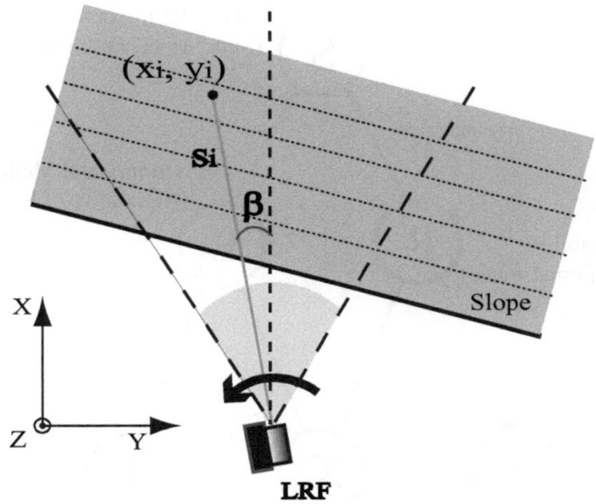

Fig. 8.6 Measuring slope 2

8.5 Stabilization Based on Internal Conditions

8.5.1 *Estimation of Probability*

The uncertainty shown in INTRODUCTION is involved in robot's stability of walking. In this research, uncertainty in locomotion motion is dealt with to evaluate the stability. There are many kinds of uncertain parameters which have various dimensions, so it is difficult to deal with them uniformly. Then, these parameters are integrated into the risk of falling down as belief with Bayesian Network. The Bayes theory assumes that parameters have distributions individually, and posterior probability is induced formally by conditional probability. Bayesian Network is the model which describes relations among phenomenon using probability. We describe the causality between the risk of falling down and the uncertain parameters.

In this research, Bayesian Network shown in Fig. 8.7 is used to estimate the risk of falling down. First, Bayesian Network estimates Robot Model Reliability "R" and Environmental Model Reliability "E". Reliability of a Robot Model R show how ideal the robot motion is, and describes the capacity of moving. Reliability of a Environmental Model E is a index which shows how correctly the robot perceive the dynamics between the environment and the robot. Secondly, R and E are induced the risk of falling down "S". "$S = 1$" shows falling down, and "$S = 0$" shows not falling down. Probability variables R and E have classes 0, 1, 2 in more reliable order. Then conditional probability $P(S \mid R,E)$ reflects the performance of the robot, and the designer arranges this probability subjectively. The evaluating parameters X_1, X_2, X_3 shown below are observed at real time. Then probability variables from 0 to 4 based on uncertainty which the parameters have input the Bayesian Network. When the probability variable is 0, the situation is most stabile. The calculation of

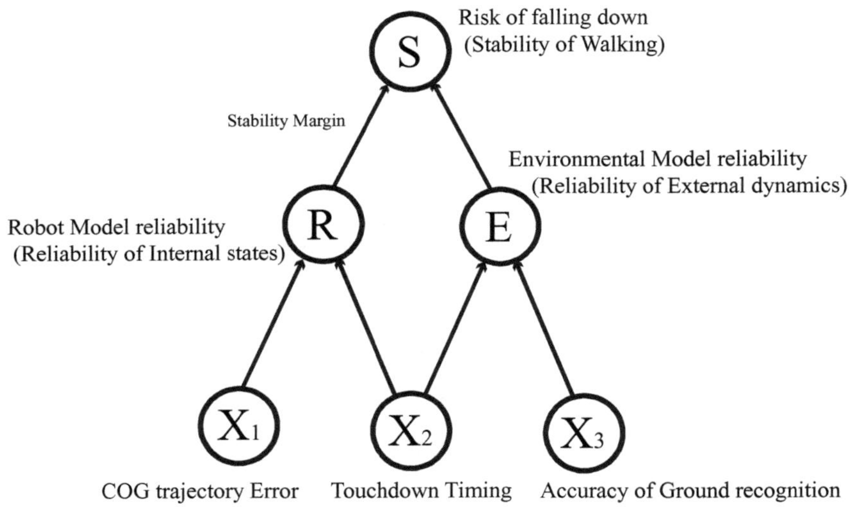

Fig. 8.7 Bayesian Network for locomotion Stabilization

Bayesian Network uses the enumeration method shown by (8.3).

$$
\begin{aligned}
P(S=1) &= \frac{\sum\limits_{R=0}^{2}\sum\limits_{E=0}^{2} P(S=1,R,E)}{\sum\limits_{S=0}^{1}\sum\limits_{R=0}^{2}\sum\limits_{E=0}^{2} P(S,R,E)} \\
&= \frac{\sum\limits_{R=0}^{2}\sum\limits_{E=0}^{2} P(S=1\mid R,E)P(R\mid X_1,X_2)P(E\mid X_2,X_3)}{\sum\limits_{S=0}^{1}\sum\limits_{R=0}^{2}\sum\limits_{E=0}^{2} P(S\mid R,E)P(R\mid X_1,X_2)P(E\mid X_2,X_3)}
\end{aligned} \tag{8.3}
$$

The evaluating parameters X_1, X_2, X_3 are always observed, so each probability $P(X_1),P(X_2),P(X_3)$ is set 1.

8.5.1.1 COG Trajectory Error X_1

The position of the center of gravity is measured by the force sensor which the robot put on its four legs. In biped posture, outputs which come from the sixth axis force sensor makes ZMP. In quadruped posture, the center of gravity is calculated with the equilibrium of moments. Then the errors between the desired trajectory and the observed trajectory decides the probability variable X_1.

8.5.1.2 Touchdown Timing X_2

The touchdown timing shows differences between the landing and the ground surface actually. When the robot is thrown off balance, or when the recognition is inadequate and the ground is higher than measured point, then the touchdown timing is earlier than the planed timing. In the robot moving, the probability variable X_2 is renewed at every landing.

8.5.1.3 Accuracy of Ground Recognition X_3

This parameter evaluates the performance of the recognition which the robot has. This shows how much information the robot attain with some sensors, and how abstracted the environmental model which the robot has is. The laser range finder has effective ranges, so over this ranges there is much uncertainty. Then the two-dimension recognition and the approximate algorithm have the uncertainty.

8.5.2 Consideration of Stability Margin

The conditional probability $P(S \mid R,E)$ describes the influence which Reliability of a Robot Model R have with the Risk of falling down S. Then when the stability margin is enough large compared with the COG errors, the influence is little even if R goes down. In reverse, when the stability margin is small, R has a big influence on S. Therefore $P(S \mid R,E)$ is decided based on the stability margin. For example, a stability margin in biped posture is smaller than one in quadruped posture, so $P(S \mid R,E)$ in biped posture is bigger than in quadruped posture.

8.5.3 Shift of Locomotion Mode

The evaluating parameters X_1, X_2, X_3 are observed at real time, and the probability of falling down is estimated. The conditional probabilities used in Bayesian Network are arranged by the subjective judgments of the designer. Therefore, when the robot falls down, the probability of falling down is not always 1.0. So we pay an attention to the fluctuation of the probability. That is, when the robot move in biped posture and the risk of falling down increases, then it has the transition motion from biped to quadruped posture and go quadruped walking. Contrarily the risk decreases in quadruped walking, the robot stands up and go biped walking.

8.6 Experiments

8.6.1 Experimental Conditions

In this experiment, the robot measures the landform with The laser range finder at starting point, and in walking, it get the gait based on the risk of falling down estimated by Bayesian Network shown in Fig. 8.8. When the risk is more than β (0.7) in biped posture, the robot squats to get quadruped posture. And when the risk is less than α (0.3) in quadruped posture, it standups. Then the robot in biped posture has three patterns of biped walking a_1, a_2, a_3 which have different efficiency. If the risk decreases, the robot get more efficient gait. In this research, this efficiency is the walking velocity, then a_1, a_2, a_3 are respectively 8.67, 6.67, 4.67[cm/sec] acquired by stride widths changed and the quadruped walking velocity is 3.00[cm/sec]. Both the standup motion and the squat motion take 10[sec] to action. Modifications of its gait are conducted in every walking cycle. The robot aims at minimizing the risk and maximizeing the efficiency all the time.

8.6.2 Experimental Result

By three experiments, we show the effectiveness of proposed method.

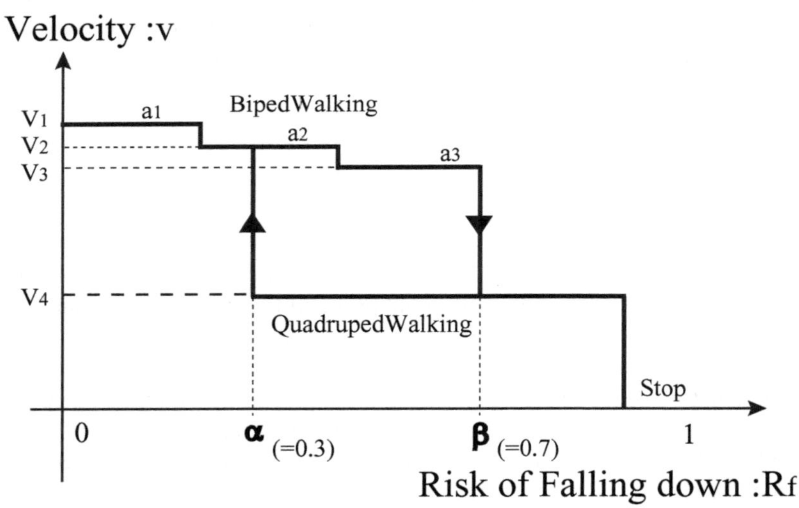

Fig. 8.8 Velocity - Risk of falling down

8.6.2.1 Experiment 1 (Planning Based on Recognition)

In this experiment, the robot walks from flat to upslope. This slope is 15[deg] and impossible for the robot to walk in biped posture. At first, it estimates the slope information in biped posture at the starting point. Based on it, the numbers of steps in biped walking are decided. Then, the robot starts biped walk and stops it at described steps to transit to quadruped posture. Finally the robot climbs up the upslope in quadruped posture. Fig. 8.9 is the slope information acquired by the laser range finder. The gradient error is 1.5[deg]. Snapshots of Experiment 1 are shown in Fig. 8.10.

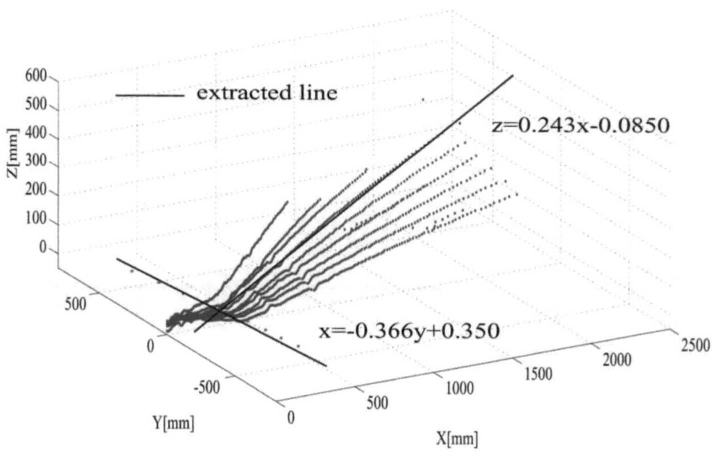

Fig. 8.9 Data for Recognition of slope

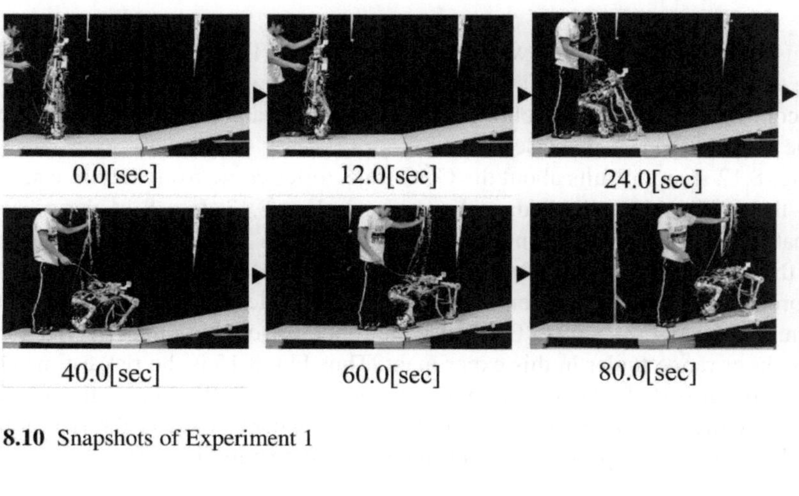

Fig. 8.10 Snapshots of Experiment 1

8.6.2.2 Experiment 2 (Transition Based on Risk)

In experiment 2, the robot walks in biped posture on flat. Fig. 8.11 is the risk of falling down. The biped action a_1, a_2 or a_3 is selected by the average of the risk during one period (1.5[sec]) in every step. We can confirm a_1, a_2 or a_3 is adjusted by the risk.

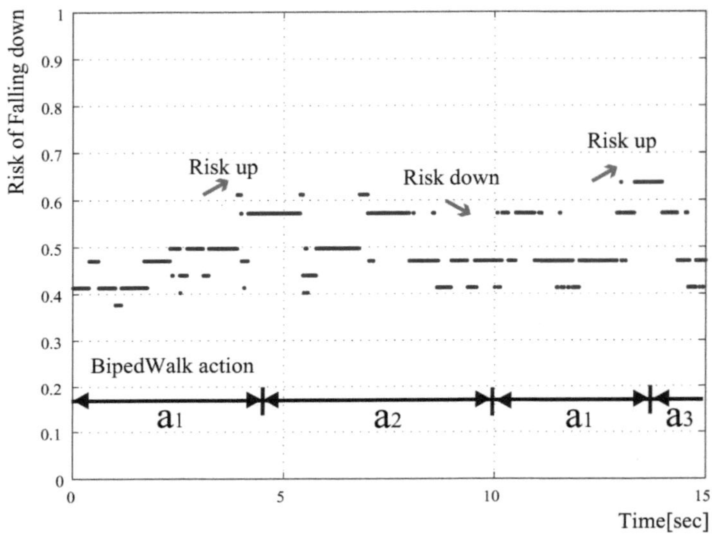

Fig. 8.11 Risk of Falling down 1

8.6.2.3 Experiment 3 (Transition Based on Risk)

In this experiment, the robot walks on rough ground. There are inequalities which have the maximum height, 5[mm]. This is not recognized by the robot on purpose. We confirmed whether the robot in biped posture changes the gait to quadruped mode because the risk increases.

Fig. 8.12 shows results about the COG trajectories come from the force sensors. And the COG trajectories induce X_1 shown in Fig. 8.13. Fig. 8.14 describes the probability variable X_2. The numbers in these figures are the threshold to apportion the probability variable. In this experiment the node X_1,X_2 have 0, 1, 2, 3, 4 as the probability variables. When the probability variable is 4, the robot almost falls down. The node X_3 is always 0 because the robot move within the effective ranges of the laser range finder in this experiment. Thus Fig. 8.15 is the risk estimated by Bayesian Network. In the transition motion, the risk is 0.0. We can see the transition caused by the risk increasing. Before the robot conducts a squat, the risk is more than β (0.7). And snapshots of Experiment 3 are shown in Fig. 8.16.

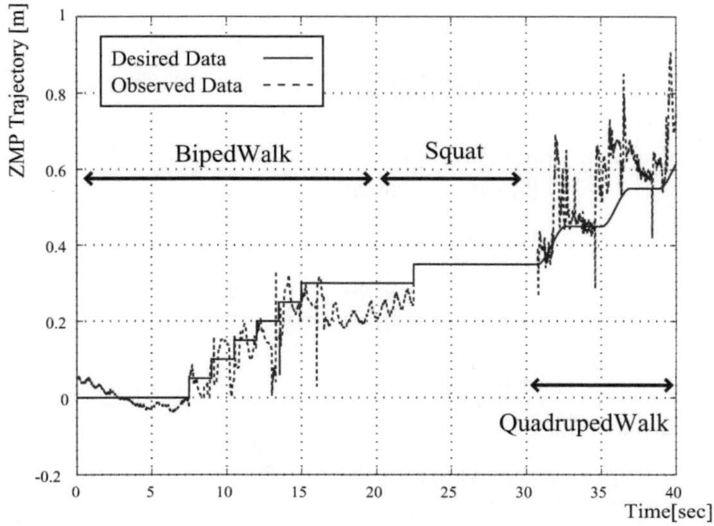

Fig. 8.12 Comparison between desired and actual COG trajectory

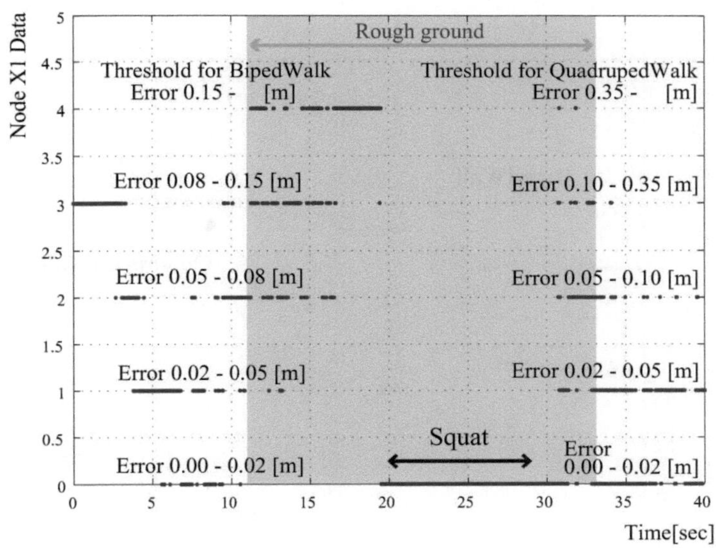

Fig. 8.13 Experimental Data of node X1

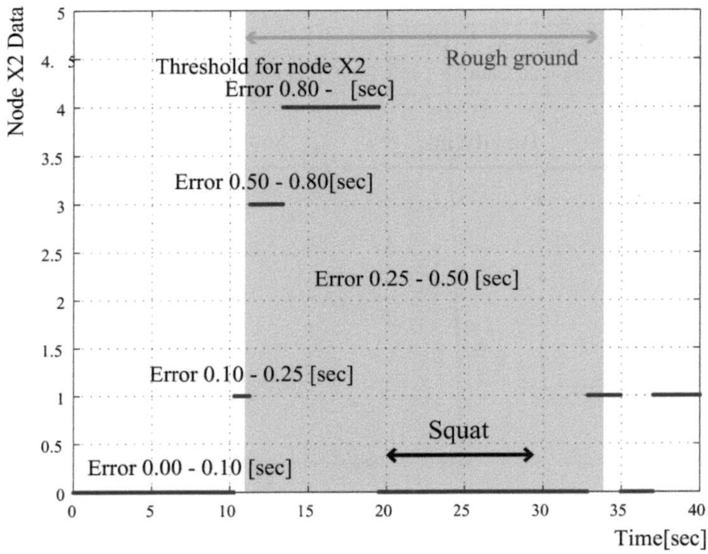

Fig. 8.14 Experimental Data of node X2

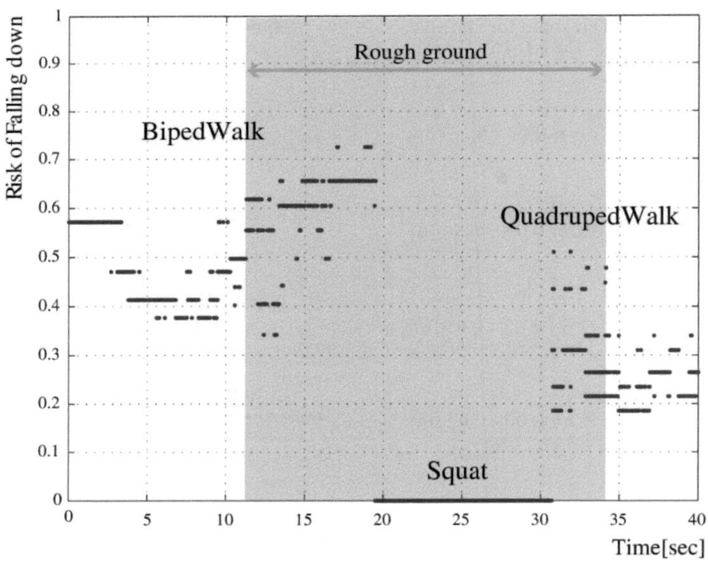

Fig. 8.15 Risk of Falling down 2

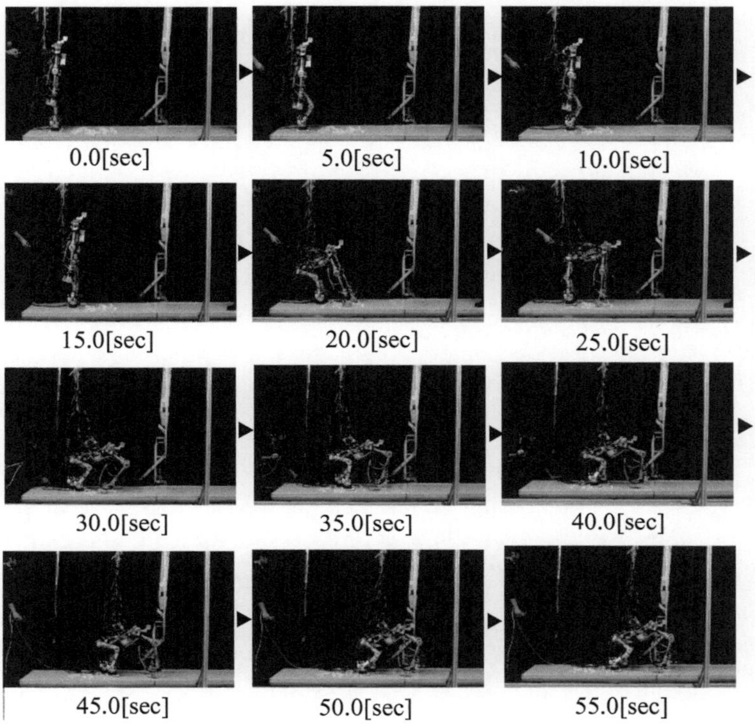

Fig. 8.16 Snapshots of Experiment 3

8.7 Summary

In this chapter, we made following two propositions. The first one was a method to recognize an unknown test course with the laser range finder and to plan for locomotion. The second one was the way of estimating the uncertainty as the risk of falling down. And the robot decides its gait based on the recognition and the risk. By experiments, we verified the proposed methods can be applied to various cases, and showed that stable locomotion with transition between biped and quadruped walk have been realized.

Although we dealt with only biped walk and quadruped walk in this chapter, we will deal with other locomotion modes such as brachation and ladder climbing for transition. Furthermore, we will add the Bayesian Network for the risk of falling down to more parameters, and diagnose the causes which give the robot system the uncertainty mainly.

Chapter 9
Application of the Passive Dynamic Autonomous Control (PDAC)

9.1 Lateral Motion Control with PDAC

Miura *et al.*[183] and Furusho *et al.*[75] 3-D motion was divided into a pair of planes—lateral plane and sagittal plane—and built the controller based on the lateral and sagittal dynamics. The lateral plane is defined as an Addison's clinical plane passing vertically on either side through a point on the interspinal halfway between the anterior portion of the iliac crest and the median plane. Meanwhile, the sagittal plane is defined as a longitudinal plane that divides the body of a bilaterally symmetrical animal into right and left sections. This dividing is based on the following two assumptions: 1) The motion in each plane is small—i.e. the amplitude of lateral motion is small and step-length in the sagittal plane is relatively short—enough that the effects of each plane on each other is low. 2) The motion in each plane is stable against disturbance generated by motion in other plane.

In this section, 3-D dynamics is divided into lateral and sagittal plane similarly to previous work [183, 75]. The controller of 3-D dynamic biped walking based on a pair of 2-D dynamics is designed by means of PDAC. In addition to the assumptions described above, it is needed to synchronize motion period in both plane in order to achieve this control framework. Thus, in this chapter, firstly PDAC is applied to the lateral motion and secondly the control to stabilize walking by adjusting the landing position at foot-contact is proposed. Finally, the motion period control by varying the amplitude of lateral sway motion is built.

9.1.1 Lateral Motion and Dynamics

9.1.1.1 Motion Design

In this section, the lateral motion is composed by means of PDAC. Many researchers investigated and proposed the lateral motion control[249, 104, 158]. In this section,

T. Fukuda et al.: Multi-Locomotion Robotic Systems, STAR 81, pp. 187–290.
springerlink.com © Springer-Verlag Berlin Heidelberg 2012

we design the lateral motion by means of PDAC as depicted in Fig. 9.1. In phase(A), a robot starts to turn over toward its swing-leg-side and is accelerated by gravitation from the tilting position at a standstill on the stance-leg-side to foot-contact. In phase(B), after foot-contact, a robot is got up toward the tilting position at a standstill by the energy obtained in phase(A). Since the mechanical energy is lost at foot-contact, a robot compensates for the lost energy by lifting up its pelvis in both phases in order to continue the side-to-side rocking motion. It is possible to consider the lifts-up of pelvis as the change of pendulum length approximately. The following model shown in Fig. 9.2 is used as the model of the lateral motion: two inverted pendulums which are opposite each other continue to rock, iterating the collision between them. Right figure in Fig. 9.2 shows the trajectory of COG and two coordinate systems, Σ^R and Σ^L, that correspond to right- and left-leg-support period respectively.

9.1.1.2 Phase Portrait Coalescence

Next, lateral motion is considered in terms of dynamics. Fig. 9.3(a) shows the phase portrait of an inverted pendulum in the coordinate system Σ_R and Σ_L.

Lateral motion continues by switching these coordinate systems at foot-contact. Coalescence of the phase portrait in Σ_R and Σ_L yields that of lateral motion as shown in Fig. 9.3(b). As can be seen in this figure, in the gray tetragon surrounded by the

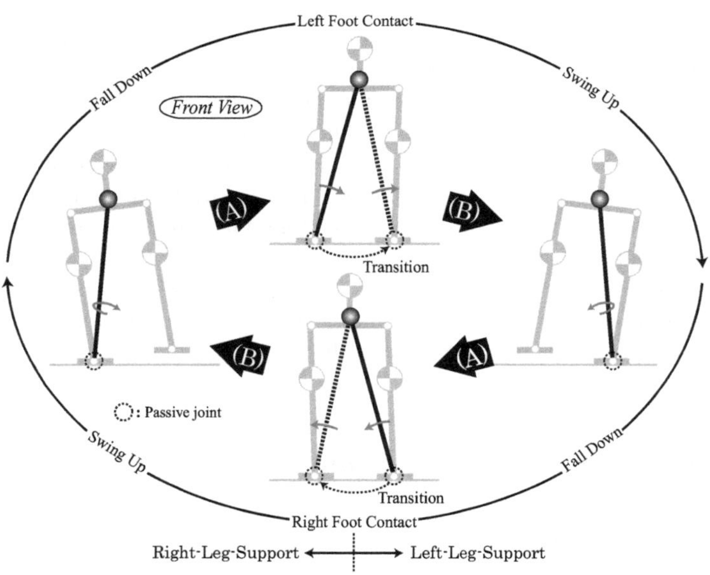

Fig. 9.1 The lateral motion of lateral-based walk (front view). The inverted pendulum falls off in phase(A) and swing up in phase(B)

pair of separatrixes, dynamics has the property to rotate, which implies the lateral sway motion. The stable lateral motion is realized only in this zone. That is, if the actual dynamics is out of it, an inverted pendulum—i.e. a robot—falls down outside and the stable motion can not be continued unless some compensation is conducted.

Next, the effect of the pelvis-lift-up on lateral dynamics is explained by use of the virtual COG trajectory shown in Fig. 9.4(a). In this trajectory, the pendulum length is switched virtually from l_0 to $l_0+\Delta l$ instantaneously at the shift from phase (B) to (A), which is unrealizable definitely. It is kept at $l_0+\Delta l$ in phase (A) and l_0 in phase (B). If $l=l_0+\Delta l$, the absolute value of the separatrixes' gradient near the origin is lower than that of $l=l_0$, and if $l=l_0-\Delta l$, the gradient is higher (see Fig. 9.4(b)). This deformation of the solution trajectory makes it possible to realize the continuous side-to-side rocking motion.

9.1.1.3 Virtual Holonomic Constraint of Lateral Joints

Here, we decide the *Virtual constraint*, that is, pendulum length l is described as the function of ϕ. It is clear that the right side of Eq. (9.12) can be integrated if $f(\phi)$ is a polynomial equation. Thus in this section, $f(\phi)$ is decided as follows:

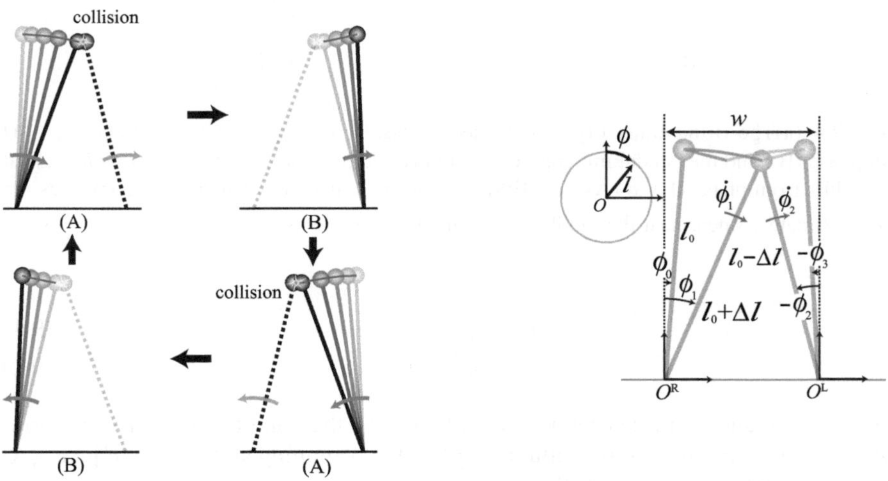

Fig. 9.2 (Left) Motion of CIPM. The collision between the foot and the ground is regarded as that between two pendulums. (A) and (B) correspond to those in Fig. 9.1. (Right) Trajectory of COG and polar coordinate systems Σ^R, Σ^L. l and ϕ denote the length and the angle of an inverted pendulum. (l_0,ϕ_0) and $(l_0+\Delta l,\phi_1)$ are the coordinates in Σ^R at the beginning and ending of phase (A), $(l_0-\Delta l,\phi_2)$ and (l_0,ϕ_3) is that of Σ^L of phase (B) respectively. $\dot{\phi}_1$, $\dot{\phi}_2$ denotes the angular velocity at the end of phase (A) and at the beginning of phase (B).

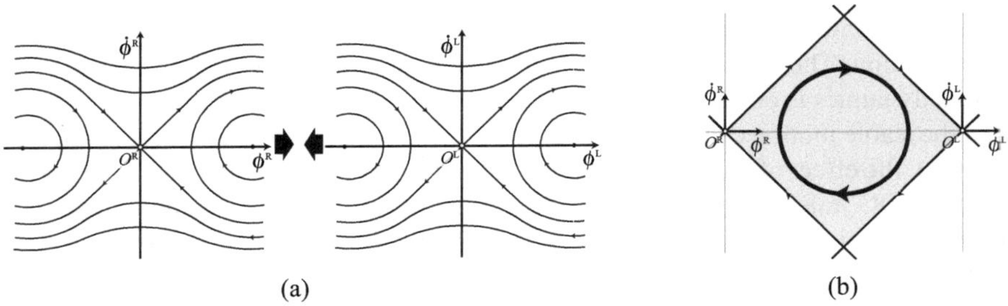

Fig. 9.3 (a) Phase portraits of ϕ^R and ϕ^L (b) Coalescence of the phase portrait in Σ_R and Σ_L

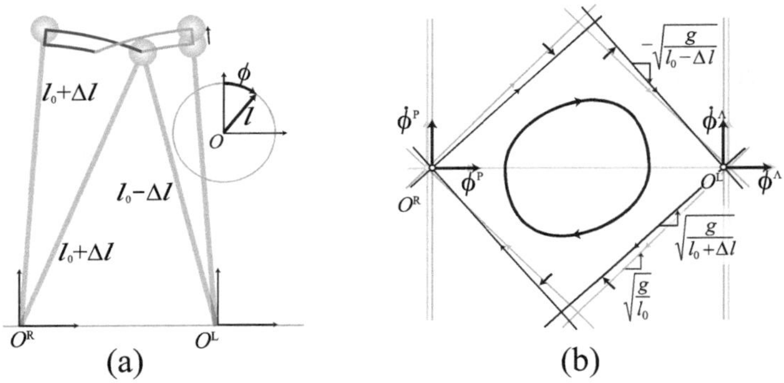

Fig. 9.4 (a)The virtual trajectory designed as follows: $l = l_0 + \Delta l$ in phase (A) and $l = l_0 - \Delta l$ in phase (B), and the pendulum lengthens instantly and discontinuously $l_0 - \Delta l \rightarrow l_0 + \Delta l$ at the shift from phase (B) to (A). (b)CIPM map of this trajectory. Pendulum separatrixes are approximately linear near the saddle points, and their gradients are $\pm\sqrt{\frac{g}{l}}$.

$$l = f(\phi) \tag{9.1}$$

$$= a\phi^2 + b\phi + c \tag{9.2}$$

where a, b, and c are determined so as to satisfy the conditions described below. At first, the conditions of pendulum length at the beginning and ending of phase(A) and phase(B) introduce the following four equations:

$$f^A(\phi_0) = l_0 \tag{9.3}$$

$$f^A(\phi_1) = l_0 + \Delta l \tag{9.4}$$

$$f^B(-\phi_2) = l_0 - \Delta l \tag{9.5}$$

$$f^B(-\phi_3) = l_0 \tag{9.6}$$

where upper-suffixes denote the differentiation of phases.

In addition, the pendulum motion is designed so that the angular velocity of robot joints is not discontinuous, that is, the velocity along pendulum is zero,

$$\frac{\partial f^A}{\partial \phi}(\phi_1) = 0 \tag{9.7}$$

$$\frac{\partial f^B}{\partial \phi}(-\phi_2) = 0. \tag{9.8}$$

From Eqs. (9.3)-(9.8), a, b, and c in each phase are decided.

9.1.1.4 Converged Dynamics

The dynamic equation of an inverted pendulum is described as follows:

$$\frac{d}{dt}\left((ml^2+J)\dot{\phi}\right) = mgl\sin\phi. \tag{9.9}$$

Multiplying both side of this equation by $(ml^2+J)\dot{\phi}$ and integrating with respect to time yields the following equations,

$$\left((ml^2+J)\dot{\phi}\right)\frac{d}{dt}\left((ml^2+J)\dot{\phi}\right) = mgl(ml^2+J)\dot{\phi}\sin\phi \tag{9.10}$$

$$\Longleftrightarrow \frac{1}{2}\left((ml^2+J)\dot{\phi}\right)^2 = \int mgl(ml^2+J)\dot{\phi}\sin\phi\, dt \tag{9.11}$$

$$\Longleftrightarrow \dot{\phi} = \frac{1}{ml^2+J}\sqrt{2\int mgl(ml^2+J)\dot{\phi}\sin\phi\, dt}. \tag{9.12}$$

The phase around contact point (phase of passive joint) is obtained from Eqs. (9.12) and (9.1) as follows:

$$\dot{\phi} = \frac{1}{mf(\phi)^2+J}\sqrt{2\int mgf(\phi)\left(mf(\phi)^2+J\right)\sin\, d\phi} \tag{9.13}$$

$$:= \frac{1}{M_l(\phi)}\sqrt{2\left(D_l(\phi)+C_l\right)} \tag{9.14}$$

$$:= F(\phi) \tag{9.15}$$

Finally, the value to lift up pelvis, Δl, is determined. Δl is necessary to be decided so that Eq. (9.15) satisfies the initial condition of phase(A) and the end condition of phase(B), that is,

$$F_l^A(\phi_0) = F_l^B(-\phi_3) = 0 \tag{9.16}$$

$$\Longleftrightarrow \frac{\sqrt{2\left(D_l^A(\phi_1)-D_l^A(\phi_0)\right)}}{M_l^A(\phi_1)}\cos(\phi_1+\phi_2) = \frac{\sqrt{2\left(D_l^B(-\phi_2)-D_l^B(-\phi_3)\right)}}{M_l^B(-\phi_2)} \tag{9.17}$$

where upper suffixes denote the differentiation of phases. Δl is so small that it is possible to find the appropriate value satisfying Eq. (9.17) by use of the quadratic approximation.

9.1.1.5 Simulation

Fig. 9.5 shows the simulation results of the above-mentioned motion design under the condition $\phi_0 = \phi_3$. Although the motion is attenuated without lengthening the pendulum as shown in Fig. 9.5(a), the periodic motion is generated with lift-up as shown in Fig. 9.5(b).

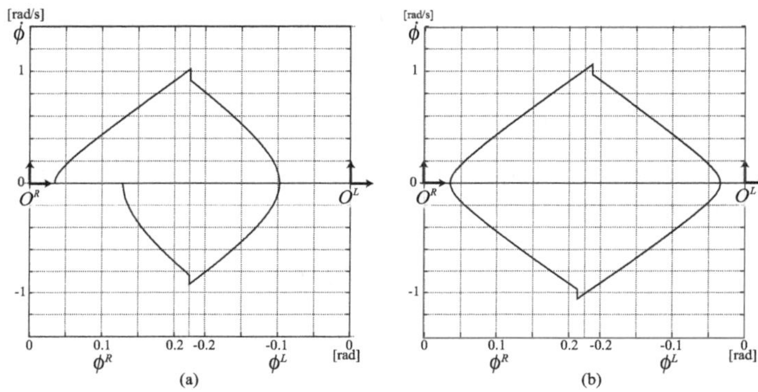

Fig. 9.5 Phase Portrait of the lateral motion (simulation, $l_0 = 0.46$[m], $\phi_0 = \phi_3 = 0.028$[rad], $w = 0.2$[m]). (a) with no lift-up (b) with lift-up

9.1.2 *Control of Lateral Period*

Next we design the period controller of the lateral motion described in the previous subsection. The period of lateral motion is decided by the amplitude of pendulum motion, that is, the period is long if the amplitude is large and it is short if the amplitude is small. In this section, the desired period is realized by controlling the lateral amplitude.

Assuming that the pendulum angle at the transition from phase(B) to phase(A) is ϕ_3, the motion period T can be found properly by the following calculation

$$\int_{-\phi_3}^{-\phi_2} \frac{1}{F_B(\phi)} d\phi + \int_{\phi_3}^{\phi_1} \frac{1}{F_A(\phi)} d\phi = T. \tag{9.18}$$

However, it is not easy to solve this equation for ϕ_3. The pendulum extension is so small that the desired amplitude is decided approximately by use of the model of inverted pendulum, length of which is not variable, as follows:

$$\phi_3 = \frac{\phi_c}{\cosh\left(\sqrt{\frac{g}{l_0}}\frac{T}{2}\right)} \tag{9.19}$$

where, ϕ_c is the pendulum angle in the standing posture, i.e. the pendulum angle at the foot-contact under the condition of $\Delta l = 0$.

9.1.3 *Stabilization*

9.1.3.1 Landing Position Control

Various methods to stabilize the lateral motion were proposed (Kuo[158], Miura *et al.*[183], Sano *et al.*[249], Hemami *et al.*[104]). In this section, we propose the stabilizing method to adjust the pendulum length and angle at foot-contact according to the error between phase around contact point derived by PDAC and actual motion. As for the robot motion, as shown in Fig. 9.6, the state at foot-contact is varied by opening or closing the swing leg in phase(A). This motion has two following effects:

- Change of the pendulum angle of swing leg at foot-contact, ϕ_2.
- Variation of the pendulum length on the side of swing leg at foot-contact, $l_0 - \Delta l$.

Strictly speaking, the pendulum angle on the side of stance leg at foot-contact, ϕ_1, is also changed by this motion, however its effect is so small that we do not give consideration to it. Hence, in this section, it is assumed that ϕ_1 does not vary.

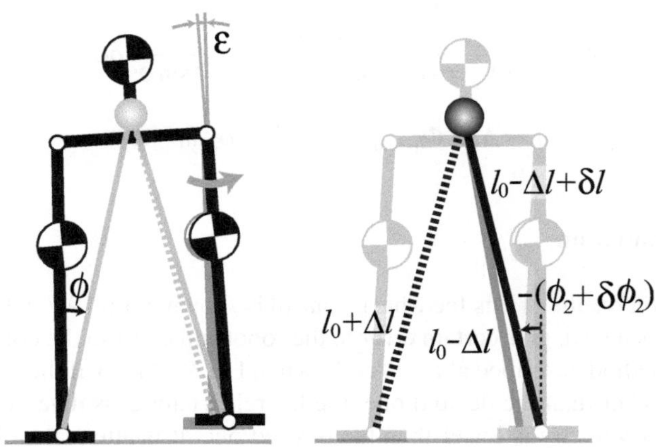

Fig. 9.6 Stabilization of lateral motion

We can describe the condition that the pendulum angle at the end of phase(B) is ϕ_3 as below:

$$\frac{2\left(D_l^A(\phi_1)+C_l\right)}{\left(M_l^A(\phi_1)\right)^2} = \Phi(\phi_2, l_0 - \Delta l) \tag{9.20}$$

where, from Eq. (9.17), the right hand side of this equation is described as

$$\Phi(\phi_2, l_0 - \Delta l) = \frac{2\left(D_l^B(-\phi_2) - D_l^B(-\phi_3)\right)}{(l_0 - \Delta l)^2 \cos^2(\phi_1 + \phi_2)}. \tag{9.21}$$

Here, we assume that the actual C_l has the error between desired value and actual one, δC_l, and determine the landing position of swing leg. In order to stabilize the motion, the robot is necessary to satisfy the desired state at the end of phase(B), that is, $F_l^B(\phi_3) = 0$. Assuming that the robot opens its swing leg by ε and consequently ϕ_2, $l - \Delta l$ are changed into $\phi_2 + \delta\phi_2$, $l_0 - \Delta l + \delta l$ respectively, Eq. (9.20) can be rewrite as follows:

$$\frac{2\left(D_l^A(\phi_1)+C_l+\delta C_l\right)}{\left(M_l^A(\phi_1)\right)^2} = \Phi(\phi_2 + \delta\phi_2, l_0 - \Delta l + \delta l) \tag{9.22}$$

In addition, from the geometric condition, the following two equations are obtained,

$$(l_0 - \Delta l)\cos\phi_2 = (l_0 - \Delta l + \delta l)\cos(\phi_2 + \delta\phi_2) \tag{9.23}$$

$$(l_0 - \Delta l + \delta l)\sin(\phi_2 + \delta\phi_2) - (l_0 - \Delta l)\sin\phi_2 = L\varepsilon \tag{9.24}$$

ε is decided from Eqs. (9.20)-(9.24) by assuming that it is possible to neglect the squared term of δ because of its smallness.

$$\varepsilon = \frac{(l_0 - \Delta l)\cos\phi}{L}\left(\tan(\phi_2 + \delta\phi_2) - \tan\phi_2\right) \tag{9.25}$$

$$\delta\phi_2 = \frac{\cos\phi_2}{\Phi_{\phi_2}\cos\phi_2 + \Phi_{l_0-\Delta l}(l_0 - \Delta l)\sin\phi_2}\delta C \tag{9.26}$$

where, $\Phi_{\phi_2} = \frac{\partial\Phi}{\partial\phi_2}(\phi_2, l_0 - \Delta l)$, $\Phi_{l_0-\Delta l} = \frac{\partial\Phi}{\partial(l_0-\Delta l)}(\phi_2, l_0 - \Delta l)$.

9.1.3.2 Simulation

The simulation results of this feed-back control is shown in Fig. 9.7, where $\phi_0 = \phi_3$. In this simulation, we give certain error to the robot and confirm the convergency by the control method described above. As shown in Fig. 9.7(a)(c), if the actual angular velocity is higher than the desired one, the lateral dynamics is diverged, and if the actual angular velocity is lower than the desired one, it is attenuated. However, as shown in Fig. 9.7(b)(d), the above-mentioned control make both states including error converged on the desired state after foot-contact.

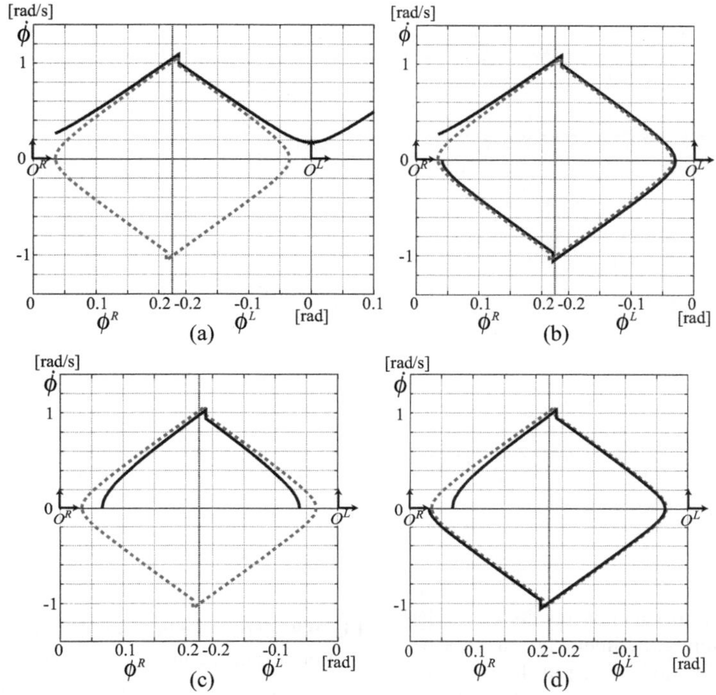

Fig. 9.7 Simulation results of the feed-back of lateral motion. (a) $\delta\dot{\phi} > 0$ without feed-back (b) $\delta\dot{\phi} > 0$ with feed-back (c) $\delta\dot{\phi} < 0$ without feed-back (d) $\delta\dot{\phi} < 0$ with feed-back

9.1.4 Experimental Results

In order to confirm the validity of proposed control of lateral motion, the following four experiments are conducted.

1) To input the disturbance—the effects of the ground variation or sagittal motion—by letting a robot step forward and to confirm the stability of lateral motion
2) To check whether autonomous dynamics of PDAC is realized or not by suspending and restarting by hand during walking
3) To ascertain that the motion realized by proposed method is energy efficient.
4) To confirm achievement of the desired period by the amplitude control of lateral sway motion

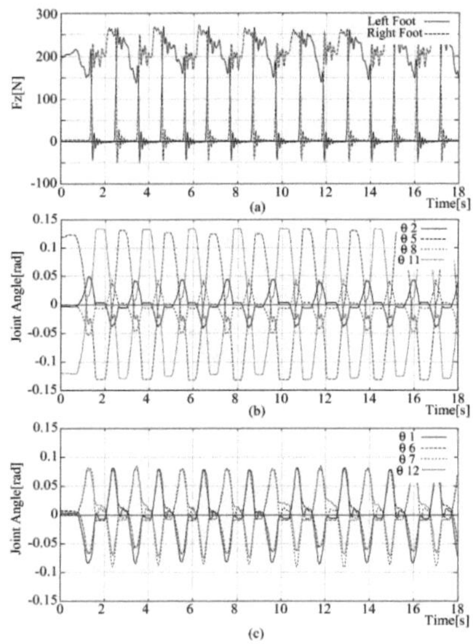

Fig. 9.8 Experimental results of the lateral motion (experiment(1)). (a) time vs. Fz (Fz is the axial force of the foot force sensor.) (b) time vs. joint angle in the lateral plane (c) time vs. joint angle in the sagittal plane

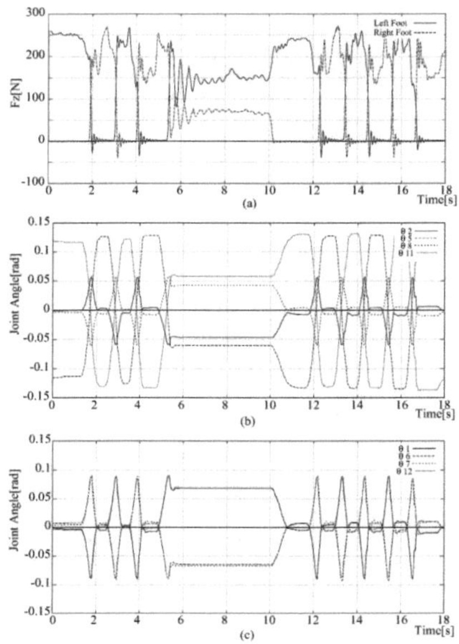

Fig. 9.9 Experimental results of the lateral motion (experiment(2)). (a) time vs. Fz (Fz is the axial force of the foot force sensor.) (b) time vs. Joint angle in the lateral plane (c) time vs. Joint angle in the sagittal plane (t=5.6[s] to 11.6[s]: suspended.)

9.1.4.1 Experiment 1)

In experiment 1), it is ascertained that the proposed lateral motion can be continued to perform stably even if disturbance is input. We generate disturbance by letting a robot step forward. Sagittal motion was designed to be associated with the lateral inverted pendulums. Its association is a simple linear function as follows:

$$Phase(A) : \psi = \frac{\phi - \phi_0}{\phi_1 - \phi_0}\psi_1 \tag{9.27}$$

$$Phase(B) : \psi = -\frac{\phi - \phi_0}{\phi_1 - \phi_0}\psi_1 \tag{9.28}$$

where ψ is the pitch angle of the support leg's ankle and ψ_1 is its angle at the foot-contact.

In this experiment, step-length was set at about 0.07[m] and walking velocity was 0.065[m/s]. The parameters was employed in the experiment as follows: l_0=0.47[m] and $\phi_0 = \phi_3$=0.037[rad], Δl=0.1[m]. Fig. 9.8 shows the data of force sensor of foot

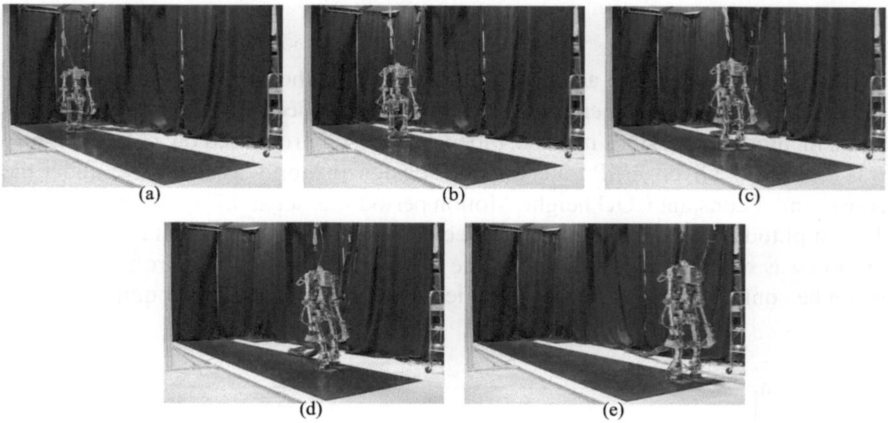

Fig. 9.10 Snapshots of lateral-based walk (a) at 2nd step (b) at 11th step (c) at 21st step (d) at 31st step (e) at 43rd step

and joint angles. From these figures, it can be seen that the quasi-periodic stable motion is realized. Fig. 9.10 shows the snapshots of the experiment.

In this experiment, since the sagittal motion was not designed based on the sagittal dynamics, small step-length was employed. Providing step-length was increased in the same algorithm, the sagittal motion lost stability and the robot fell down. As a result, it was difficult to continue walking stably. It is conceivable that longer step-length can be achieved by designing the sagittal controller based on the sagittal motion. This is described in the next chapter.

The upper bound of stability against irregularity of the ground was approximately $|0.5|$[m]. This limit implies that the proposed stabilizing method cannot accommodate to the large error of the ground height since its convergency of dynamics is dependent on the lost energy at foot-contact on the flat ground. Robustness against the ground irregularity is future work.

9.1.4.2 Experiment 2)

Next, experiment 2) is conducted in order to ascertain that the whole robot dynamics becomes time-dependent autonomous system. The graph of experimental results is shown in Fig. 9.9. We suspended the robot motion by hand from 5.6[s] to 10.1[s], then released at 11.6[s] in the initial position ($\phi = \phi_0$). The figure indicates that the robot could restart motion after the motion was completely suspended. From these results, it can be confirmed that the dynamics of PDAC was realized.

9.1.4.3 Experiment 3)

Thirdly, to investigate the advantage of proposed method in terms of energy efficiency, we compare the energy consumption of proposed method with that of previous method. As previous method, biped walking control based on ZMP trajectory was employed. Given ZMP trajectory was designed simply as sine wave under the constraint of constant COG height. Motion period was set at 0.7[s] in both methods. The amplitude of ZMP trajectory was decided experimentally so that the smoothest motion was realized. Fig. 9.11 shows the total power of all joints. From this figure, it can be confirmed that the energy efficiency of proposed method is quite high.

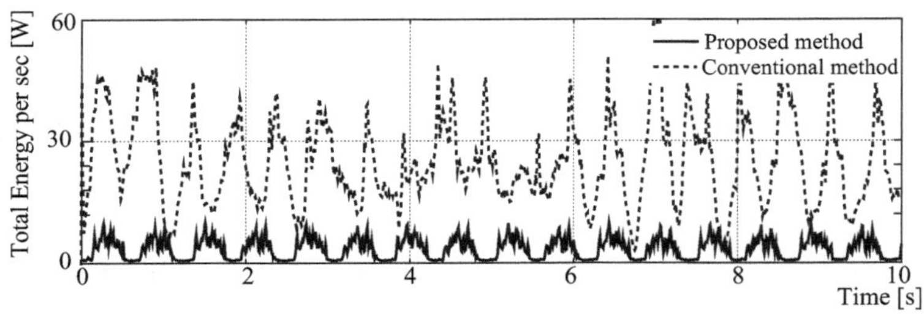

Fig. 9.11 Comparative results of total Energy per second

9.1.4.4 Experiment 4)

Finally, the validity of motion period control was tested. In this experiment, at first, by increasing and decreasing the amplitude, we checked that the period could be controlled by amplitude variation. Fig. 9.12 shows the angle and angular velocity graph of the inverted pendulum and left figure in Fig. 9.13 shows the phase portrait. The change of the phase portrait size indicates that of the period. Since the robot needs the potential energy to start the walking, we lifted up the lateral pendulum to the position at the beginning of phase (A) and released.

Next, we confirmed the validity of the period controller by giving it desired walking period from 0.45[s] to 0.8[s] in increments of 0.05[s]. Right figure in Fig. 9.13 depicts the graph of the actual period vs. desired period. From this figure, it can be seen that the desired period is nearly achieved.

9.1.5 Summary of This Section

It is possible to divide 3-D dynamics into a pair of 2-D dynamics—lateral and sagittal plane if the following condition is held: 1) The motion in each plane is small—i.e. the amplitude of lateral motion is small and step-length in the sagittal plane

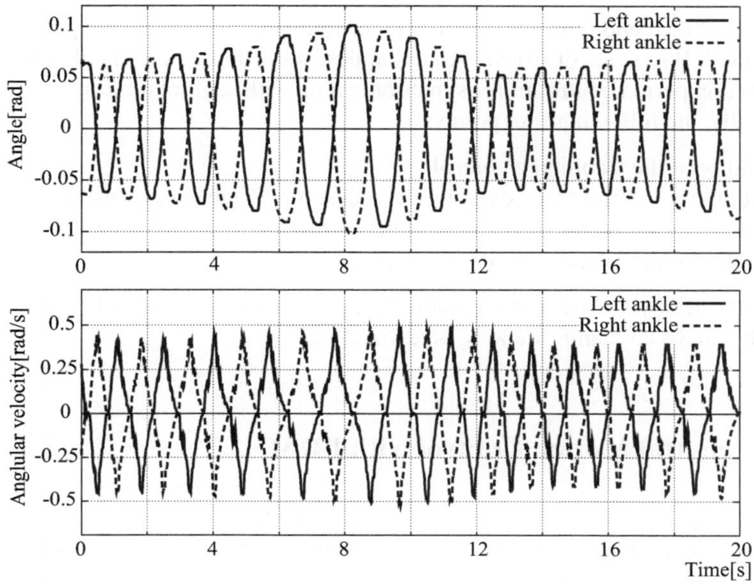

Fig. 9.12 Angle and angular velocity of the ankle joint (experiment 4))

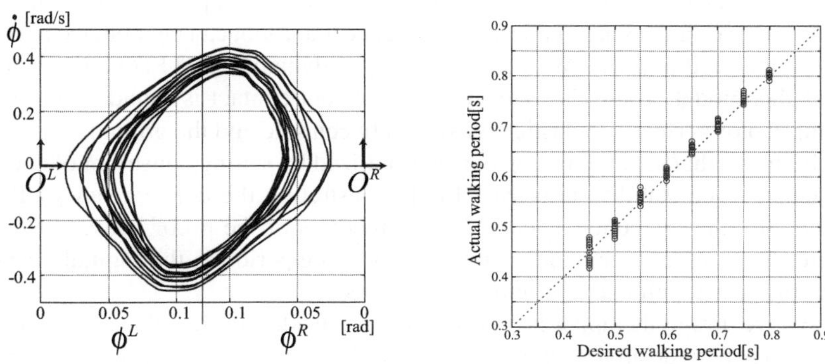

Fig. 9.13 (Left)Phase Portrait of the inverted pendulums (experiment (3)). (Right)Experimental result of the period control of the lateral motion

is relatively short—enough that the effects of each plane on each other is low. 2) The motion in each plane is stable against disturbance generated by motion in other plane. In this chapter and next, the controller of 3-D dynamic walking is built by dividing 3-D motion into lateral and sagittal motion.

In this chapter, lateral motion of 3-D dynamic walking was designed by means of PDAC. The robot was modeled as an inverted pendulum, length of which is variable, in the lateral plane. Then, the pendulum length was designed to express the function of the pendulum angle (Virtual constraint) and lateral Converged dynamics was found. By means of Converged dynamics, we proposed the stabilizing method and motion period controller for synchronization with sagittal motion.

Finally, the following four points:

1. Validity of the stabilizing control
2. Autonomy of Converged dynamics and designed controller by means of PDAC
3. Superiority of the proposed method in terms of energy efficiency
4. Validity of the lateral motion period control

are confirmed by the experiment. In the next chapter, the controller of sagittal motion is built and 3-D dynamic walking is realized.

9.2 Sagittal Motion Control with PDAC

As described in section 9.1, the controller of 3-D dynamic walking is designed by dividing 3-D motion into lateral and sagittal motion. In section 9.1, the lateral motion controller was built based on lateral dynamics. Also, stabilization and motion period control were realized.

Thus, in this chapter, the sagittal motion controller based on sagittal dynamics is designed by means of PDAC. In order to build the basic simple controller, arms are not moved during walking in this section. Hence, the robot can be modeled as 3-link robot in the sagittal plane. We derive Converged dynamics and PDAC Constant by use of this model. Then, discrete dynamics at foot-contact is found based on the assumption of perfectly inelastic collision between sole and the ground.

It is shown that sagittal motion can be stabilized by setting constraint of constant step-length. The stability is proved by demonstrating the convergency of PDAC Constant by the Lyapunov Theory, and correctness of the proof is ascertained by the numerical simulation. Additionally, the walking period in the sagittal plane for synchronization with lateral motion is calculated.

Next, we design the upper layer controller to decide step-length and walking period according to desired walking velocity so that energy consumption per traveled distance is minimized. By use of this controller, the robot decides step-length and walking period autonomously according to the desired walking velocity.

Finally, the validity of proposed control architecture is tested by experiment and 3-D dynamic walking is realized.

9.2.1 Sagittal Motion and Dynamics

9.2.1.1 3-Link Model

For the sake of simplicity, in this section upper body of a robot is not moved, hence the 3-link model as shown in Fig. 9.14 is employed. The dynamic equation of this model is described as Eq. (2.13) and that of the ankle joint of the stance leg is Eq. (2.18) where $n = 3$. The left side of Eq. (2.18) is described as follows:

$$\begin{aligned}
M_{11}(\Theta) &= J_1 + J_2 + J_3 + m_1 a_1^2 + m_2 l_1^2 + m_2 a_2^2 \\
&\quad -2m_2 a_2 l_1 \cos\theta_2 + m_3 l_1^2 + m_3 a_3^2 \\
&\quad +2m_3 a_3 l_1 \cos\theta_3 \\
M_{12}(\Theta) &= -J_2 - m_2 a_2^2 + m_2 a_2 l_1 \cos\theta_2 \\
M_{13}(\Theta) &= J_3 + m_3 a_3^2 + m_3 a_3 l_1 \cos\theta_3 \\
G_1(\Theta) &= (m_1 a_1 + m_2 l_1 + m_3 l_1) g \sin\theta_1 \\
&\quad + m_2 g a_2 \sin(\theta_2 - \theta_1) \\
&\quad + m_3 g a_3 \sin(\theta_1 + \theta_3)
\end{aligned}$$

where, $m_1(\Theta) = [M_{11}(\Theta), M_{12}(\Theta), M_{13}(\Theta)]$.

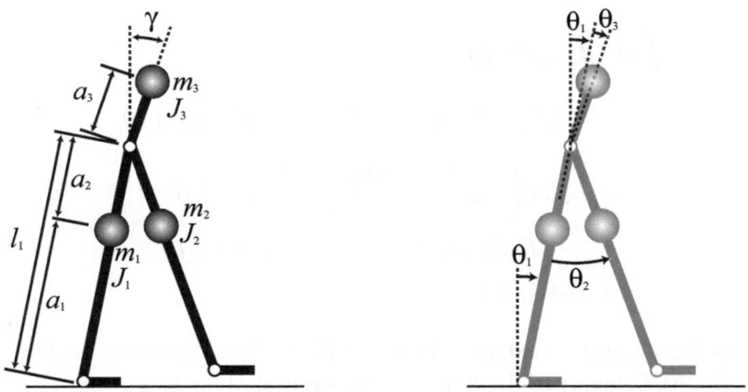

Fig. 9.14 3-link model in the sagittal plane. m_i, J_i, l_i and a_i are the mass, the moment of inertia, the length of link and the distance from the joint to the link COG of link i respectively. γ is the angle of the forward tilting. In the right figure, θ_1, θ_2 and θ_3 are the ankle angle of the stance leg, the angle from the stance leg to the the swing leg, the relative angle between the stance-leg and trunk respectively.

9.2.1.2 Converged Dynamics

Constraints are simply designed as follows:

- The angle of the torso is constant.
- The swing leg is symmetrical to the stance leg.

That is,

$$\theta_1 = f_1(\theta) = \theta \tag{9.29}$$
$$\theta_2 = f_2(\theta) = 2\theta \tag{9.30}$$
$$\theta_3 = f_3(\theta) = -\theta + \gamma \tag{9.31}$$

From Eqs. (9.29)-(9.31) and (2.11), Eq. (2.20) is

$$\begin{aligned} M_s(\theta) &= J_1 - J_2 + m_1 a_1^2 + m_2 l_1^2 - m_2 a_2^2 \\ &\quad + m_3 l_1^2 + m_3 a_3 l_1 \cos(\gamma - \theta) \\ &:= E_1 + E_2 \cos(\gamma - \theta) \end{aligned} \tag{9.32}$$

$$\begin{aligned} G_s(\theta) &= (m_1 a_1 + m_2 l_1 + m_2 a_2 + m_3 l_1) g \sin\theta \\ &\quad + m_3 g a_3 \sin\gamma \\ &:= E_3 + E_4 \sin\theta. \end{aligned} \tag{9.33}$$

Thus,

$$\begin{aligned} &\int M_s(\theta) G_s(\theta) d\theta \\ &= \int (E_1 + E_2 \cos(\gamma - \theta))(E_3 + E_4 \sin\theta) d\theta \\ &= E_2 E_4 \left(\frac{\sin\gamma}{2}\theta - \frac{\cos(2\theta - \gamma)}{4} \right) + E_1 E_3 \theta \\ &\quad + E_2 E_3 \sin(\theta - \gamma) - E_1 E_4 \cos\theta + C_s \\ &:= D_s(\theta) + C_s \end{aligned} \tag{9.34}$$

where C_s is the integral constant, which is PDAC Constant of the sagittal motion. From Eq. (2.26), Converged dynamics in the sagittal plane is

$$\dot{\theta} = \frac{1}{M_s(\theta)} \sqrt{2(D_s(\theta) + C_s)} \tag{9.35}$$
$$:= F_s(\theta). \tag{9.36}$$

Note that it is necessary that γ is decided so that $M_s(\theta) > 0$ in order to avoid singular point. Generally speaking, as for humanoid robots and biped robots, $E_1 > E_2$ since $l_1 > a_3$. Thus, we assume $M_s(\theta) > 0$ below.

9.2.1.3 Foot-Contact Model

Regarding foot-contact, it is assumed that the ground is perfectly inelastic collision and occurred for a moment similarly to previous work[87, 297, 158, 84]. That is, the angular momentum around the contact point is conserved before and after foot-contact.

Fig. 9.15 shows the angle and length of the inverted pendulum at foot-contact. Here, consider the foot-contact at the end of kth step, i.e. at the beginning of $k+1$th step. Denoting the angular velocity of ankle joint of the rear leg at foot-contact as $\dot{\theta}_e[k]$, the following equation is derived from Eq. (2.26):

$$\dot{\theta}_e[k] = \frac{1}{M_s(\theta_e[k])}\sqrt{2\big(D_s(\theta_e[k])+C_s[k]\big)} \tag{9.37}$$

where, $C_s[k]$ denotes PDAC Constant of kth step.

Since the torso angle is constant and COG is not rotated, the angular velocity of ankle joint of the fore leg at foot-contact, $P_i[k+1]$, is described as follows:

$$\begin{aligned} P_i[k+1] &= m_t l_e[k] l_i[k+1]\dot{\theta}_e[k] \\ &\quad \cdot\cos(\xi_e[k]+\xi_i[k+1]) \\ &= \frac{m_t l_e[k] l_i[k+1]\cos(\xi_e[k]+\xi_i[k+1])}{M_s(\theta_e[k])} \\ &\quad \cdot\sqrt{2\big(D_s(\theta_e[k])+C_s[k]\big)} \\ &:= h[k]\sqrt{2\big(D_s(\theta_e[k])+C_s[k]\big)} \end{aligned} \tag{9.38}$$

where, $m_t = m_1+m_2+m_3$.

Since the angular velocity around the passive joint is

$$P = M_s(\theta)\dot{\theta},$$

PDAC Constant after foot-contact, $C_s[k+1]$, is represented as

$$\begin{aligned} C_s[k+1] &= \frac{1}{2}P_i[k+1]^2 - D_s(\theta_i[k]) \\ &= h[k]^2 C_s[k] \\ &\quad + h[k]^2 D_s(\theta_e[k]) - D_s(\theta_i[k]) \\ &:= s_1[k]C_s[k] + s_2[k]. \end{aligned} \tag{9.39}$$

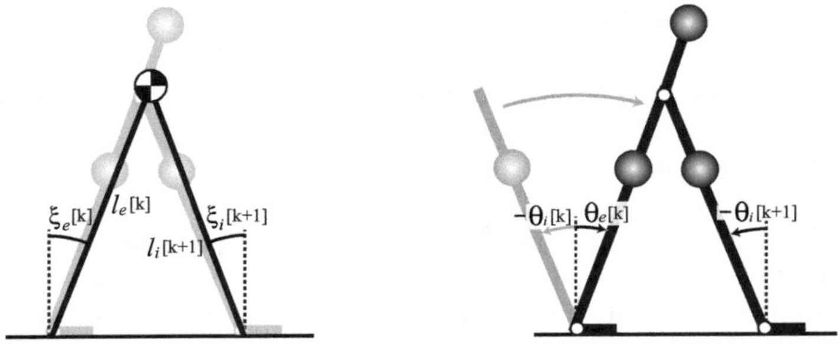

Fig. 9.15 Parameters at foot-contact. $l_e[k]$ and $\xi_e[k]$ are the length and inclination of the inverted pendulum which connects the ankle of support leg and robot COG before impact at the end of kth step. $l_i[k+1]$ and $\xi_i[k+1]$ are those after impact. $\theta_e[k]$ and $\theta_i[k+1]$ are the angles around the contact point before and after impact.

9.2.2 *Stabilization*

9.2.2.1 Constraint of Constant Step-Length

In order to stabilize the sagittal dynamics, constraint of constant step-length is set on the sagittal motion. Under such condition, it is clear that the following is held:

$$\theta_e[k] = -\theta_i[k] = \arcsin\left(\frac{\lambda_d}{2l_1}\right) := \theta_c = \text{const} \tag{9.40}$$

where $k \in \mathbf{N}$ and $0 \leq \theta_c < \frac{\pi}{2}$. Since the torso angle, γ, is kept constant, $\xi_e[k]$ and $\xi_i[k]$, $l_e[k]$, $l_i[k]$ are also all constant similarly. Hence, in Eq. (9.38),

$$h[k] := H = \text{const}$$

is held. Besides, in Eq. (9.39),

$$\begin{aligned} s_1[k] &= h[k]^2 = H^2 := S_1 = \text{const} \\ s_2[k] &= h[k]^2 D_s(\theta_e[k]) - D_s(\theta_i[k]) \\ &= H^2 D_s(\theta_c) - D_s(-\theta_c) := S_2 = \text{const} \end{aligned}$$

are also held.

9.2.2.2 Requisite to Perform Walking Continuously

We consider the requisite to generate walking. In terms of practicality, let γ be $0 < \gamma < \frac{\pi}{2}$. In order to perform walking continuously, $\dot{\theta} > 0$ is required at all times. Considering $M_s(\theta) > 0$, this condition is equivalent to $P = M_s(\theta)\dot{\theta} > 0$. Since

$$\begin{aligned}\frac{dD_s(\theta)}{d\theta} &= M_s(\theta)G_s(\theta)\\ &= \left(E_1 + E_2\cos(\gamma - \theta)\right)\left(E_3 + E_4\sin\theta\right),\end{aligned}$$

and also since $E_3 > 0$ and $E_4 > 0$, in can be seen that $D_s(\theta)$ i.e. the angular velocity of passive joint, P, is minimum when

$$\theta = \arcsin\left(-\frac{E_3}{E_4}\right) := \hat{\theta}.$$

and that it decreases monotonically on $-\frac{\pi}{2} < \theta < \hat{\theta}$ and increases monotonically on $\hat{\theta} < \theta < \frac{\pi}{2}$. Since $\frac{1}{2}P^2 = D_s(\theta) + C_s$, the condition discussed above, $P > 0$, is described as below,

$$P > 0 \Longleftrightarrow C_s > -D_s(\hat{\theta}) := \hat{C}_s. \tag{9.41}$$

Therefore, from Eq. (9.39), the requisite to perform walking continuously is found as follows:

$$S_1\hat{C}_s + S_2 > \hat{C}_s. \tag{9.42}$$

Next, we argue the state that dynamics of walking is converged on a sole trajectory, i.e. the equilibrium state. On such condition, PDAC Constant of every step is converged on constant value. That is,

$$C_s[k] = C_s[k+1] := C_s^*.$$

Hence, from Eq. (9.39),

$$\begin{aligned} C_s^* &= S_1C_s^* + S_2\\ \Longleftrightarrow S_2 &= C_s^*(1 - S_1)\end{aligned} \tag{9.43}$$

is held. Substituting Eq. (9.43) into Eq. (9.42),

$$(1 - S_1)(C_s^* - \hat{C}_s) > 0 \tag{9.44}$$

is obtained. From Eq. (9.41), it is clear that

$$C_s^* > \hat{C}_s. \tag{9.45}$$

From Eqs. (9.44) and (9.45), the requisite to perform walking continuously, Eq. (9.42), is described as follows:

$$1 - S_1 > 0. \tag{9.46}$$

As for our robot (Gorilla Robot II: Fig. 3.7, Table 9.2), Eq. (9.46) is the range shown in Fig. 9.16. Although we assume $\gamma < \frac{\pi}{2}$ in Fig. 9.16, actual upper bound is decided according to the limitation of robot's specification such as the torque to swing a leg forward or to keep a torso angle at constant value.

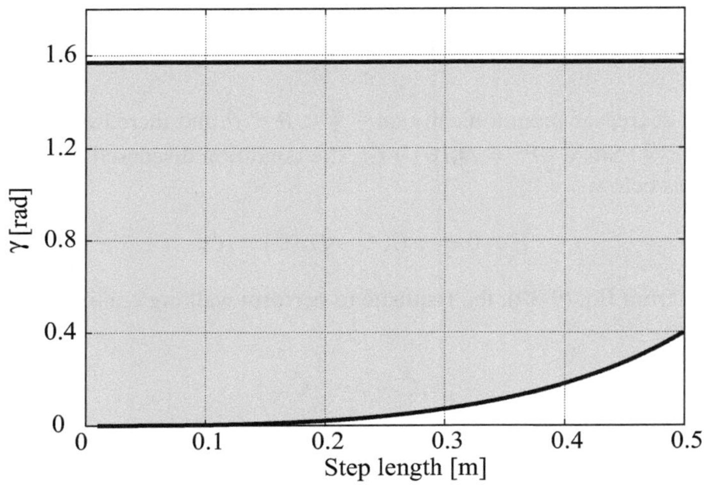

Fig. 9.16 Condition of θ and γ in order to generate continual walking

9.2.2.3 Proof of Stability

Lastly, we prove the stability by Lyapunov Theory using the conditions found above. The error between actual C_s and convergent value is defined as

$$\delta C_s := C_s^* - C_s. \tag{9.47}$$

The following positive definite function, V, is defined,

$$V = (\delta C_s)^2.$$

Since $V(0) = 0$ and $V > 0$ $(\delta C_s \neq 0)$, it is apparent that V is positive definite. From Eqs. (9.39) and (9.43), (9.47), finite difference of V is

$$\begin{aligned}
\Delta V &= V[k+1] - V[k] \\
&= \delta C_s[k+1]^2 - \delta C_s[k]^2 \\
&= (\delta C_s[k+1] + \delta C_s[k])(\delta C_s[k+1] - \delta C_s[k]) \\
&= -(2C_s^* - C_s[k+1] - C_s[k]) \\
&\qquad \cdot (C_s[k+1] - C_s[k]) \\
&= -(2C_s^* - S_1 C_s[k] - S_2 - C_s[k]) \\
&\qquad \cdot (S_1 C_s[k] + S_2 - C_s[k]) \\
&= -\left(2C_s^* - S_1 C_s[k] - (1 - S_1)C_s^* - C_s[k]\right) \\
&\qquad \cdot \left(S_1 C_s[k] + (1 - S_1)C_s^* - C_s[k]\right) \\
&= -(1 + S_1)(1 - S_1)(C_s^* - C_s[k])^2 \\
&= -(1 + S_1)(1 - S_1)\Delta C_s[k]^2.
\end{aligned} \tag{9.48}$$

Since it is clear that $1 + S_1 = 1 + H^2 > 0$ from Eq. (9.41) and that $1 - S_1 > 0$ from Eq. (9.46), Eq. (9.48) is

$$\Delta V = 0 \quad (\delta C_s = 0). \tag{9.49}$$

In addition,

$$\Delta V < 0 \quad (\delta C_s \neq 0) \tag{9.50}$$

is held. From Eqs. (9.49) and (9.50), ΔV is negative definite. Therefor, the equilibrium point, C_s^*, is asymptotically stable in the range shown in Fig. 9.16.

Note that this stability proved above does not depend on the form of virtual holonomic constraint. That is, as long as the parameters are set in the range shown in Fig 9.16, how to move the swing leg is of no concern to the walking stability, under the proposed framework. Thus, we can design the swing-leg motion under the constraint of constant step-length at foot-contact. In addition, it is significantly important that physical parameters such as mass distribution, link length, and leg weight do not affect the stability.

9.2.2.4 Simulation

Finally, stability proof described above is confirmed by the numerical simulation. Physical parameters employed in the simulation is described in Table 9.2.

Fig. 9.17 is the phase portrait of θ and alteration in terms of time. From these figures, the convergency of the sagittal motion can be ascertained. Fig. 9.18 shows the snapshots of the simulation. On this simulation, step-length is 0.18[m], the torso angle is γ=0.035[rad].

Fig. 9.19 depicts the return maps of PDAC Constant and the angular velocity of passive joint at foot-contact. These figures show that the sagittal dynamics has a sole stable fixed point.

Table 9.1 Link parameters of Gorilla Robot II

Mass[kg]	link1	m_1	2.618
	link2	m_2	3.451
	link3	m_3	15.143
COG position	link1	a_1	0.23
	link2	a_2	0.28
	link3	a_3	0.22
Moment of inertia	link1	J_1	0.042
	link2	J_2	0.070
	link3	J_3	0.047

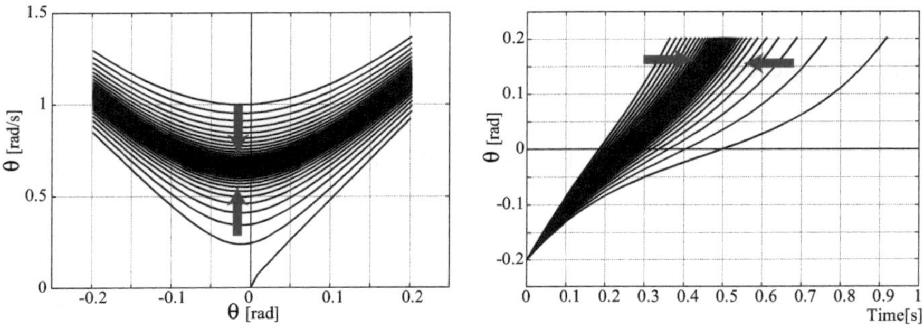

Fig. 9.17 Simulation results of sagittal stabilization. (Left) Phase portrait of the sagittal dynamics. (Right) θ trajectory vs. time.

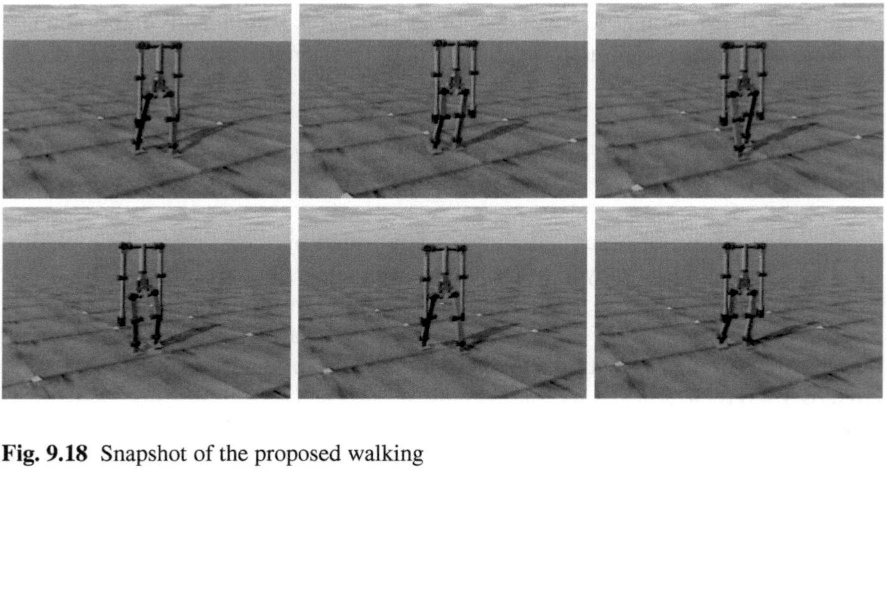

Fig. 9.18 Snapshot of the proposed walking

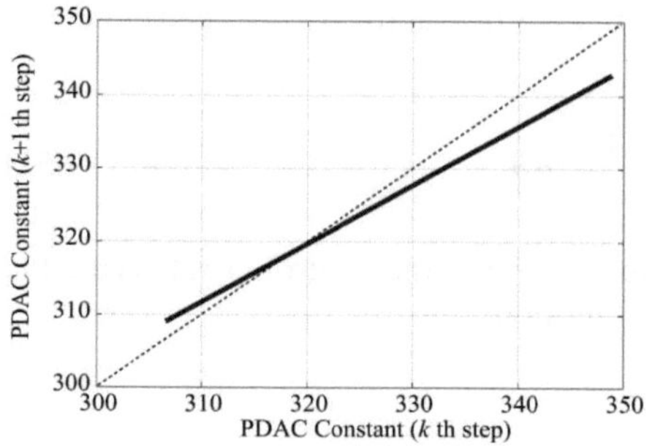

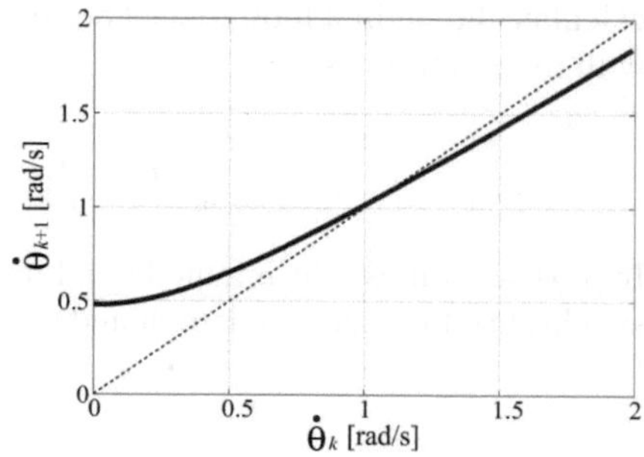

Fig. 9.19 Simulation results of sagittal stabilization. (Left) Return map of PDAC Constant. (Right) Return map of $\dot{\theta}$ at foot-contact.

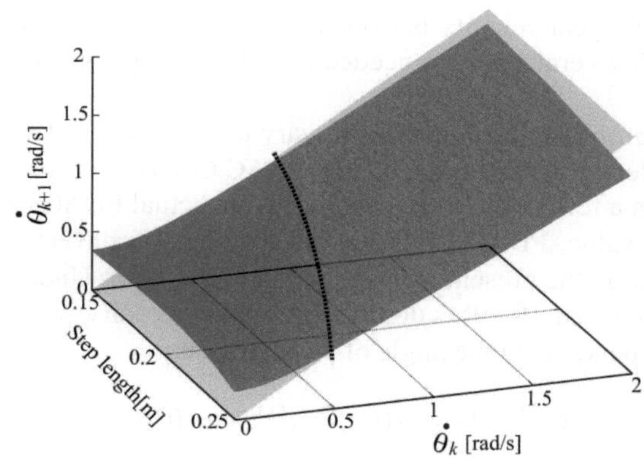

Fig. 9.20 Return map of $\dot{\theta}$ at foot-contact vs. step-length

Finally, to confirm that the stability of the sagittal motion is independent of step-length, we perform the simulation of the various step-length. Fig. 9.20 is the graph of the return map of the angular velocity of passive joint at foot-contact v.s. step-length. From this figure, it can be confirmed that the sagittal dynamics is stable regardless of step-length.

9.2.3 Sagittal Motion Period

In order to satisfy the condition of constant step-length, it is necessary to control the lateral motion so that lateral foot-contact period matches sagittal one. Since sagittal dynamics is expressed as an 1-dimensional autonomous dynamics, it is possible to calculate the sagittal foot-contact period by integrating sagittal Converged dynamics with time as follows:

$$T_s = \int_{{}_i\theta^+}^{{}_{i+1}\theta^-} \frac{1}{F_s(\theta)}\, d\theta. \tag{9.51}$$

In next section, we design the lateral motion and build the controller satisfying the synchronization between lateral and sagittal motion.

9.2.4 *Quick Convergency Method*

9.2.4.1 Variable Step-Length

The stability of above-mentioned method was proved, however it has the disadvantage that the convergent velocity is slow since the convergence is not strong. That is, in this method, several steps are needed in order to converge the dynamics on the fixed point.

To solve this problem, the controller to vary the constraint of step-length is proposed. The step-length is controlled so that PDAC Constant is converged on the desired value within a few steps. To be specific, if the actual PDAC Constant is higher than the desired value, PDAC Constant of next step is converged on the desired value by lengthening the present step according to the error. Under the condition of variable step-length, Eq. (9.39) is described as below because $s_1[k]$ and $s_2[k]$ are no more constant dependent on the angle of passive joint,

$$C_s[k+1] = s_1(\theta)[k]C_s[k] + s_2(\theta)[k]. \tag{9.52}$$

Letting C_s^* and δC_s the converged value and the error of PDAC Constant and letting $\delta\theta$ the adjusted value of passive joint angle at foot-contact respectively, Condition to converge PDAC Constant on the fixed point at next step is described as follows:

$$C_s^* = s_1(\theta_c + \delta\theta)[k](C_s^* + \delta C_s) + s_2(\theta_c + \delta\theta)[k]. \tag{9.53}$$

It is possible to converge dynamics on the fixed point by employing $\delta\theta$ derived from Eq. (9.53).

9.2.5 *Simulation*

Fig. 9.21 is the map of the adjusted value, $\delta\theta$, v.s. the error of PDAC Constant, C_s, under the condition that trunk inclination is 4.5[deg], step-length is 0.2[m]. Intersection of the curve depicted in the map with the origin indicates that $\delta\theta = 0$ if $\delta C_s = 0$. As shown in this map, if PDAC Constant is lower than desired value, there is limit

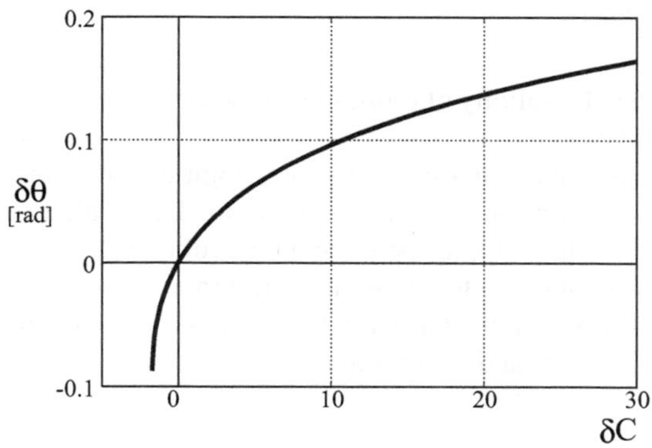

Fig. 9.21 Map of $\delta\theta$ v.s. C_s (Trunk inclination: 4.5[deg], step-length: 0.2[m])

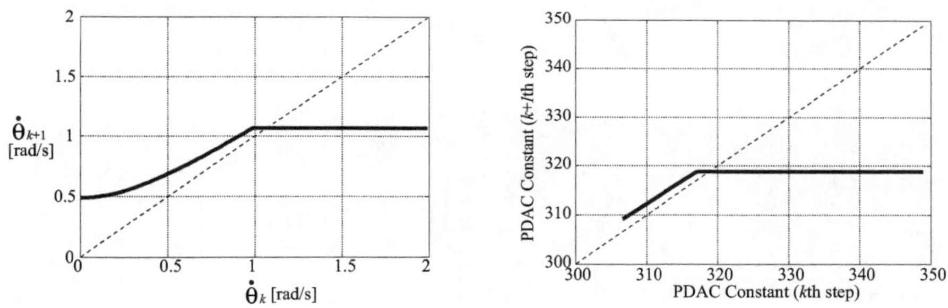

Fig. 9.22 Simulation results of quick convergence method. (Left) Return map of PDAC Constant. (Right) Return map of $\dot{\theta}$ at $\theta = 0$.

of the convergence ability. Thus, if δC_s is beyond this limit, 1-step convergence of PDAC Constant is impossible. Considering the upper limit of supplied energy and the lower limit of lost energy at foot-contact, this result is the natural.

Fig. 9.22 shows the return map of PDAC Constant and $\dot{\theta}$ at $\theta = 0$. If the PDAC Constant error is positive i.e. in the right area from the fixed point, it is possible to converge it on the desired value at next step. However, in the area of negative PDAC Constant error i.e. in the left area from the fixed point, there is a limit as mentioned above. If the PDAC Constant error is within the limit, the dynamics can be converged within a step. Meanwhile, outside the limit, PDAC Constant is converged within a few steps.

9.2.6 Experiment

In order to ascertain the validity of proposed method, the experiment of 3-D dynamic walking was conducted.

In this experiment, the robot bends its knee joint of the swing leg so as to prevent the foot being in friction with the ground immediately after foot-contact on the assumption that the effect of knee bending on the robot dynamics can be neglected. The foot of the swing leg is actuated so as to be kept parallel to the ground. The experiment was performed on the basically level and flat ground which has maximum 1.0[cm] irregularity without discontinuities.

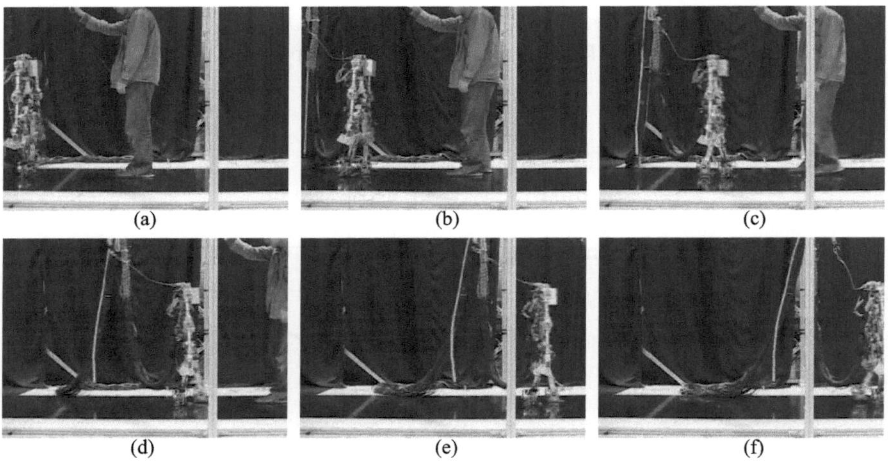

Fig. 9.23 Snapshots of the walking experiment. Each figure shows the snapshots at (a)1st (b)5th (c)9th (d)12th (e)14th (f)16th step.

In this section, the desired step-length is given to be gradually increased within initial 5 steps up to 0.15[m] and after 5 steps it is fixed at 0.15[m]. As a result of experiment, the dynamic and natural walking is realized over 25 steps. The step-length is about 0.15[m] and the walking velocity is about 0.23[m/s]. Fig. 9.23 shows the snapshots of the PDAC walking at 1st, 7th, 12th, 16th, 19th, 22nd step respectively. The angle and angular velocity of the lower body joints are depicted in Fig. 9.24 and Fig. 9.25. As shown in these figures, the smooth dynamic motion is realized periodically.

9.2.7 Upper Layer Controller

To generate 3-D dynamic walking by means of the proposed control architecture, it is necessary to input both the desired step-length and walking period. Considering the usability of controller, it is desirable that the controller decides step-length and walking period by just inputting the desired walking velocity. However, there are a infinite number of combination of them that satisfies the desired walking velocity. Thus, in this section, we design the upper layer controller to decide step-length and walking period so that the energy consumption per traveled distance is minimized.

Fig. 9.26 shows the overview of upper layer controller. PDAC Controller was built above, hence the gray box, which is the upper layer controller, is needed to be designed. Energy consumption is calculated by the following equation:

$$E = \frac{\int_0^{T^d} \tau^T \dot{\Theta} dt}{\lambda} = \frac{\int_0^{T^d} \tau^T d\Theta}{\lambda} \tag{9.54}$$

where T^d denotes walking period and λ is step-length. Note that this calculation is conducted under the following conditions: 1) No viscosity and no friction of all joints 2) Flat and uniform ground 3) perfectly inelastic collision between sole and the ground.

As a result, the energy consumption map shown in Fig. 9.27 is acquired. As shown in this figure, there is a valley in this map, which is depicted by a thick line. The dotted line drawn on the bottom is the projection of valley onto the bottom surface. By means of this line, it is possible to decide the step-length and walking period

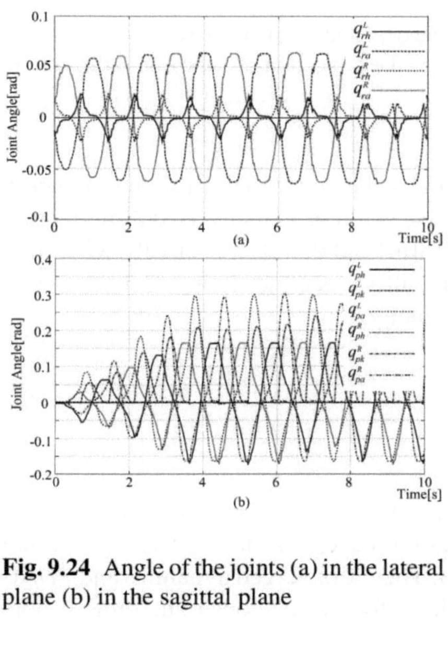

Fig. 9.24 Angle of the joints (a) in the lateral plane (b) in the sagittal plane

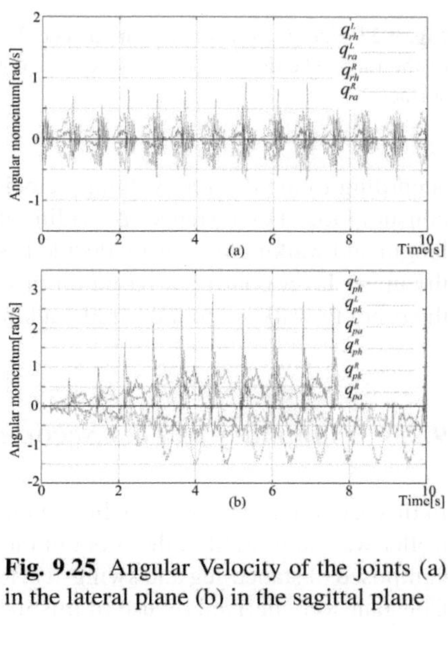

Fig. 9.25 Angular Velocity of the joints (a) in the lateral plane (b) in the sagittal plane

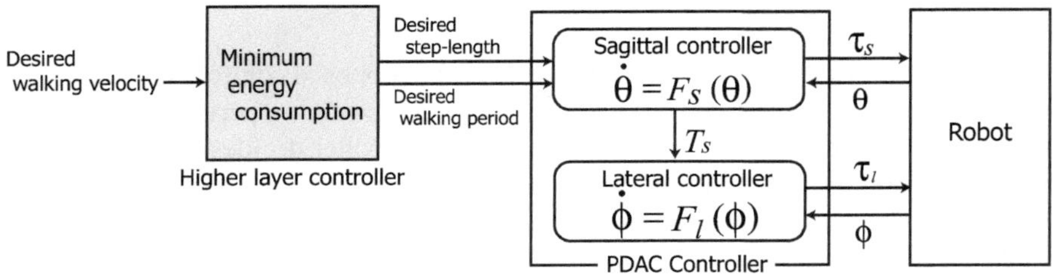

Fig. 9.26 Upper layer controller

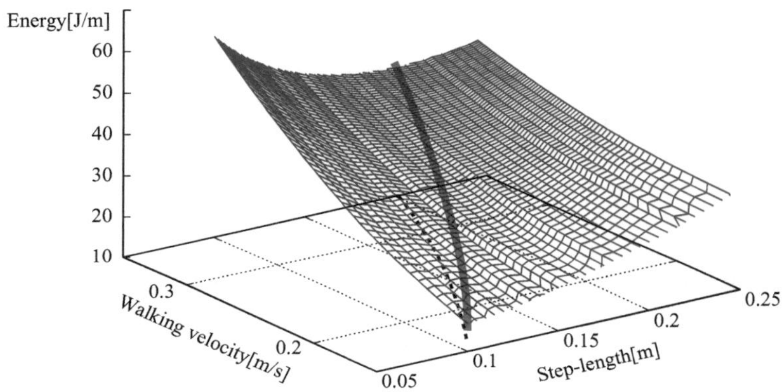

Fig. 9.27 Energy consumption map. The thick line depicted on the map expresses the valley of the map. The dotted line drawn on the bottom is the projection of valley onto the bottom surface.

according to the desired walking velocity under the condition of minimum energy consumption. For instance, providing desired walking velocity is 0.3[m/s], step-length and walking period are decided as 0.18[m] and 0.6[s] respectively. Mounting the upper layer controller based on this map, all we need to do in order to operate the robot to walk is input desired walking velocity.

9.2.8 Summary of This Section

In this section, PDAC was applied to the sagittal dynamics and sagittal motion controller was built. Firstly, the robot in the sagittal plane was modeled as 3-link robot composed of stance-leg and swing-leg, torso. Then, Converged dynamics and PDAC Constant were derived by use of this model. As for Virtual constraint, quite simple constraint was employed as follows: 1) Constant torso inclination 2) The swing leg is symmetrical to the stance leg.

Thirdly, discrete dynamics at foot-contact was found based on the assumption of perfectly inelastic collision between sole and the ground. In this model, before and after foot-contact, the angular momentum is conserved, however in the meanwhile, the mechanical energy is not conserved. This energy loss at foot-contact is important for stabilization. The higher the mechanical energy is, the higher the energy loss is. Thus, sagittal motion can be stabilized by setting constraint of constant step-length. We proved the stability by the Lyapunov Theory. In addition, by the numerical simulation, it was ascertained that the proof was correct. Also, the walking period in the sagittal plane for synchronization with lateral motion is calculated.

Next, we design the upper layer controller that determines step-length and walking period according to the desired walking velocity so as to minimize the energy consumption per traveled distance. The upper layer controller was based on the energy consumption map associating combination of walking velocity and step-length with the consumption energy.

Finally, the experiment of 3-D dynamic walking was conducted and the validity of proposed control architecture was confirmed.

9.3 Heel-off Walking Control with PDAC

In this section, we design the planar biped walking with heel-off motion, which makes it possible to avoid the problem for the impact shock at foot-contact. The 3-D motion is separated into lateral and sagittal motion under the assumption that each motion is approximately independent each other. The stable lateral motion to achieve the desired motion period in section 9.1. Thus, in this section, the sagittal motion including heel-off motion is presented.

Sagittal motion consists of two following phases: 1) The center of rotation (contact point) is ankle joint. 2) The center of rotation is toe. In phase 2), heel-off motion is designed to diminish the impact shock. The approximate 1-mass model is employed and COG trajectory, which is not the trajectory with respect to time but the COG path independent of time, is designed. By applying PDAC to the sagittal dynamics, phase around the contact point can be derived based on the desired COG trajectory and point-contact condition. Based on this phase, simple stabilizing controller by varying COG trajectory in heel-off phase is proposed and the motion period necessary to synchronize the lateral motion with sagittal motion is calculated.

By numerical simulation, it is confirmed that walking dynamics is converged on certain state and that heel-off motion reduced impact force at foot-contact. Finally, we examine energy efficiency of heel-off walking and propose the mechanical system to improve the energy efficiency.

9.3.1 Sagittal Motion Design

As mentioned previously, heel-off motion is introduced at the end of the stance-leg phase. Heel-off motion has the function to decrease the impact shock at foot-contact and to make walking smooth. In this section, two kinds of COG trajectory are given: 1) designed trajectory that the COG moving direction is parallel to the ground at foot-contact (Trajectory 1). 2) Human COG trajectory rescaled in accordance with the robot size (Trajectory 2).

In order to apply PDAC, it is necessary that pendulum length is expressed as the function of pendulum angle. In this section, the following equation is employed as the function:

$$l = a\theta^3 + b\theta^2 + c\theta + d. \tag{9.55}$$

a, b, c and d are decided so as to satisfy the desired COG trajectory, Trajectory 1 and 2. We derive the phase around contact-point by applying PDAC and design the stabilizing control.

Regarding dynamics, the following assumptions are employed in this section: 1) The effects of legs on the dynamics are neglected since these masses are quite light compared to the upper body as for our robot, "Gorilla Robot III". 2) A perfectly inelastic collision is occurred at foot-contact and angular momentum is conserved before and after foot-contact.

9.3.1.1 Trajectory 1

Fig. 9.28 shows Trajectory 1 and parameters, variables models. Σ_p is the polar coordinate system around the ankle joint and Σ_h is that around the toe. Since the knee joint is not bended, the length of inverted pendulum in Σ_p is constant. On the other hand, that of the inverted pendulum in Σ_h is varied by the torque of ankle joint. The designed walking has the two following phases:

- $2 \rightarrow 0 \rightarrow 1$: Constant pendulum length. center of rotation is the ankle joint. The coordinate system is Σ_p. (Stance phase)
- $1 \rightarrow 2$: Variable pendulum length. center of rotation is the toe. The coordinate system is Σ_h. (Heel-off phase)

[**Stance phase**]
In Stance phase, pendulum length is constant. As shown in Fig. 9.28,

$$l_{p0} = l_{p1} = l_{p2} = L_2 + L_3 := L. \tag{9.56}$$

Hence, $a = 0, b = 0, c = 0, d = L$

[**Heel-off phase**]
The heel-off motion is designed to satisfy the following conditions at the end and beginning of Heel-off phase:

- The pendulum length is continuous at the beginning.
- The COG vector is continuous at the beginning.
- The COG is located at the desired position at the end.
- The COG vector is parallel to the ground at the end.

a, b, c, and d are decided so as to satisfy the 4 conditions described above,

$$a = -\frac{2(l_{h2} - l_{h1})}{(\theta_{h2} - \theta_{h1})^3} + \frac{s_2 + s_1}{(\theta_{h2} - \theta_{h1})^2} \tag{9.57}$$

$$b = \frac{s_2 - s_1}{2(\theta_{h2} - \theta_{h1})} - \frac{3a}{2}(\theta_{h2} + \theta_{h1}) \tag{9.58}$$

$$c = s_1 - 3a\theta_{h1}{}^2 - 2b\theta_{h1} \tag{9.59}$$

$$d = l_{h1} - a\theta_{h1}{}^3 - b\theta_{h1}{}^2 - c\theta_{h1} \tag{9.60}$$

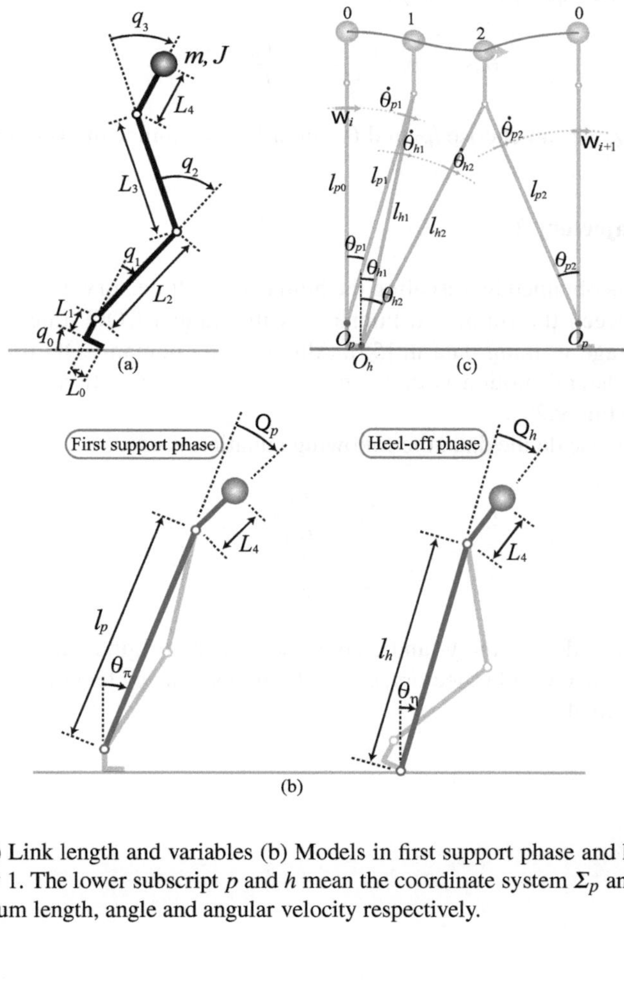

Fig. 9.28 (a) Link length and variables (b) Models in first support phase and heel-off phase (c)Trajectory 1. The lower subscript p and h mean the coordinate system Σ_p and Σ_h. l, θ and $\dot{\theta}$ are pendulum length, angle and angular velocity respectively.

where,

$$s_1 = -l_{h1}\tan(\theta_{p1} - \theta_{h1}) \tag{9.61}$$
$$s_2 = l_{h2}\tan(\theta_{h2}). \tag{9.62}$$

Human has Achilles' tendon, thus it is seemed that when it extends completely, the ankle joint is locked and the heel-off phase starts. Based on this assumption, we decide the ankle joint at the beginning of heel-off phase at constant value, $\theta_{p1} := \theta_{ac}$=10.0[deg]. Hence, l_{h1} and θ_{h1} are determined as follows:

$$l_{h1} = \sqrt{(L\sin\theta_{ac} - L_0)^2 + (L\cos\theta_{ac} + L_1)^2} \tag{9.63}$$
$$\theta_{h1} = \tan^{-1}\left(\frac{L\sin\theta_{ac} - L_0}{L\cos\theta_h + L_1}\right) \tag{9.64}$$

θ_{p2} is decided uniquely according to the state at foot-contact as below,

$$\theta_{p2} = \cos^{-1}\left(\frac{l_{h2}\cos\theta_{h2} - L_1}{L}\right). \tag{9.65}$$

Consequently, we can choose l_{h2} and θ_{h2} in order to stabilize the walking.

9.3.1.2 Trajectory 2

Trajectory 2 is obtained by rescaling the human COG trajectory in accordance with the ratio between the robot and human. As the human COG trajectory, we employ the average walking data in [55]. Although the human COG trajectory is 3-dimensional, lateral motion is quite small, hence the x-direction is neglected approximately (Fig. 9.29).

In order to rescale the data, the following equation is used,

$$Y_g = \frac{L_r}{L_m} y_g \tag{9.66}$$
$$Z_g = \frac{L_r}{L_m} z_g \tag{9.67}$$

where y_g and z_g denote the y- and z-coordinate of the original data. Y_g and Z_g are that of the rescaled COG trajectory. L_r and L_m denote the leg length of the robot and human respectively.

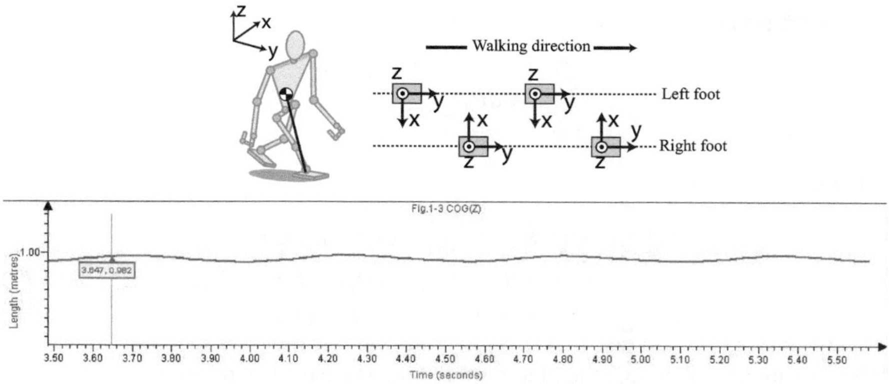

Fig. 9.29 Coordinate system definition. Lower figure is the human COG trajectory data in yz plane.

9.3.2 *Converged Dynamics*

[**Stance phase**]
Dynamic equation of the model shown in Fig. 9.28 is

$$\frac{dP_p}{dt} = G_p \tag{9.68}$$

where,

$$\begin{aligned} P_p(\theta_p) =& m\big(l^2\dot{\theta}_p + L_4^2(\dot{\theta}_p + \dot{\Theta}_p) \\ &+ lL_4(2\dot{\theta}_p + \dot{\Theta}_p)\cos\Theta_p\big) + J(\dot{\theta}_p + \dot{\Theta}_p) \end{aligned} \tag{9.69}$$

$$G_p(\theta_p) = mg\big(l\sin\theta_p + L_4\sin(\theta_p + \Theta_p)\big). \tag{9.70}$$

Thus, by means of PDAC, the phase around toe is derived as follows:

$$\dot{q}_1 = \frac{1}{M_p(\theta_p)}\sqrt{2\int M_p(\theta_p)G_p(\theta_p)d\theta_p} \tag{9.71}$$

$$:= \frac{1}{M_p(\theta_p)}\sqrt{2\big(D_p(\theta_p) + C_p\big)} \tag{9.72}$$

$$:= F_p(\theta_p) \tag{9.73}$$

where,

$$M_p(\theta_p) = m(L^2 + L_4L\cos\theta_p) \tag{9.74}$$

$$G_p(\theta_p) = mgL\sin\theta_p. \tag{9.75}$$

[**Heel-off phase**]

$$\frac{dP_h}{dt} = G_h \tag{9.76}$$

where,

$$P_h(\theta_h)=m\big(l^2\dot{\theta}_h + L_4^2(\dot{\theta}_h + \dot{\Theta}_h) - L_4\dot{l}\sin\Theta_h \\ + lL_4(2\dot{\theta}_h + \dot{\Theta}_h)\cos\Theta_h\big) + J(\dot{\theta}_h + \dot{\Theta}_h) \tag{9.77}$$

$$G_h(\theta_h) = mg\big(l\sin\theta_h + L_4\sin(\theta_h + \Theta_h)\big). \tag{9.78}$$

Thus, by means of PDAC, the phase around toe is derived as follows:

$$\dot{\theta}_h = \frac{1}{M_h(\theta_h)}\sqrt{2\int M_h(\theta_h)G_h(\theta_h)d\theta_h} \tag{9.79}$$

$$:= \frac{1}{M_h(\theta_h)}\sqrt{2\big(D_h(\theta_h)+C_h\big)} \tag{9.80}$$

$$:= F_h(\theta_h) \tag{9.81}$$

where,

$$M_h(\theta_h)=m\big(l^2 + L_4 f'(\theta_h)\sin\theta_h + lL_4\cos\theta_h\big)\dot{\theta}_h \tag{9.82}$$

$$G_h(\theta_h)=mgl\sin\theta_h. \tag{9.83}$$

9.3.3 Stabilization

Since in order to stabilize the walking, it is possible to choose the values of l_{h2} and θ_{h2}, we can design the controller to satisfy another condition such as desired step-length or desired walking velocity. However, as the first step of the heel-off walking, we employ the simple stabilizing control that θ_{h2} is fixed at constant value and l_{h2} is adjusted according to PDAC Constant of stance-leg phase as follows:

$$l_{h2} = l_{h0} + k_{l_h}(C_p^d - C_p) \tag{9.84}$$

where, l_{h2} is l_{h2} at stationary state, k_{l_h} is the feedback gain, and C_p^d is the desired PDAC Constant of the phase of Σ_p, i.e. C_p.

9.3.4 Foot-Contact

9.3.4.1 Foot-Contact Dynamics

Similarly to the previous chapter, we assume perfectly inelastic collision at foot-contact in heel-off walking. That is, the angular momentum around contact-point is conserved before and after impact. Considering the COG trajectory and moving direction of COG immediately before foot-contact, the discrete dynamics of foot-contact is described as follows:

$$\frac{ml_{p2}l_{h2}\dot{\theta}_{h2}\cos\theta_{p2}}{\cos\theta_{h2}} = ml_{p2}{}^2\dot{\theta}_{p2}$$
$$\Longleftrightarrow \dot{\theta}_{p2} = \frac{l_{h2}\cos\theta_{p2}}{l_{p2}\cos\theta_{h2}}\dot{\theta}_{h2}. \tag{9.85}$$

9.3.4.2 Foot-Contact Period

In order to realize 3-D walking, it is necessary that the periods from foot-contact to next foot-contact (foot-contact period) in each plane are identical. As for the lateral motion, we realized the stable lateral motion to achieve the desired motion period in Chapter 9.1. Thus, it is necessary that the sagittal foot-contact period is calculated, and then the lateral motion is adjusted so as to satisfy its period.

The sagittal foot-contact period is found by integrating the phase of sagittal motion as follows:

$$T_s = \int_{\theta_{p0}}^{\theta_{p1}} \frac{1}{F_p(\theta_p)}\, d\theta_p + \int_{\theta_{h1}}^{\theta_{h2}} \frac{1}{F_h(\theta_h)}\, d\theta_h. \tag{9.86}$$

9.3.5 Simulation

9.3.5.1 Simulation Setup

Table 9.2 is the robot parameters. The model is faithfully based on Gorilla Robot III. Note that, in the following simulation, it was assumed that the ground friction is strong enough to prevent slipping.

9.3.5.2 Sagittal Stabilization

At first, the stability of sagittal motion was investigated. Fig. 9.30 shows the simulation results under the condition that l_2^0=0.532[m], k_{l_2}=0.01, $\dot{\Theta}_1^p$=0.96. Left figure depicts the phase portrait around the toe. The center of rotation is switched from the

Table 9.2 Link parameters of Gorilla Robot III

Mass[kg]	m	24.0
Moment of Inertia[kgm^2]	J	24.0
Link length[m]	L_0	0.08
	L_1	0.087
	L_2	0.23
	L_3	0.225
	L_4	0.2

ankle joint to the toe at the beginning of the heel-off phase. The reason why the angular acceleration is discontinuous at the point $\theta_b = 0.03$ is because contact-point is switched from the ankle joint to toe at this point. As shown in Fig. 9.30, both if the energy is higher than desired state and if it is lower, the walking status is converged on the desired state. Right figure shows the return map of the angular velocity. From this figure, it can be confirmed that the dynamics of proposed walking has the stable fixed point.

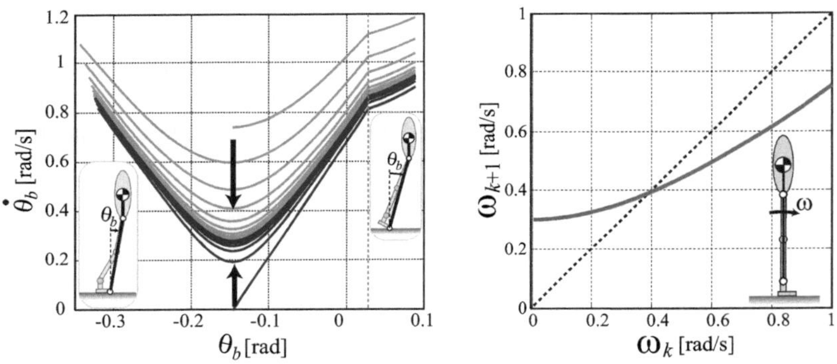

Fig. 9.30 Simulation results of the simple feedback control. (Left) Phase portrait around the toe. θ_b is the inclination of the line drawn from toe to the hip joint. (Right) Return map. ω is the angular velocity of ankle joint at 0[rad].

9.3.5.3 Comparison with Non-heel-off Walking

Next, we applied the non-heel-off method proposed in Chapter 9.2 to the simple model employed in this section (Fig. 9.28) and compared proposed heel-off walking with previous method regarding the impact force and energy loss at foot-contact, energy efficiency of walking.

At first, the previous method is explained briefly. In the previous method, point-contact and no knee-bending are adopted, hence $\tau_1 = 0$ and $q_2 = 0$, $\dot{q}_2 = 0$. In addition, the inclination of upper body is kept at constant value, γ, that is, Virtual Constraint is $q_3 = -q_1 + \gamma$. Thus, the energy is supplied by the hip torque, τ_3, which supplies mechanical energy to compensate for energy loss at foot-contact. Fig. 9.31 depicts the brief overview and phase portrait of previous walking.

By applying PDAC, it is possible to express the robot dynamics as the phase around ankle joint as follows:

$$\dot{q}_1 = \frac{1}{M_s(q_1)}\sqrt{2\int M_s(q_1)G_s(q_1)dq_1} \tag{9.87}$$

$$:= \frac{1}{M_s(q_1)}\sqrt{2(D_s(q_1)+C_s)} \tag{9.88}$$

$$:= F_s(q_1) \tag{9.89}$$

where,

$$M_s(q_1) = m(L^2 + L_4 L\cos(\gamma - q_1)) \tag{9.90}$$

$$G_s(q_1) = mg(L\sin q_1 + L_4\sin\gamma). \tag{9.91}$$

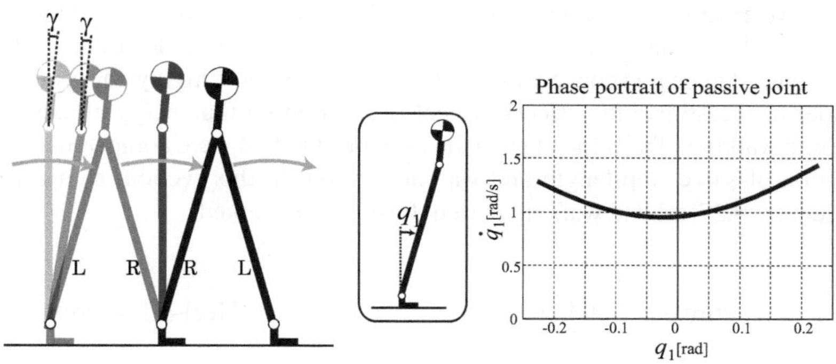

Fig. 9.31 Brief overview and phase portrait of previous walking (step-length=0.2[m], γ=0.075[rad])

Impact Force

The impact force and energy loss at foot-contact were checked. We compared heel-off walking with previous method and examined the effects of heel-off motion. Table 9.3 is the simulation results of slow walking (walking velocity: 0.35[m/s], step-length: 0.18[m]) and Table 9.4 is that of fast walking (walking velocity: 0.55[m/s], step-length: 0.27[m]). These simulation results indicate that the proposed heel-off motion has the effects to reduce the impact shock and energy loss at foot-contact.

Table 9.3 Impact force and energy loss at foot-contact (Slow walking: Walking velocity is about 0.35[m/s] and step-length is about 0.18[m])

	Previous method	Trajectory 1	Trajectory 2
Heel-off	×	○	○
Impact force[N]	5.062	0.551	0.023
Energy loss[J]	0.533	0.050	0.001

Table 9.4 Impact force and energy loss at foot-contact (Fast walking: Walking velocity is about 0.55[m/s] and step-length is about 0.27[m])

	Previous method	Trajectory 1	Trajectory 2
Heel-off	×	○	○
Impact force[N]	11.27	3.656	0.252
Energy loss[J]	2.651	0.287	0.021

Energy Consumption

Finally, we examined the energy efficiency of heel-off walking. The simulation results of Table 9.3 and 9.4 shows that the heel-off motion reduced energy loss at foot-contact, hence it is conceivable that heel-off walking is highly energy efficient. In order to investigate the effects of heel-off motion on the energy efficiency, the proposed walking (Trajectory 1) and the previous method were compared under the condition of same step-length and walking period. In this section, as a example, step-length: 0.27[m] and walking period: 0.68[s] are adopted.

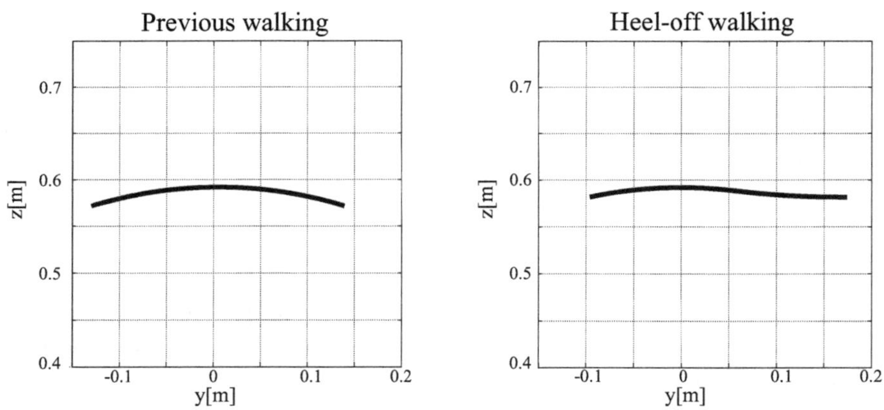

Fig. 9.32 COG trajectory. (Left) Previous walking (Right) Heel-off walking

Fig. 9.32 depicts COG trajectory of both walking. In this figure, it is seen that the heel-off motion makes COG trajectory parallel to the ground at foot-contact, which contributes low foot-impact as described in previous subsection.

Fig. 9.33 and 9.34, 9.35 shows the angle and angular velocity, torque of the robot's joints. As shown in these figures, the robot is rotated around toe in heel-off walking differently from previous walking. Note that knee joint is assumed to have the apparatus to be locked i.e. $\tau_2 = 0$.

Fig. 9.36 shows the energy consumption per second calculated by the following equation,

$$\dot{E} = \sum_{i=0}^{3} \dot{q}_i \tau_i. \tag{9.92}$$

Energy consumption per step is derived by integrating this equation from foot-contact to next foot-contact with respect to time. As a result, the energy consumption of previous method is 2.53[J] and that of the heel-off walking is 3.11[J]. Considering the low energy loss of heel-off walking at foot-contact, it is expected that the energy consumption of heel-off walking is lower than that of previous walking, however the result is different from such anticipation.

This mismatch between the anticipation and actual results is ascribed to the transition from the first support phase to the heel-off phase. As shown in Fig. 9.37(a), the robot apply torque to the ankle joint opposite of rotating direction in the beginning of the heel-off phase in order to switch center of rotation from ankle joint to toe. Hence, by use of ankle joint, the robot decreases the mechanical energy at first, then increases in the heel-off phase as shown in Fig. 9.37(b). The cause to deteriorate the energy efficiency of heel-off walking is this energy-diminishing motion. That is, in previous walk, this energy is lost by impact shock at foot-contact, whereas it is dissipated in heel-off walk by robot actuating energy, which results in low foot-contact shock.

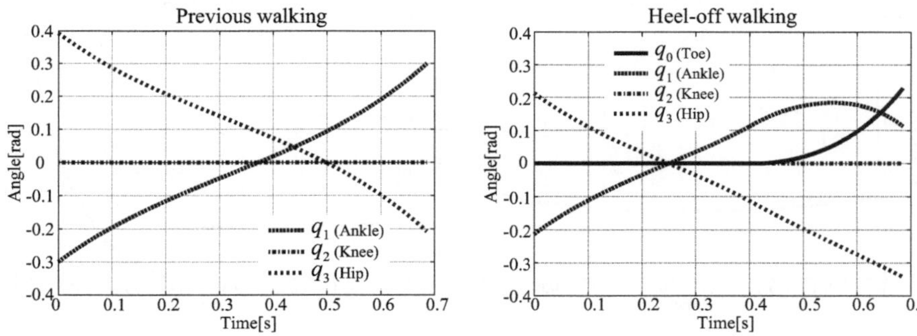

Fig. 9.33 Joint angle. (Left) Previous walking (Right) Heel-off walking. q_0 and q_1, q_2, q_3 denote the joint angle as shown in Fig. 9.28.

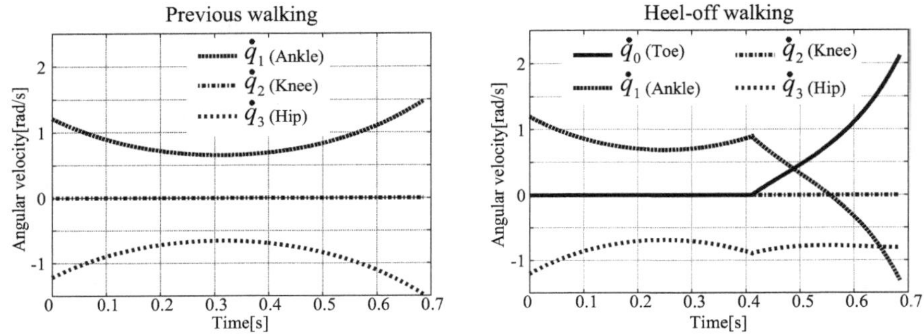

Fig. 9.34 Joint angular velocity. (Left) Previous walking (Right) Heel-off walking. $\dot{q}_0$ and $\dot{q}_1$, $\dot{q}_2$, $\dot{q}_3$ denote the angular velocity around toe and ankle, knee, hip joint.

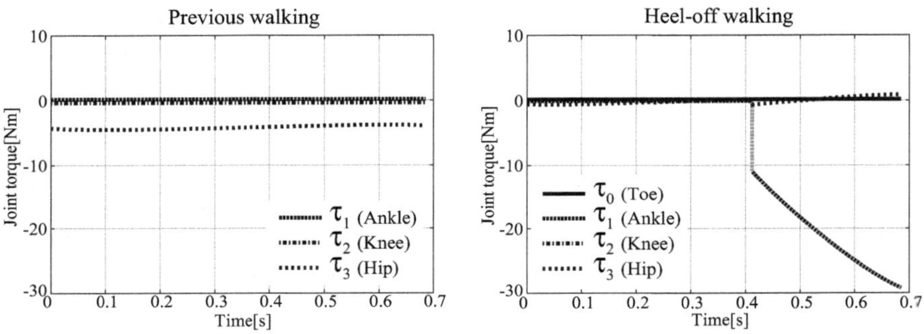

Fig. 9.35 Joint torque. (Left) Previous walking (Right) Heel-off walking. τ_0 and τ_1, τ_2, τ_3 denote the torque applied toe and ankle, knee, hip joint.

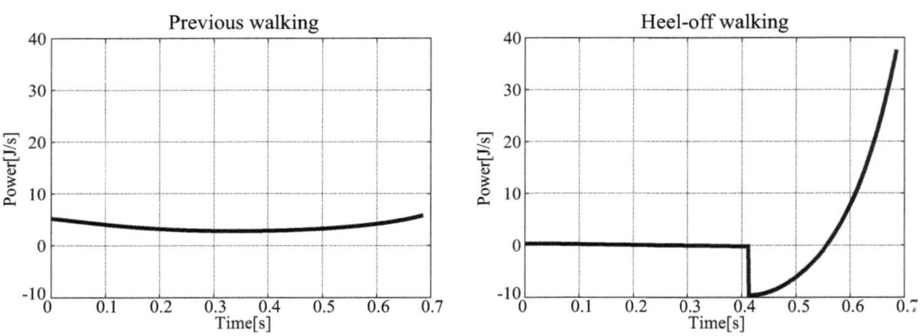

Fig. 9.36 Energy consumption per second. (Left) Previous walking (Right) Heel-off walking

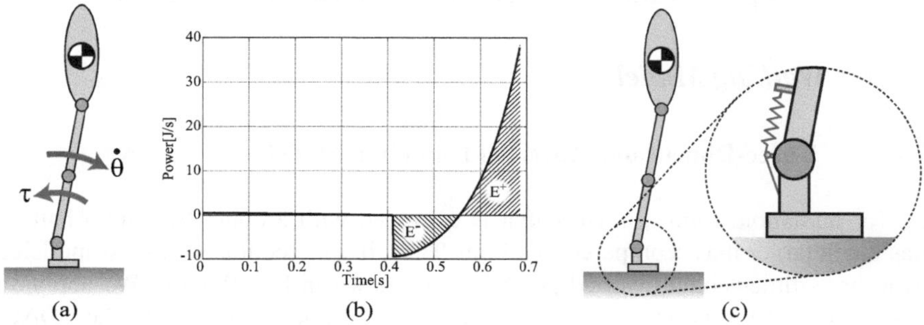

Fig. 9.37 Mechanical system to accumulate the energy

In order to improve the energy efficiency, a mechanical device to accumulate the energy such as Fig. 9.37(c) is necessary. This type of device stores the mechanical energy E^- shown in Fig. 9.37(b) at the transition from first support phase to heel-off phase and release it as E^+ in the end of heel-off phase. Assuming that the robot had this device, the energy consumption of heel-off walking was 1.04[J] as a result of simulation, which is much lower than previous walking.

9.3.6 Summary of This Section

In this section, the controller to achieve smooth biped walking by means of heel-off motion was designed. Heel-off walking makes it possible to avoid the problem for the impact shock at foot-contact to break the stability of walking. We described the following:

1. The control method of heel-off walking by means of PDAC was proposed.
2. The stabilizing control was designed and convergency of the walking dynamics was shown.
3. The proposed heel-off walking and previous method were compared and it was confirmed that heel-off motion reduced impact shock at foot-contact.
4. Energy efficiency of heel-off walking was investigated and the mechanical system to improve the energy efficiency was proposed.

9.4 3-D Biped Walking Based on 3-D Dynamics with PDAC

9.4.1 Walking Model

9.4.1.1 Three-Dimensional Inverted Pendulum Model

As for humanoid robots, a trunk that is equipped with a controller and a battery has much larger mass compared to a limb. Thus, in this section, a robot is modeled as a three-dimensional inverted pendulum as shown in Fig. 9.38(a). We apply an assumption of the point-contact to this pendulum in accordance with the PDAC, hence it is possible to choose the axes of pendulum angle around the contact point to express its motion. In this section, we utilize the polar coordinate system. The state variables and parameters are shown in Fig. 9.39(b). Angles q_1 and q_2 are the variables of the pendulum inclination around the contact point. Joint angles q_3, q_4, and q_5 decide the upper body posture. l is the variable of the pendulum length. L is the virtual value for convenience of description and equals zero. By use of the six variables q_1 to q_5 and l, it is possible to express any states of the robot.

The left-handed system is used in the left-leg supporting phase and vice versa as shown in Fig. 9.38(b), so that it is possible to describe the robot dynamics in both supporting phases as a single dynamics.

9.4.1.2 Three-Dimensional Dynamics

Let a mass of the robot be m and let a tensor of inertia be

$$I = \begin{bmatrix} I_{xx} & I_{xy} & I_{xz} \\ I_{yx} & I_{yy} & I_{yz} \\ I_{zx} & I_{zy} & I_{zz} \end{bmatrix}.$$

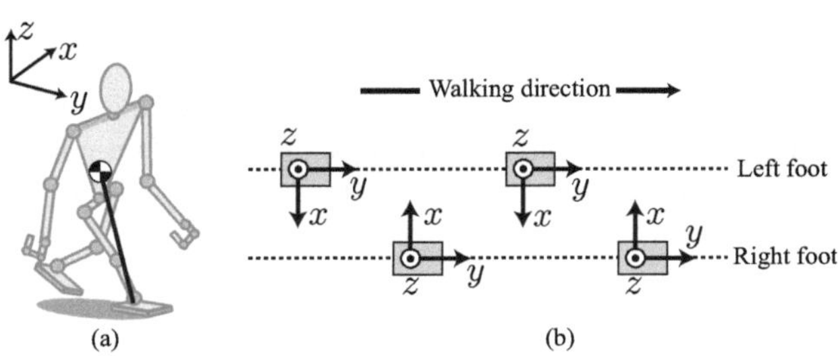

Fig. 9.38 (a) 3D inverted pendulum model. (b) Definition of coordinate system. Note that this figure shows just a coordinate system definition and doesn't mean that foot placement is in alignment.

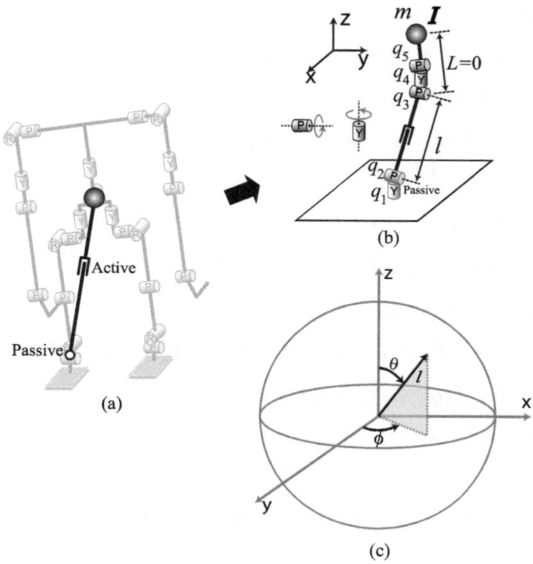

Fig. 9.39 (a) Passive joints (point-contact) and active pendulum length actuation (b) Parameters and variables of the 3D inverted pendulum model. (c) Polar coordinate system around contact point

The configuration of humanoid robots is basically symmetrical, hence generality of an assumption that all products of inertia are zero is not lost, i.e. $I_{xy} = I_{yx} = I_{yz} = I_{zy} = I_{zx} = I_{xz} = 0$,

$$\begin{aligned} I &= diag(I_{xx}, I_{yy}, I_{zz}) \\ &:= diag(I_x, I_y, I_z). \end{aligned} \tag{9.93}$$

Note that this tensor of inertia is parameter in the local coordinate system that is attached to the robot body, not in the global one.

The dynamic equations with respect to q_1, q_2, and l are obtained respectively as follows:

$$\frac{d}{dt}\Big(M_1(q)\dot{q}_1\Big) = H_1(q) + \tau_1, \tag{9.94}$$

$$\frac{d}{dt}\Big(M_2(q)\dot{q}_2\Big) = H_2(q) + \tau_2, \text{ and} \tag{9.95}$$

$$\frac{d}{dt}\Big(m\dot{l}\Big) - ml(\dot{q}_1^2 \sin^2 q_2 + \dot{q}_2^2) - mg\cos q_2 = f, \tag{9.96}$$

where

$$
\begin{aligned}
M_1(q) &= I_z m_{11}(q) + ml^2 S_2^2 \dot{q}_1, \\
H_1(q) &= (I_x - I_y)\big(g_{11}(q) + g_{12}(q) + g_{13}(q) + g_{14}(q)\big), \\
M_2(q) &= m_{21}(q) + m_{22}(q) + ml^2, \\
H_2(q) &= -S_1 C_1 (I_x - I_y) g_{21}(q) + ml^2 S_2 C_2 \dot{q}_1^2 + mglS_2, \\
m_{11}(q) &= \dot{q}_1 + \dot{q}_4 C_{23} + \dot{q}_5 S_{23} S_4, \\
g_{11}(q) &= \big((\dot{q}_2 + \dot{q}_3)^2 - \dot{q}_4^2 S_{23}\big) S_1 C_1, \\
g_{12}(q) &= -\dot{q}_5^2 (C_1 C_{23} S_4 + S_1 C_4)(S_1 C_{23} S_4 - C_1 C_4), \\
g_{13}(q) &= (\dot{q}_2 + \dot{q}_3)(\dot{q}_4 + \dot{q}_5)(S_1^2 - C_1^2), \\
g_{14}(q) &= \dot{q}_4 \dot{q}_5 S_{23} \big((S_1^2 - C_1^2) C_4 + 2 S_1 C_1 C_{23} S_4\big)(I_x S_1^2 + I_y C_1^2 - I_z) g_{22}(q) - I_z \dot{q}_1 \dot{q}_4 S_{23}, \\
m_{21}(q) &= (I_x S_1^2 + I_y C_1^2)\big((\dot{q}_2 + \dot{q}_3) + \dot{q}_5 C_4\big), \\
m_{22}(q) &= S_1 C_1 (I_x - I_y)(-\dot{q}_4 S_{23} + \dot{q}_5 C_{23} C_4), \\
g_{21}(q) &= \dot{q}_4 (\dot{q}_2 + \dot{q}_3) C_{23} + \dot{q}_5 (\dot{q}_2 + \dot{q}_3) S_{23} S_4 + \dot{q}_4 \dot{q}_5 C_{23} C_4 + \dot{q}_5^2 S_{23} S_4 C_4 \\
g_{22}(q) &= \dot{q}_4^2 S_{23} C_{23} - \dot{q}_4 \dot{q}_5 C_{23}^2 S_4 + \dot{q}_4 \dot{q}_5 S_{23}^2 S_4 - \dot{q}_5^2 S_{23} C_{23} S_4^2.
\end{aligned}
$$

In the above equations, τ_1 and τ_2 are torques of joints q_1 and q_1 respectively. f is a thrust force to change the pendulum length. S_a denotes $\sin q_a$, C_a denotes $\cos q_a$, S_{ab} denotes $\sin(q_a + q_b)$, and C_{ab} denotes $\cos(q_a + q_b)$.

9.4.1.3 Derivation of Converged Dynamics

Tasks of humanoid robots or other humanoid-like robots are not only to walk, but to do various work by use of their hands. Thus the swinging of the upper body is necessary to be restrained. In this section, the trunk posture is kept in the vertical direction and the upper body does not rotate around yaw-axis, that is,

$$q_3 = -q_2, \tag{9.97}$$

$$q_4 = -q_1, \tag{9.98}$$

$$q_5 = 0. \tag{9.99}$$

In order to control the three-dimensional inverted pendulum by means of the PDAC, the assumption of point-contact is applied, i.e.

$$\tau_1 = \tau_2 = 0. \tag{9.100}$$

For simplicity of description, we describe q_1 and q_2 as ϕ and θ respectively in the below. From Eqs. (9.97)-(9.100), Eqs. (9.94) and (9.95) are expressed as follows:

$$\frac{d}{dt}\left(ml^2 \sin^2\theta\dot{\phi}\right) = 0, \tag{9.101}$$

$$\frac{d}{dt}\left(ml^2\dot{\theta}\right) = ml^2\dot{\phi}^2 \sin\theta\cos\theta + mgl\sin\theta. \tag{9.102}$$

By multiplying both sides of Eq. (9.101) by $ml^2 \sin^2\theta\dot{\phi}$, and by integrating with respect to time, the following equation is obtained,

$$\dot{\phi} = \frac{\sqrt{2C_1}}{ml^2\sin^2\theta} \tag{9.103}$$

$$:= F_1(\theta), \tag{9.104}$$

where C_1 is the integral constant which is determined by an initial state immediately after a foot-contact. Substituting Eq. (9.103) into Eq. (9.102) results in

$$\dot{\theta} = \frac{1}{ml^2}\sqrt{2\int\left(\frac{2C_1\cos\theta}{\sin^3\theta} + m^2gl^3\sin\theta\right)d\theta} \tag{9.105}$$

$$:= \frac{1}{M(\theta)}\sqrt{2(D(\theta)+C_2)} \tag{9.106}$$

$$:= F_2(\theta). \tag{9.107}$$

Next, in accordance with PDAC, the pendulum length is described as the function of θ,

$$l := \lambda(\theta). \tag{9.108}$$

In this section, for simplicity, λ is defined as the following function of θ,

$$\lambda(\theta) =: \sqrt[3]{p_1\theta^3 + p_2\theta^2 + p_3\theta^+ p_4} \tag{9.109}$$

$$=: \sqrt[3]{f(\theta)}. \tag{9.110}$$

By substituting this equation into Eq. (9.106), the converged dynamics is derived,

$$M(\theta) = mf(\theta)^{2/3}, \tag{9.111}$$

$$D(\theta) = -\frac{C_1}{\sin^2\theta} - m^2g\Big((f(\theta) - f''(\theta))\cos\theta - (f'(\theta) - f'''(\theta))\sin\theta\Big). \tag{9.112}$$

9.4.1.4 Design of Walking Cycle

In this subsection, the actual motion of the robot is designed. Figure 9.40 shows the schematics of the pendulum motion and the COG trajectory. The continuous line shows a trajectory of the COG in the right-leg support phase and the dotted line shows in the left-leg support phase. The dot on the edge of both the continuous line and the dotted one means a foot-contact. Figure 9.41 shows the parameters and variables. S_0 and S_2 denote moments right before and after a foot-contact, and S_1 is a moment at $\dot{\theta} = 0$. θ_i, ϕ_i, and l_i denote the roll angle, yaw angle, and pendulum length at S_i $(i = 0, 1, 2)$ respectively. During a cycle of walking motion, ϕ is monotonically increasing. Meanwhile, θ decreases at first, and then increases, after posing for a moment at θ_1. Thus, we compartmentalize a walking cycle from a foot-contact to the next foot-contact into two phases—Phase (A): from S_0 to S_1 $(\dot{\theta} < 0)$, Phase (B): from S_1 to S_2 $(\dot{\theta} > 0)$. In the phase (A), the pendulum length is not varied, thus

$$p_1 = p_2 = p_3 = 0 \tag{9.113}$$

$$p_d = l_0^3. \tag{9.114}$$

In the phase (B), the coefficients p_1-p_4 are decided so that the following four conditions are satisfied,

$$f(\theta_1) = l_1^3, \tag{9.115}$$

$$f(\theta_2) = l_2^3, \tag{9.116}$$

$$f'(\theta_2) = 0, \text{ and} \tag{9.117}$$

$$-f''(\theta_1)\cos\theta_1 + \big(-f'(\theta_1) + f'''(\theta_1)\big)\sin\theta_1 = 0. \tag{9.118}$$

Eqs. (9.115) and (9.116) signify the condition of pendulum length continuity, and Eq. (9.117) is the condition that the velocity of pendulum along l is 0 at a foot-contact. The objective of Eq. (9.118) is to match PDAC constants of the phase (A)

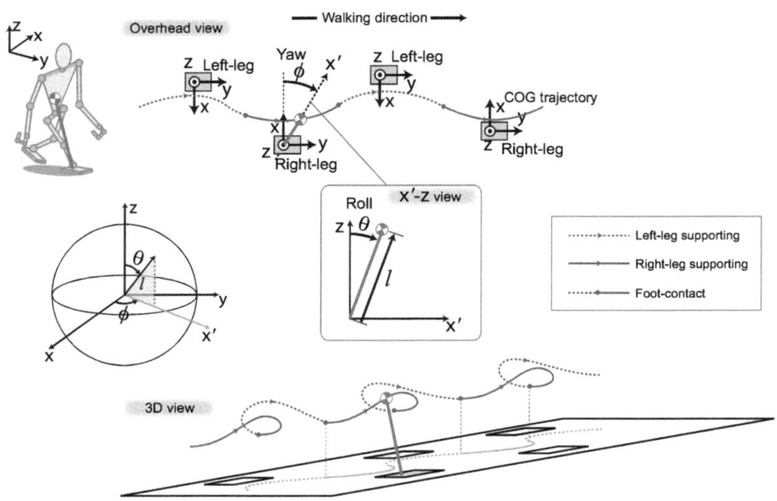

Fig. 9.40 Motion of a 3D inverted pendulum.

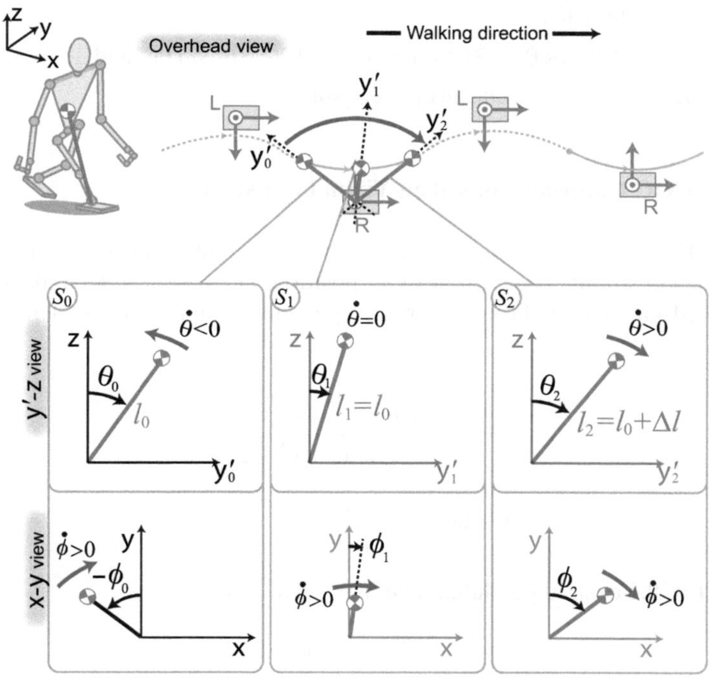

Fig. 9.41 Parameters and variables of dynamic walking based on 3D inverted pendulum model.

and (B). In the phase (A), conditions of $f'(\theta)=0$, $f''(\theta)=0$, and $f'''(\theta)=0$ are satisfied because the pendulum length is the constant value, i.e. $f(\theta)=l_0^3$. Thus, if Eq. (9.118) is held at the moment of $\theta=\theta_1$, Eq. (9.112), that is converged dynamics, is continual during a walking cycle and PDAC constants become constant during a step.

From Eqs. (9.115)-(9.118), the coefficients p_1-p_4 are derived as follows:

$$p_1 = -\frac{l_2^3 - l_0^3}{(\theta_2-\theta_1)^2}\frac{u_3}{u_1u_3-u_2}, \tag{9.119}$$

$$p_2 = -\frac{l_2^3 - l_0^3}{(\theta_2-\theta_1)^2}\frac{u_2}{u_1u_3-u_2}, \tag{9.120}$$

$$p_3 = -3p_1\theta_2^2 - 2p_2\theta_2, \text{ and} \tag{9.121}$$

$$p_4 = l_2^3 - p_1\theta_2^3 - p_2\theta_2^2 - p_3\theta_2, \tag{9.122}$$

where

$$u_1 = 2\theta_2 + \theta_1, \tag{9.123}$$
$$u_2 = -6\theta_1 \cos\theta_1 - 3\theta_1^2 \sin\theta_1 + 6\sin\theta_1 + 3\theta_2^2 \sin\theta_1, \text{ and} \tag{9.124}$$
$$u_3 = -2\cos\theta_1 - 2\theta_1 \sin\theta_1 + 2\theta_2 \sin\theta_1. \tag{9.125}$$

9.4.1.5 COG Controller Based on Virtual Constraint

Figure 9.42 shows the block diagram of the robot joint control based on virtual holonomic constraint. The current COG position, (x_g, y_g, z_g) is derived by solving the forward kinematics. From this result, pendulum angles can be determined as below,

$$\theta = \cos^{-1}\left(\frac{z_g}{\sqrt{x_g^2 + y_g^2 + z_g^2}}\right), \tag{9.126}$$
$$\phi = \tan^{-1}\left(\frac{y_g}{x_g}\right). \tag{9.127}$$

By use of θ, the desired pendulum length is calculated,

$$l^d = \lambda(\theta). \tag{9.128}$$

Consequently, the desired COG position is derived as follows:

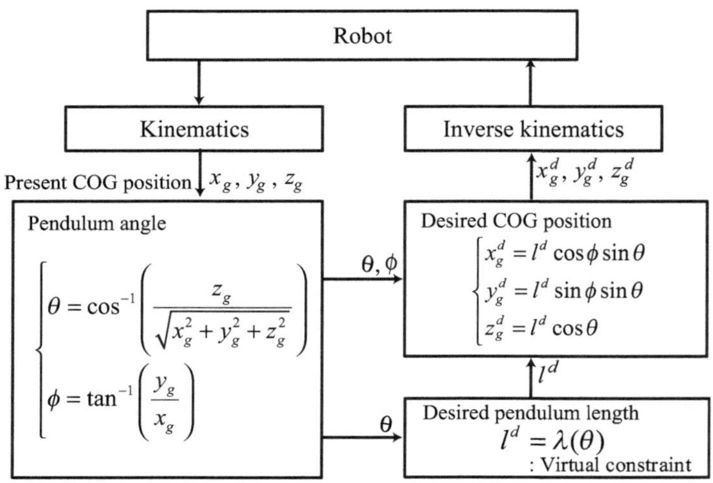

Fig. 9.42 COG controller based on virtual constraint

$$x_g^d = l^d \cos\phi \sin\theta, \tag{9.129}$$

$$y_g^d = l^d \sin\phi \sin\theta, \text{ and} \tag{9.130}$$

$$z_g^d = l^d \cos\theta. \tag{9.131}$$

The desired joint angles are decided by solving the inverse kinematics with reference to the desired COG position.

9.4.2 *Foot-Contact Model*

9.4.2.1 Discrete Dynamics at Foot-Contact

In this section, the impact between the foot of a swing-leg and the ground is assumed to be perfectly inelastic. Thus, the velocity in the parallel direction to the swing-leg is diminished and vertical one has no effect. Assuming that ϕ_0 is the angle of ϕ immediately after foot-contact, the vector of the pendulum after impact, L is

$$L = \begin{bmatrix} l_0 \cos\phi_0 \sin\theta_0 \\ l_0 \sin\phi_0 \sin\theta_0 \\ l_0 \cos\theta_0 \end{bmatrix} := \begin{bmatrix} l_x \\ l_y \\ l_z \end{bmatrix} \tag{9.132}$$

where ϕ_0, θ_0 are in the coordinate system of next step (See Fig. 9.43).

The vector of velocity immediately prior to foot-contact, V_1, is calculated as follows:

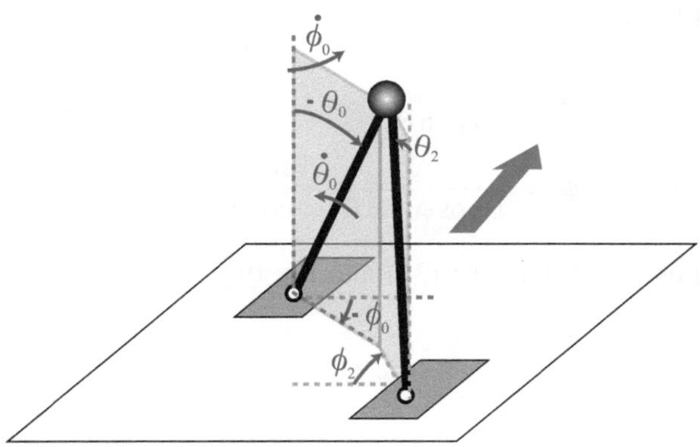

Fig. 9.43 Variables of the inverted pendulum at foot contact

$$V_1 = \begin{bmatrix} l_2(\dot{\phi}_2 \cos\phi_2 \sin\theta_2 + \dot{\theta}_2 \sin\phi_2 \cos\theta_2) + \dot{l}_2(\sin\phi_2 \sin\theta_2) \\ l_2(-\dot{\phi}_2 \sin\phi_2 \sin\theta_2 + \dot{\theta}_2 \cos\phi_2 \cos\theta_2) + \dot{l}_2(\cos\phi_2 \sin\theta_2) \\ -l_2\dot{\theta}_2 \sin\theta_2 + \dot{l}_2(\cos\theta_2) \end{bmatrix} \quad (9.133)$$

$$:= [v_x^-, v_y^-, v_z^-]^T \quad (9.134)$$

where ϕ_2 is the angle of ϕ before impact.

The velocity vector after impact, V_0, is found by the following calculation,

$$V_0 = \frac{V_1 \cdot (L \times (V_1 \times L))}{|L \times (V_1 \times L)|}(L \times (V_1 \times L)) \quad (9.135)$$

$$= \frac{L \times (V_1 \times L)}{l^2} \quad (9.136)$$

$$:= [v_x^+, v_y^+, v_z^+]^T. \quad (9.137)$$

where,

$$\begin{bmatrix} v_x^+ \\ v_y^+ \\ v_z^+ \end{bmatrix} = \frac{1}{l_0^2} \begin{bmatrix} l_y(v_x^- l_y - v_y^- l_x) - l_z(v_z^- l_x - v_x^- l_z) \\ l_z(v_y^- l_z - v_z^- l_y) - l_x(v_x^- l_y - v_y^- l_x) \\ l_x(v_z^- l_x - v_x^- l_z) - l_y(v_y^- l_z - v_z^- l_y) \end{bmatrix}. \quad (9.138)$$

Note that in the above calculation of V_0, V_1 must be treated as $V_1 = [-v_x, v_y, v_z]$ since left- and right-handed systems are switched at foot-contact.

9.4.2.2 Alteration of Angular Velocity Around Yaw and Roll Axis

From Eq. (9.137), $\dot{\theta}_0$ and $\dot{\phi}_0$ are decided,

$$\dot{\theta}_0 = -\frac{v_z^+}{l_0 \sin\theta_0} \quad (9.139)$$

$$\dot{\phi}_0 = \frac{v_y^+}{l_0 \cos\phi_0 \sin\theta_0} - \frac{\sin\phi_0 \cos\theta_0}{\cos\phi_0 \sin\theta_0}\dot{\theta}. \quad (9.140)$$

Substituting Eqs. (9.132) and (9.134) into Eq. (9.137), v_z^+ and v_y^+ are derived as follows:

$$v_z^+ = \frac{1}{l_0^2}\Big(l_x(v_z^- l_x - v_x^- l_z) - l_y(v_y^- l_z - v_z^- l_y)\Big) \tag{9.141}$$

$$= \frac{1}{l_0^2}\Big(v_z^-(l_x^2 + l_y^2) - v_x^- l_x l_z - v_y^- l_y l_z\Big) \tag{9.142}$$

$$\begin{aligned} &= \frac{1}{l_0^2}\Big(- l_2\dot{\theta}_2 \sin\theta_2 (l_0^2 \sin^2\theta_0) \\ &\quad + l_2(-\dot{\phi}_2 \sin\phi_2 \sin\theta_2 + \dot{\theta}_2 \cos\phi_2 \cos\theta_2) l_0^2 \cos\phi_0 \cos\theta_0 \sin\theta_0 \\ &\qquad - l_2\Big(\dot{\phi}_2 \cos\phi_2 \sin\theta_2 + \dot{\theta}_2 \sin\phi_2 \cos\theta_2\Big) l_0^2 \sin\phi_0 \cos\theta_0 \sin\theta_0\Big) \end{aligned} \tag{9.143}$$

$$\begin{aligned} &= l_2\Big(\dot{\theta}_2 \sin\theta_0(-\sin\theta_2 \sin\theta_0 + \cos\phi_0 \cos\theta_0 \cos\phi_2 \cos\theta_2 \\ &\quad - \sin\phi_0 \cos\theta_0 \sin\phi_2 \cos\theta_2) \\ &\qquad + \dot{\phi}_2 \cos\theta_0 \sin\theta_0 \sin\theta_2(-\sin\phi_2 \cos\phi_0 - \cos\phi_2 \sin\phi_0)\Big) \end{aligned} \tag{9.144}$$

$$\begin{aligned} &= l_2 \sin\theta_0 \Big(\dot{\theta}_2\big(- \sin\theta_2 \sin\theta_0 + \cos\theta_0 \cos\theta_2 \cos(\phi_0 + \phi_2)\big) \\ &\qquad - \dot{\phi}_2 \cos\theta_0 \sin\theta_2 \sin(\phi_0 + \phi_2)\Big) \end{aligned} \tag{9.145}$$

$$v_y^+ = \frac{1}{l_0^2}\Big(l_z(v_y^- l_z - v_z^- l_y) - l_x(v_x^- l_y - v_y^- l_x)\Big) \tag{9.146}$$

$$= \frac{1}{l_0^2}\Big(v_y^-(l_x^2 + l_z^2) - v_x^- l_x l_y - v_z^- l_y l_z\Big) \tag{9.147}$$

$$\begin{aligned} &= l_2\Big((\dot{\phi}_2 \cos\phi_2 \sin\theta_2 + \dot{\theta}_2 \sin\phi_2 \cos\theta_2)(\cos^2\theta_0 + \cos^2\phi_0 \sin^2\theta_0) \\ &\quad + \dot{\theta}_2 \sin\theta_2 \cos\theta_0 \sin\theta_0 \sin\phi_0 \\ &\qquad + (-\dot{\phi}_2 \sin\phi_2 \sin\theta_2 + \dot{\theta}_2 \cos\phi_2 \cos\theta_2) \cos\phi_0 \sin\phi_0 \sin^2\theta_0\Big) \end{aligned} \tag{9.148}$$

$$\begin{aligned} &= l_2\Big(\dot{\theta}_2(\cos^2\theta_0 \sin\sin\phi_2 \cos\theta_2 + \cos^2\phi_0 \sin^2\theta_0 \sin\phi_2 \cos\theta_2 \\ &\quad + \cos\theta_0 \sin\theta_0 \sin\phi_0 \sin\theta_2 \\ &\qquad + \cos\phi_0 \sin\phi_0 \sin^2\theta_0 \cos\phi_2 \cos\theta_2) + \dot{\phi}_2(\cos^2\theta_0 \cos\phi_2 \sin\theta_2 \\ &\qquad + \cos^2\phi_0 \sin^2\theta_0 \cos\phi_2 \sin\theta_2 - \cos\phi_0 \sin\phi_0 \sin^2\theta_0 \sin\phi_2 \sin\theta_2) \end{aligned} \tag{9.149}$$

Consequently, $\dot{\theta}_0$ and $\dot{\phi}_0$ are found as below,

$$\dot{\theta}_0 = -\frac{v_z^+}{l_0 \sin\theta_0} \tag{9.150}$$

$$= \frac{l_2}{l_0}\Big(\dot{\theta}_2\big(\sin\theta_2\sin\theta_0 - \cos\theta_0\cos\theta_2\cos(\phi_0+\phi_2)\big) + \dot{\phi}_2\cos\theta_0\sin\theta_2\sin(\phi_0+\phi_2)\Big) \tag{9.151}$$

$$\begin{aligned}\dot{\phi}_0 = \frac{l_2}{l_0\cos\phi_0\sin\theta_0}\Big(&\dot{\theta}_2(\cos^2\theta_0\sin\phi_2\cos\theta_2 + \cos^2\phi_0\sin^2\theta_0\sin\phi_2\cos\theta_2\\ &+\cos\phi_0\sin\phi_0\sin^2\theta_0\cos\phi_2\cos\theta_2 + \cos^2\theta_0\cos\phi_0\sin\phi_0\cos\theta_2\cos\phi_2\\ &-\cos^2\theta_0\sin^2\phi_0\cos\theta_2\sin\phi_2) + \dot{\phi}_2(\cos^2\theta_0\cos\phi_2\sin\theta_2\\ &+\cos^2\phi_0\sin^2\theta_0\cos\phi_2\theta_2 - \cos\phi_0\sin\phi_0\sin^2\theta_0\sin\phi_2\sin\theta_2\\ &+\cos^2\theta_0\sin\theta_0\sin^2\phi_0\sin\theta_2\cos\phi_2 + \cos^2\theta_0\sin\theta_0\sin\phi_0\cos\phi_0\sin\theta_2\sin\phi_2\Big)\end{aligned} \tag{9.152}$$

$$\begin{aligned}= \frac{l_2}{l_0\cos\phi_0\sin\theta_0}\Big(&\dot{\theta}_2(\cos^2\phi_0\sin\phi_2\cos\theta_2 + \cos\phi_0\sin\phi_0\cos\theta_2\cos\phi_2)\\ &+\dot{\phi}_2(\cos^2\phi_0\cos\phi_2\sin\theta_2 - \cos\phi_0\sin\phi_0\sin\phi_2\sin\theta_2)\Big)\end{aligned} \tag{9.153}$$

$$= \frac{l_2}{l_0\sin\theta_0}\Big(\dot{\theta}_2\cos\theta_2\sin(\phi_0+\phi_2) + \dot{\phi}_2\sin\theta_2\cos(\phi_0+\phi_2)\Big) \tag{9.154}$$

9.4.2.3 Energy Loss

In addition, the energy loss at foot-contact is derived. The potential energy of a robot does not vary before and after foot-contact, hence the energy loss is dependent on only kinetic energy. Energy of movement immediately after foot-contact is described as

$$\frac{2l_0^4E_k^+}{m} = l_0^4({v_x^+}^2 + {v_y^+}^2 + {v_z^+}^2) \tag{9.155}$$

$$\begin{aligned}&= l_y^2(v_x^-l_y - v_y^-l_x)^2 + l_z^2(v_z^-l_x - v_x^-l_z)^2 - 2l_yl_z(v_x^-l_y - v_y^-l_x)(v_z^-l_x - v_x^-l_z)\\ &\quad + l_z^2(v_y^-l_z - v_z^-l_y)^2 + l_x^2(v_x^-l_y - v_y^-l_x)^2 - 2l_zl_x(v_y^-l_z - v_z^-l_y)(v_x^-l_y - v_y^-l_x)\\ &\quad + l_x^2(v_z^-l_x - v_x^-l_z)^2 + l_y^2(v_y^-l_z - v_z^-l_y)^2 - 2l_xl_y(v_z^-l_x - v_x^-l_z)(v_y^-l_z - v_z^-l_y)\end{aligned} \tag{9.156}$$

$$\begin{aligned}&= l_0^2\big({v_x^-}^2(l_y^2+l_z^2) + {v_y^-}^2(l_z^2+l_x^2) + {v_z^-}^2(l_x^2+l_y^2)\\ &\qquad -2v_x^-v_y^-l_xl_y - 2v_y^-v_z^-l_yl_z - 2v_z^-v_x^-l_zl_x\big).\end{aligned} \tag{9.157}$$

From this result, the difference of kinetic energy before and after foot-contact is found as below,

$$\frac{2(E_k^+ - E_k^-)}{m} = (v_x^{+2} + v_y^{+2} + v_z^{+2}) - (v_x^{-2} + v_y^{-2} + v_z^{-2}) \tag{9.158}$$

$$= \frac{1}{l_0^2}\Big(v_x^{-2}(l_0^2 - l_y^2 - l_z^2) + v_y^{-2}(l_0^2 - l_z^2 - l_x^2) + v_z^{-2}(l_0^2 - l_x^2 - l_y^2)$$
$$+2v_x^- v_y^- l_x l_y + 2v_y^- v_z^- l_y l_z + 2v_z^- v_x^- l_z l_x\Big) \tag{9.159}$$

$$= \frac{1}{l_0^2}\Big(v_x^{-2} l_x^2 + v_y^{-2} l_y^2 + v_z^{-2} l_z^2 + 2v_x^- v_y^- l_x l_y + 2v_y^- v_z^- l_y l_z + 2v_z^- v_x^- l_z l_x\Big) \tag{9.160}$$

$$= v_x^{-2} \cos^2\phi_0 \sin^2\theta_0 + v_y^{-2} \sin^2\phi_0 \sin^2\theta_0 + v_z^{-2}\cos^2\theta_0$$
$$+2v_x^- v_y^- \sin\phi_0 \cos\phi_0 \sin^2\theta_0 + +2v_y^- v_z^- \sin\phi_0 \sin\theta_0 \cos\theta_0$$
$$+2v_z^- v_x^- \cos\phi_0 \sin\theta_0 \cos\theta_0 \tag{9.161}$$

$$= \sin^2\theta_0 (v_x^- \cos\phi_0 + v_y^- \sin\phi_0)$$
$$+2v_z^- \sin\theta_0 \cos\theta_0 (v_x^- \cos\phi_0 + v_y^- \sin\phi_0) + v_z^{-2}\cos^2\theta_0 \tag{9.162}$$

$$= \Big((v_x^- \cos\phi_0 + v_y^- \sin\phi_0)\sin\theta_0 + v_z \cos\theta_0\Big)^2. \tag{9.163}$$

From Eq. (9.134),

$$v_x^- \cos\phi_0 + v_y^- \sin\phi_0 = -l_2(-\dot\phi_2 \sin\phi_2 \sin\theta_2 + \dot\theta_2 \cos\phi_2 \cos\theta_2)\cos\phi_0$$
$$+l_2(\dot\phi_2 \cos\phi_2 \sin\theta_2 + \dot\theta_2 \sin\phi_2 \cos\theta_2)\sin\phi_0 \tag{9.164}$$
$$= l_2\dot\phi_2 \sin\theta_2(\sin\phi_2\cos\phi_0 + \cos\phi_2 \sin\phi_0)$$
$$-l_2\dot\theta_2 \cos\theta_2(\cos\phi_2\cos\phi_0 - \sin\phi_2\sin\phi_0) \tag{9.165}$$
$$= l_2\dot\phi_2 \sin\theta_2 \sin(\phi_2+\phi_0) - l_2\dot\theta_2\cos\theta_2\cos(\phi_2+\phi_0). \tag{9.166}$$

Substituting Eq. (9.166) into Eq. (9.166), the energy loss at foot-contact is derived as follows:

$$\Delta E = \frac{ml_2^2}{2}\Big(\dot\phi_2 \sin\theta_2 \sin(\phi_2+\phi_0)\sin\theta_0 - \dot\theta_2\big(\cos\theta_2\cos(\phi_2+\phi_0)\sin\theta_0 + \sin\theta_2\cos\theta_0\big)\Big)^2. \tag{9.167}$$

9.4.3 *Stabilization*

9.4.3.1 Geometrical Constraints

In order to stabilize walking, some constraints are given. At first, the lengthening value of pendulum is fixed at constant value. In this constraint, supplied energy is nearly constant. In addition, the following two constraints at foot-contact, are designed as shown in Fig. 9.74,

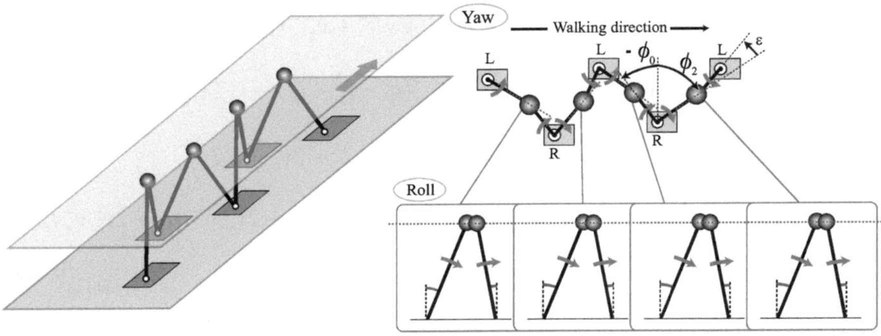

Fig. 9.44 Geometrical constraints at foot-contact

- COG height h is constant, i.e. roll angles of stance- and swing-leg are constant at foot-contact.
- Yaw angle of swing-leg is shifted by ε from the symmetrical position with stance-leg at foot-contact, i.e. it is $\phi_0[k+1] = -\phi_2[k] + \varepsilon$ where $\phi_0[k+1]$ and $\phi_2[k]$ denote ϕ_0 and ϕ_2 at $k+1$th and kth step respectively.

9.4.3.2 Zero-shift Condition

At first, we tested the condition of $\varepsilon = 0$. Fig. 9.45 shows the simulation results under zero-shift condition.

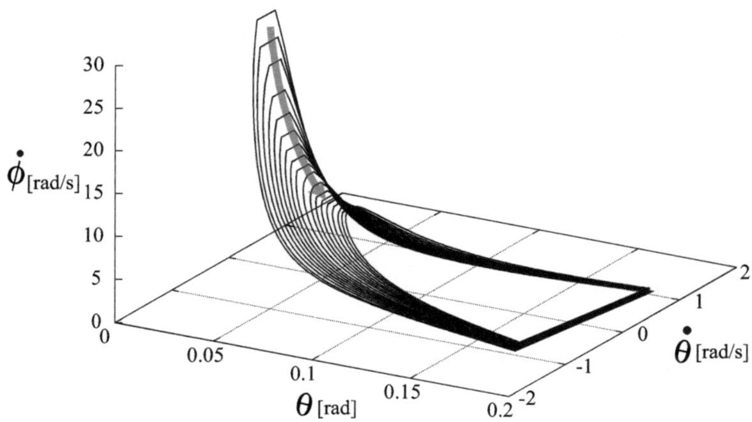

Fig. 9.45 Simulation results under the condition of $\varepsilon = 0$, l_0 =0.51[m], Δl =0.005[m].

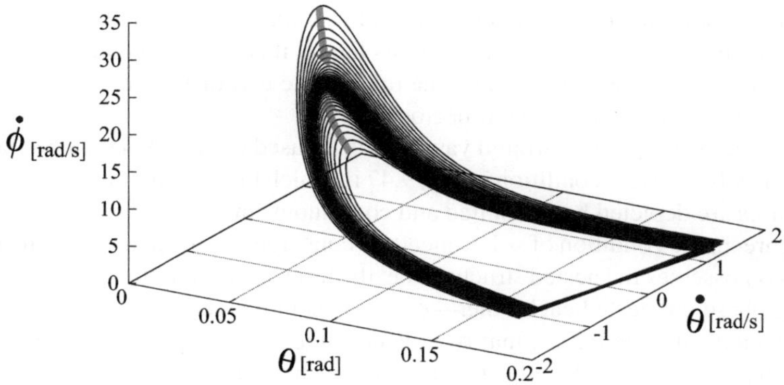

Fig. 9.46 Simulation results under the condition of ε=0.018[rad]=const, l_0 =0.51[m], Δl =0.007[m]

As can be seen in this figure, the energy around yaw axis diminishes gradually. Meanwhile, the energy around roll axis is increased. This result indicates that step-length is decreased steadily and the amplitude of lateral sway motion is increased.

The reason for this result is explicit from Eqs. (9.104) and (9.107), (9.151), (9.154). Letting $\varepsilon = 0$, Eqs. (9.151) and (9.154) are described as follows:

$$\dot{\theta}_0 = -\frac{l_2}{l_0}\dot{\theta}_2 \cos(\theta_0 + \theta_2) \tag{9.168}$$

$$\dot{\phi}_0 = \frac{l_2 \sin\theta_2}{l_0 \sin\theta_0}\dot{\phi}_2. \tag{9.169}$$

These equations indicate that the energy around roll- and yaw- axis is effected each other. It can be seen from Eq. (9.104) that the angular velocity around yaw-axis at foot-contact is decreased compared with that at the beginning of stance-leg phase. Meanwhile, we can see from Eq. (9.107) that the angular velocity around roll-axis at foot-contact is increased up to certain state compared to that at the beginning. Therefore, at each foot-contact, the energy around yaw-inies and that around roll axis increases up to certain state.

9.4.3.3 Constant-shift Condition

Next, the condition of constant shifting was tested. Fig. 9.46 depicts the simulation result of this condition. As can be seen in this figure, θ and $\dot{\theta}$, $\dot{\phi}$ are converged on a certain fixed point.

However, it can be seen in Fig. 9.47, which shows the phase portrait around yaw- and roll axis, that the trajectory is not converged on a sole one and 2-cycle trajectory appears. If 2-cycle occurs, a robot walks leftwards and cannot walk straight.

In this simulation, the initial angle around yaw-axis, ϕ_0, is varied without changing any other state and condition. From this figure, it can be seen that the converged trajectory seems to be depending on the initial state and that there is the certain set of initial state converging a sole trajectory.

Here, the phase portrait around yaw-axis is focused on. Fig. 9.48 shows the ones under the left and right condition in Fig. 9.47 in which the left- and right-leg supporting phases are depicted by the dotted and continuous line respectively. As shown in this figure, in the condition of sole-trajectory-convergence, the phase portrait around yaw axis possesses the nested structure, i.e. the trajectories in left- and right-leg supporting phase are nested each other—$\phi_0[k+1]$ is in between $\phi_0[k]$ and $\phi_0[k+2]$—and gradually attracted. Note that $\phi_0[k]$ is ϕ_0 value at kth step. Hence, it is contemplated that by the controller adjusting ε so as to achieve the nested structure of phase portrait around yaw-axis, it is possible to converge the dynamics on a sole trajectory.

9.4.3.4 Landing Position Control

In this subsection, we design the stabilizing controller that adjusts the landing position of stance-leg foot base on the nested structure.

If the present state is inside the converged trajectory, the condition to achieve the nested structure is described as

$$(\phi_0[k] > \phi_0[k+1]) \wedge (\phi_0[k+1] > \phi_0[k+2]). \tag{9.170}$$

Meanwhile, it the present state is outside, it is

$$(\phi_0[k] < \phi_0[k+1]) \wedge (\phi_0[k+1] < \phi_0[k+2]). \tag{9.171}$$

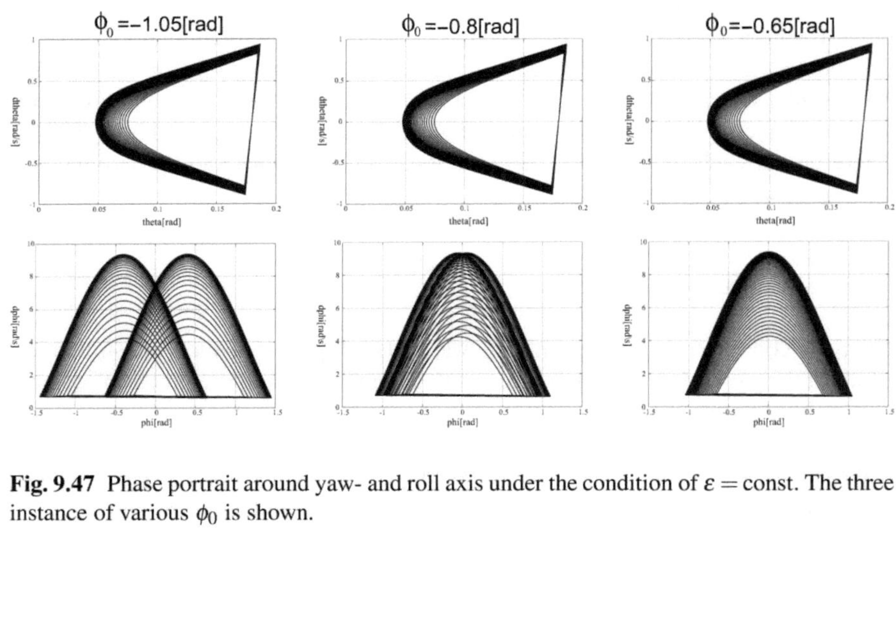

Fig. 9.47 Phase portrait around yaw- and roll axis under the condition of $\varepsilon = \text{const}$. The three instance of various ϕ_0 is shown.

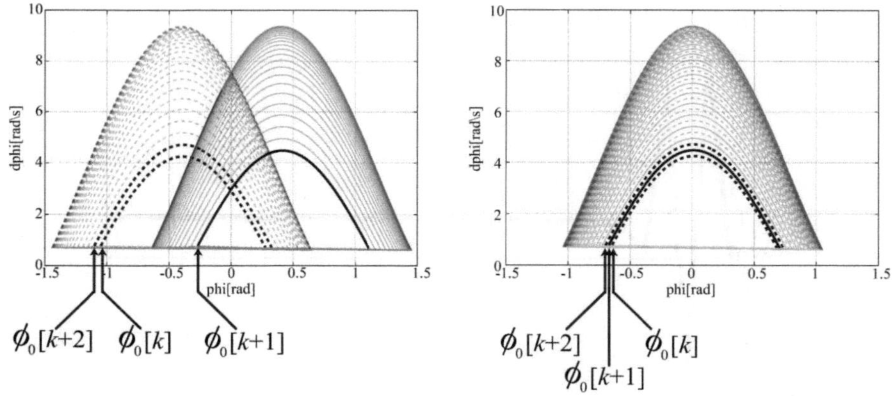

Fig. 9.48 Nested structure of the trajectory. $\phi_0[k]$ denotes the yaw angle at the beginning of stance-leg phase.

In order to build the stabilizing controller making the yaw dynamics attract to the nested structure, we define the distance between present state and the nested structure and design the stabilizing method minimizing this distance as follows:

$$\min(\sqrt{(\phi_0[k+2]-\phi_0[k+1])^2+(\phi_0[k+1]-\phi_0[k])^2}). \tag{9.172}$$

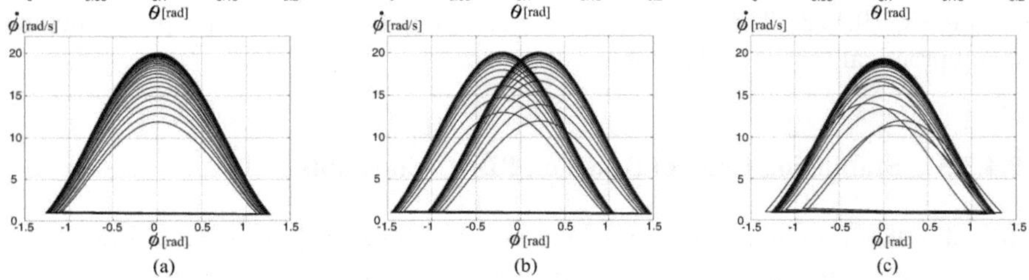

Fig. 9.49 Simulation results of the proposed stabilizing control. ε =0.018[m], l_0 =0.51[m], Δl =0.005[m].

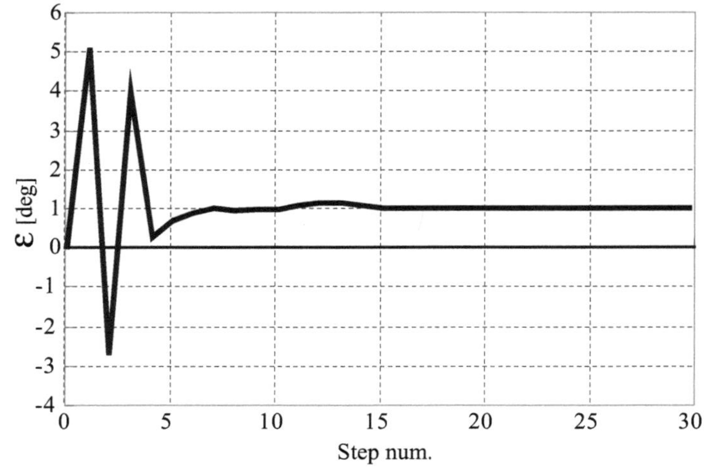

Fig. 9.50 Alteration of ε in landing position control

By adjusting the landing position according to this equation, it is conceivable that robot dynamics is attracted to the nested structure and consequently converged on a sole trajectory.

Fig. 9.49 shows the simulation results of this control. In Fig. 9.49(a) and (b), ε is constant value, 0.018[rad]. The initial condition of Fig. 9.49 is the set that the dynamics is converged on a sole trajectory without stabilization. Fig. 9.49(c) depicts the result of the controller embedded the above-mentioned stabilization. In Fig. 9.49(b) and (c), same initial condition is employed. Note that, as shown in Fig. 9.49(b), the 2-cycle trajectory appears without stabilization, however in Fig. 9.49(c), the dynamics is converged on sole trajectory with stabilization.

The proposed stabilizing method succeeded in converging the 2-cycle trajectory on a sole trajectory. However, in this method, the robot requires to search for ε satisfying Eq. (9.172). This search process needs long time, thus the proposed method is not practical for the real robot.

9.4.3.5 Stabilizing Control Based on PDAC Constants

To solve this problem, we propose the novel stabilizing control by use of PDAC Constants. In the proposed control framework, the converged 2-dimensional dynamics has two conserved quantity, i.e. PDAC Constants C_1,C_2. These two PDAC Constants decide the trajectory in the 3-dimensional space composed of θ, $\dot{\theta}$, and $\dot{\phi}$. Note that ϕ is directly independent of θ, $\dot{\theta}$, and $\dot{\phi}$ as can be seen in Eqs. (9.104) and (9.107). Thus, in order to build the controller converging the robot dynamics on sole trajectory in the 4-dimensional space composed of θ, ϕ, $\dot{\theta}$, and $\dot{\phi}$, it is necessary to design the following two controllers: 1) the stabilizing controller converging

PDAC Constants on the desired values 2) the walking-direction controller adjusting 2-cycle of ϕ phase. Hereinafter, controller 1) is described first, and then 2) is mentioned. Lastly, we shows the convergency and stability by means of return maps.

Constraint Condition Based on PDAC Constant

In order to converge PDAC Constants on the desired values, it is necessary to find the condition of l_2, h, and ε. Hence, the constraint condition satisfying the desired PDAC Constants is derived.

At first, convergent values of PDAC Constants are found. Letting PDAC Constants at kth step be $C_1[k]$ and $C_2[k]$, $C_1[k+1]$ and $C_2[k+1]$ are described as follows from Eqs. (9.104), (9.107), (9.151) and (9.154):

$$
\begin{aligned}
C_1[k+1] &= \left(R_{11}\sqrt{C_1[k]} + R_{12}\sqrt{\frac{-C_1[k]}{\sin^2\theta_2} + D(\theta_2) + C_2[k]} \right)^2 && (9.173)\\
&=: \xi_1(C_1[k], C_2[k], \theta_0, \theta_2, \varepsilon, l_2) && (9.174)\\
C_2[k+1] &= \left(R_{21}\sqrt{C_1[k]} + R_{22}\sqrt{\frac{-C_1[k]}{\sin^2\theta_2} + D(\theta_2) + C_2[k]} \right)^2 \\
&\quad + \frac{1}{\sin^2\theta_0}\left(R_{11}\sqrt{C_1[k]} + R_{12}\sqrt{\frac{-C_1[k]}{\sin^2\theta_2} + D(\theta_2) + C_2[k]} \right)^2 \\
&\quad + m^2 g l_0^3 \cos\theta_0 && (9.175)\\
&=: \xi_2(C_1[k], C_2[k], \theta_0, \theta_2, \varepsilon, l_2) && (9.176)
\end{aligned}
$$

where

$$
\begin{aligned}
R_{11} &= \frac{l_0 \sin\theta_0 \cos\varepsilon}{l_2 \sin\theta_2} && (9.177)\\
R_{12} &= \frac{l_0 \sin\theta_0 \cos\theta_2 \sin\varepsilon}{l_2} && (9.178)\\
R_{21} &= \frac{l_0 \cos\theta_0 \sin\varepsilon}{l_2 \sin\theta_2} && (9.179)\\
R_{22} &= \frac{l_0}{l_2}\left(\sin\theta_0\theta_2 - \cos\theta_0 \cos\theta_2 \cos\varepsilon \right). && (9.180)
\end{aligned}
$$

Note that l_0 is constant value.

Assuming $C_1[k] = C_1[k+1] = C_1^*$ and $C_2[k] = C_2[k+1] = C_2^*$, Eqs. (9.174) and (9.176) are described as follows:

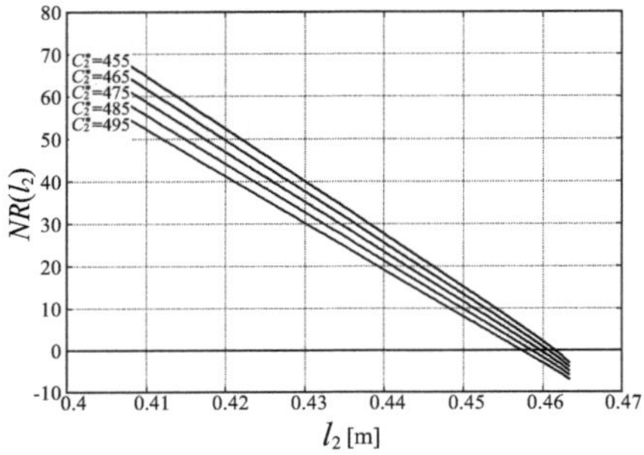

Fig. 9.51 Graph of $NR(l_2)$ vs. l_2 (C_1^*=0.02, h=0.428, l_0=0.455)

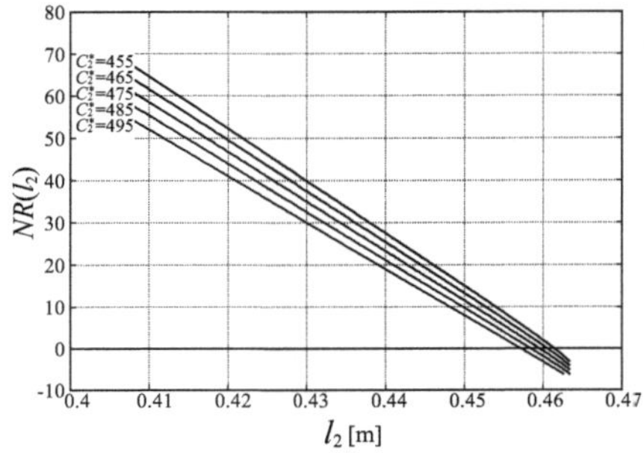

Fig. 9.52 Graph of $NR(l_2)$ vs. l_2 (C_1^*=0.05, h=0.428, l_0=0.455)

$$C_1^* = \left(R_{11}\sqrt{C_1^*}+R_{12}\sqrt{\frac{-C_1^*}{\sin^2\theta_2}+D(\theta_2)+C_2^*}\right)^2 \tag{9.181}$$

$$=: \xi_1(C_1^*,C_2^*,\theta_0,\theta_2,\varepsilon,l_2) \tag{9.182}$$

$$C_2^* = \left(R_{21}\sqrt{C_1[k]}+R_{22}\sqrt{\frac{-C_1^*}{\sin^2\theta_2}+D(\theta_2)+C_2^*}\right)^2$$

$$\frac{1}{\sin^2\theta_0}\left(R_{11}\sqrt{C_1[k]}+R_{12}\sqrt{\frac{-C_1^*}{\sin^2\theta_2}+D(\theta_2)+C_2^*}\right)^2+m^2gl_0^3\cos\theta_0 \tag{9.183}$$

$$=: \xi_2(C_1^*,C_2^*,\theta_0,\theta_2,\varepsilon,l_2). \tag{9.184}$$

Thus, the convergent PDAC Constants C_1^* and C_2^* are derived as below,

$$C_1^* = -\frac{m^2gl_0^3\cos\theta_0+D(\theta_2)}{\left(R_{21}+R_{22}\sqrt{\frac{1}{\sin^2\theta_0}-\frac{1}{\sin^2\theta_2}+\frac{(1-R_{11})^2}{R_{12}^2}}\right)^2}+\frac{(1-R_{11})^2}{R_{12}^2} \tag{9.185}$$

$$=: \eta_1(\theta_0,\theta_2,\varepsilon,l_2) \tag{9.186}$$

$$C_2^* = D(\theta_2)-\left(\frac{1}{\sin^2\theta_0}+\frac{(1-R_{11})^2}{R_{12}^2}\right)C_1^* \tag{9.187}$$

$$=: \eta_2(\theta_0,\theta_2,\varepsilon,l_2). \tag{9.188}$$

The geometrical constraint h represents the relationship of θ_0, θ_2, and l_2, i.e.

$$h = l_0\cos\theta_0 \tag{9.189}$$

$$= l_2\cos\theta_2. \tag{9.190}$$

Hence, from Eqs. (9.186)-(9.190), it is possible to decide θ_0, θ_2, ε, and l_2.

However, it is difficult to solve Eqs. (9.186)-(9.190) for θ_0, θ_2, ε, and l_2. Thus, in this section, we derive the constraint equation of l_2 from Eqs. (9.186)-(9.190), and find the solution numerically by Newton-Raphson method. The constraint equation of l_2 is

$$NR(l_2) := \left(Q_1 + \frac{\sin\theta_2 \cos\theta_2 \sin\theta_0}{\cos\theta_0} * Q_2\right)^2 + \frac{\cos^2\theta_0}{\sin^2\theta_0} C_1^* \tag{9.191}$$

$$-\frac{\cos^2\theta_2}{\sin^2\theta_2} C_1^* - \cos^4\theta_2 Q_2^2 = 0 \tag{9.192}$$

where

$$\theta_0 = \arccos\left(\frac{h}{l_0}\right) \tag{9.193}$$

$$\theta_2 = \arccos\left(\frac{h}{l_2}\right) \tag{9.194}$$

$$Q_1 = -\frac{C_1^*}{\sin^2\theta_0} - m^2 g l_0^3 \cos\theta_0 + C_2^* \tag{9.195}$$

$$Q_2 = D(\theta_2) + C_2^*. \tag{9.196}$$

To find solution online by Newton-Raphson method, it is required to inquire the search speed. Fortunately, the relationship between $NR(l_2)$ and l_2 is nearly linear as shown in Fig. 9.51 and 9.52. As a result of simulation, $NR(l_2)$ is converged on 0 within 5 steps of Newton-Raphson searching in various condition.

By employing θ_0, θ_2, ε, and l_2 satisfying Eqs. (9.186) and (9.188), if $C_1[k] = C_1^*$ and $C_2[k] = C_2^*$, $C_1[k+1] = C_1^*$ and $C_2[k+1] = C_2^*$. Thus, if C_1^* and C_2^* exist and are unique, and if $|\frac{\partial \xi_1}{\partial C_1}| < 1$ and $|\frac{\partial \xi_2}{\partial C_2}| < 1$ in the vicinity of (C_1^*, C_2^*), PDAC Constants are converged on a fixed point. Fig. 9.53 shows constant C_2 manifold and constant C_1 manifold in 3-dimensional space composed of θ, $\dot{\theta}$, and $\dot{\phi}$. If C_2 is converged on C_2^*=456, the trajectory in 3-dimensional space composed of $(\theta, \dot{\theta}, \dot{\phi})$ is attracted to the manifold depicted in the left figure of Fig. 9.53. Similarly, If C_1 has the convergency on C_1^*=456, the trajectory is attracted to the manifold shown in

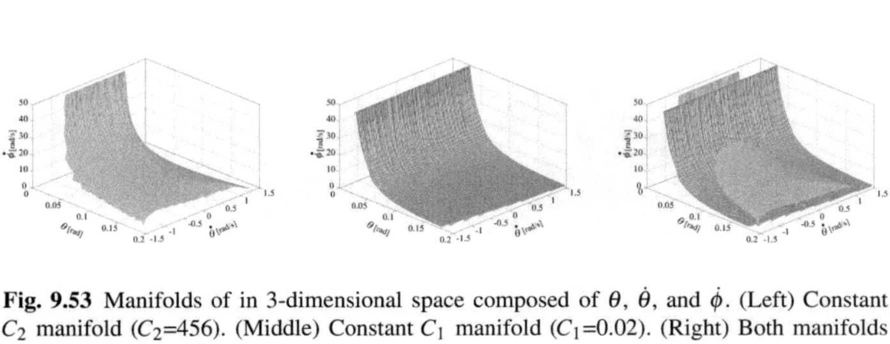

Fig. 9.53 Manifolds of in 3-dimensional space composed of θ, $\dot{\theta}$, and $\dot{\phi}$. (Left) Constant C_2 manifold (C_2=456). (Middle) Constant C_1 manifold (C_1=0.02). (Right) Both manifolds compose the trajectory.

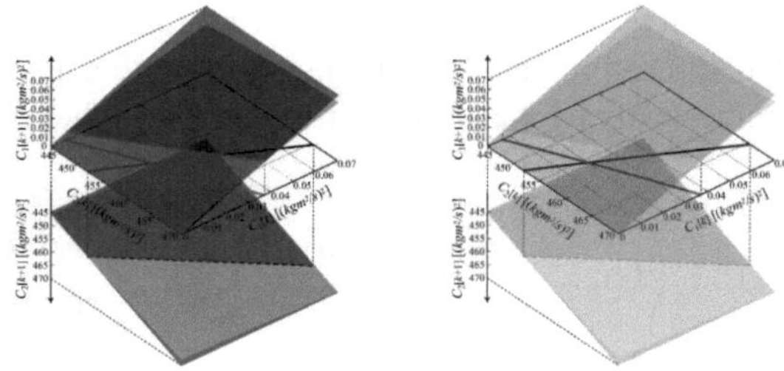

Fig. 9.54 Return map of C_1 and C_2 (C_1^*=0.02, C_2^*=456.0, h=0.428, l_0=0.455)

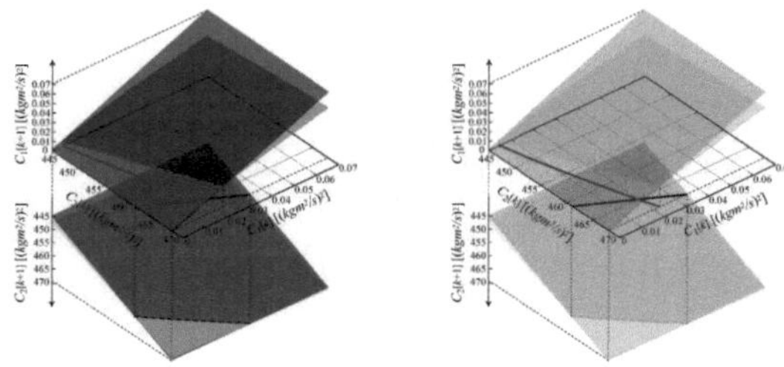

Fig. 9.55 Return map of C_1 and C_2 (C_1^*=0.02, C_2^*=465.0, h=0.428, l_0=0.455)

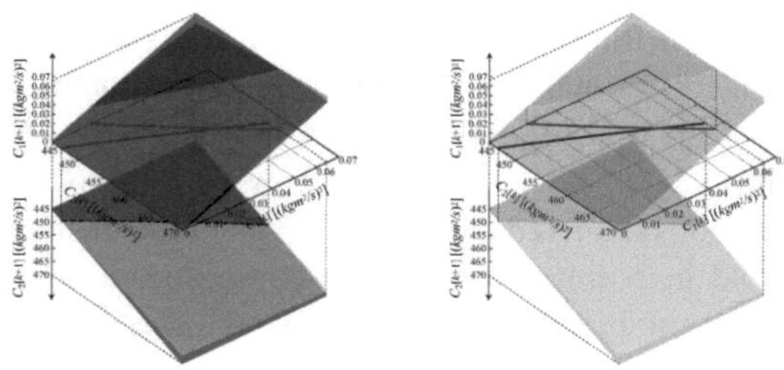

Fig. 9.56 Return map of C_1 and C_2 (C_1^*=0.05, C_2^*=456.0, h=0.428, l_0=0.455)

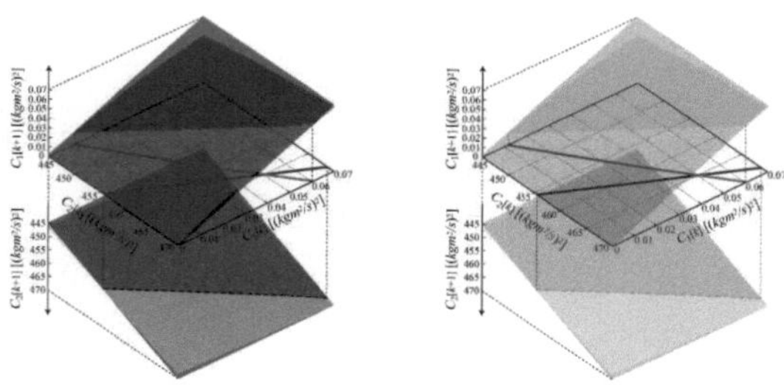

Fig. 9.57 Return map of C_1 and C_2 (C_1^*=0.05, C_2^*=465.0, h=0.428, l_0=0.455)

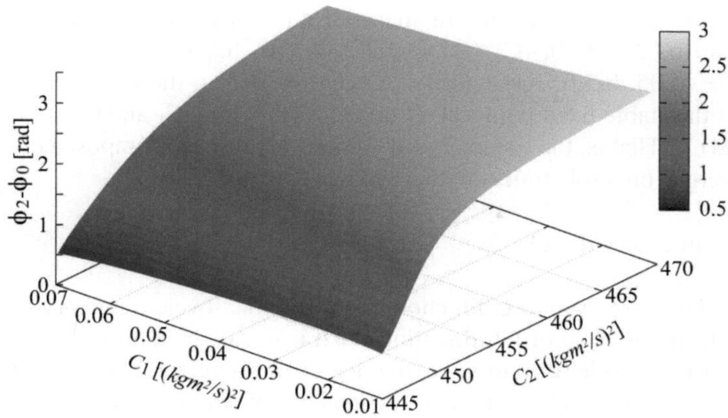

Fig. 9.58 3D map of C_1^*, C_2^*, and $\phi_2 - \phi_0$

Fig. 9.53(Middle). Consequently, if C_1 and C_2 are converged on (C_1^*, C_2^*), the robot state is attracted to the trajectory composed of both manifolds as shown in the right figure of Fig. 9.53.

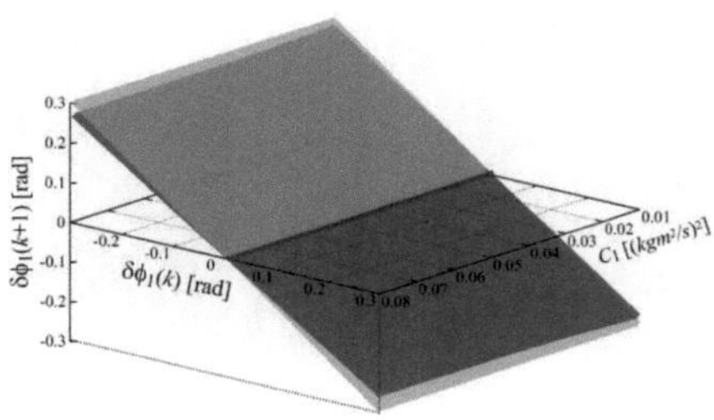

Fig. 9.59 Return map of $\delta\phi_1$ with respect to C_1^* (C_2^*=456.0, g_ϕ=0.3, h=0.428, l_0=0.455)

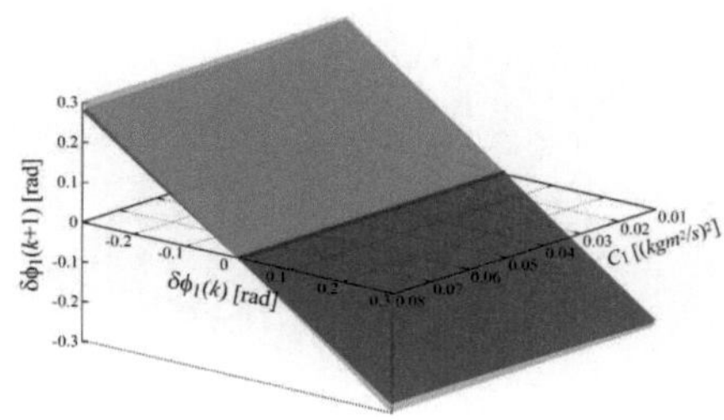

Fig. 9.60 Return map of $\delta\phi_1$ with respect to C_1^* (C_2^*=465.0, g_ϕ=0.3, h=0.428, l_0=0.455)

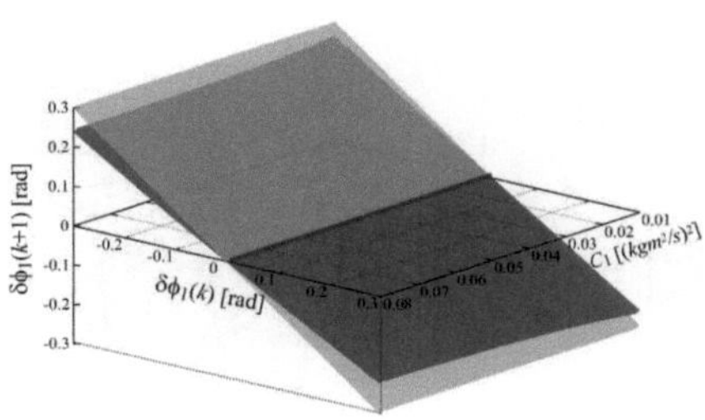

Fig. 9.61 Return map of $\delta\phi_1$ with respect to C_1^* (C_2^*=456.0, g_ϕ=0.6, h=0.428, l_0=0.455)

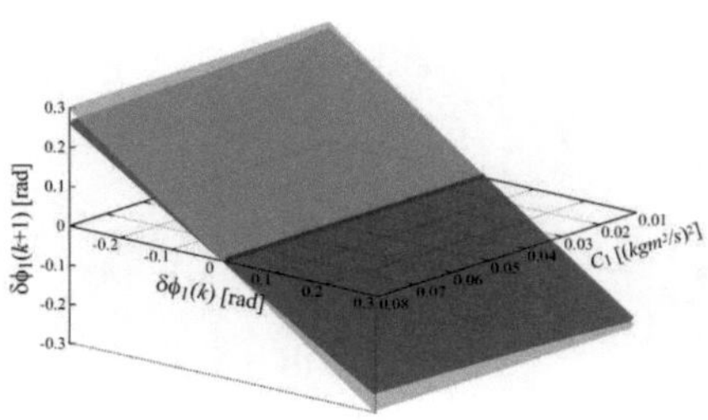

Fig. 9.62 Return map of $\delta\phi_1$ with respect to C_1^* (C_2^*=465.0, g_ϕ=0.6, h=0.428, l_0=0.455)

Fig. 9.54-9.57 show the return maps of C_1 and C_2 under the condition of $(C_1^*,C_2^*) = (0.02,456.0)$, $(C_1^*,C_2^*) = (0.02,465.0)$, $(C_1^*,C_2^*) = (0.05,456.0)$, and $(C_1^*,C_2^*) = (0.05,465.0)$ respectively. As can be seen in these figures, C_1 and C_2 possesses the stable fixed points at C_1^* and C_2^*. Therefore, C_1 and C_2 are converged on C_1^* and C_2^*. That is, the trajectory in 3-dimensional space composed of θ, $\dot{\theta}$, and $\dot{\phi}$ is converged on a sole trajectory determined by C_1^*, C_2^*, and h.

Walking Direction Control

Next, we design the walking direction controller. If 2-cycle trajectory occurs, the robot walks to the right or left direction. That is to say, by adjusting the degree of the error from 1-cycle trajectory, i.e. the degree of 2-cycle, it is possible to control the walking direction. The walking direction control requires to be able to generate desired degree of 2-cycle in the phase ϕ in order to walk in the desired direction, or converge 2-cycle trajectory on a sole trajectory in order to walk straight.

Fig. 9.58 shows the relationship between PDAC Constants, C_1^*, C_2^*, and the variation of ϕ for a step, $\phi_2 - \phi_0$. As can be seen in this figure, the variation of ϕ is monotonically increased with respect to C_2^*, and monotonically decreased with respect to C_1^*. Besides, it can be seen from this figure that the effect of C_2^* on the

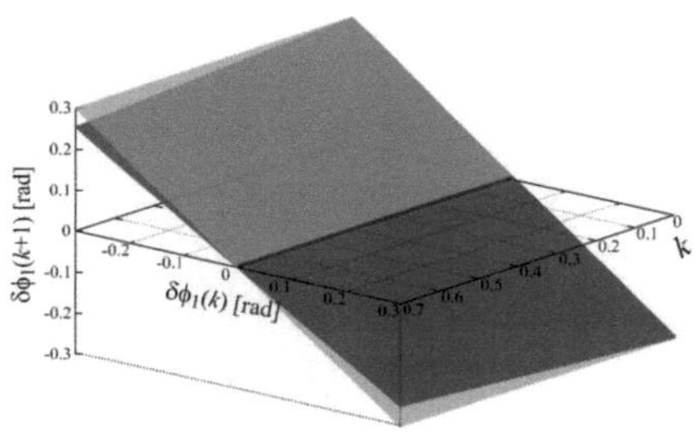

Fig. 9.63 Return map of $\delta\phi_1$ with respect to g_ϕ (C_1^*=0.02, C_2^*=456.0, h=0.428, l_0=0.455)

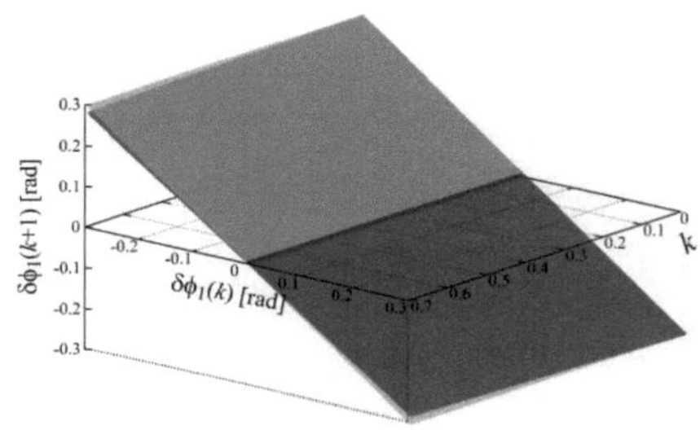

Fig. 9.64 Return map of $\delta\phi_1$ with respect to g_ϕ (C_1^*=0.02, C_2^*=465.0, h=0.428, l_0=0.455)

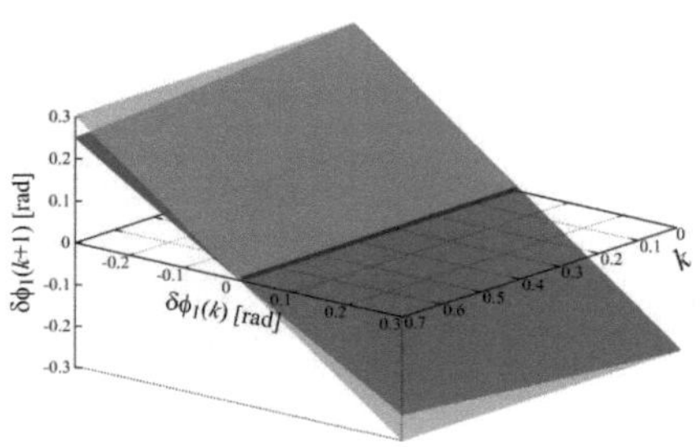

Fig. 9.65 Return map of $\delta\phi_1$ with respect to g_ϕ (C_1^*=0.05, C_2^*=456.0, h=0.428, l_0=0.455)

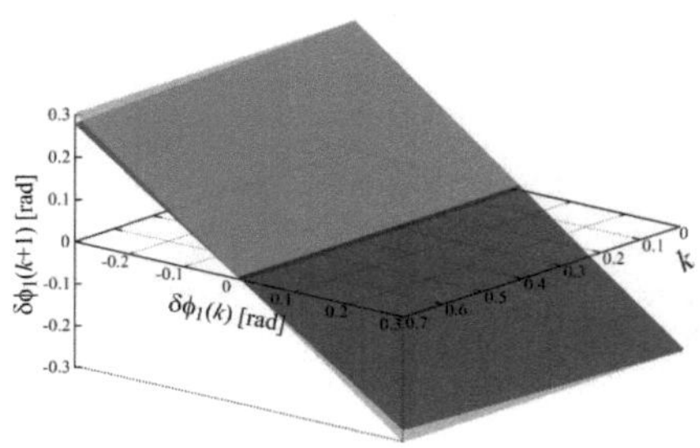

Fig. 9.66 Return map of $\delta\phi_1$ with respect to g_ϕ (C_1^*=0.05, C_2^*=465.0, h=0.428, l_0=0.455)

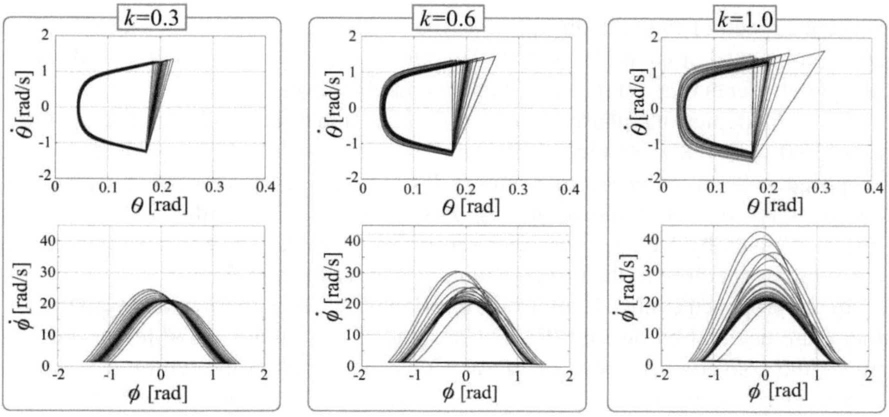

Fig. 9.67 Phase portrait of θ and ϕ (C_1^*=0.02, C_2^*=456.0, h=0.428, l_0=0.455, ϕ_1^d=0). (Left) g_ϕ=0.3 (Middle) g_ϕ=0.6 (Right) g_ϕ=1.0

variation of ϕ is large, whereas that of C_1^* is quite small. Thus, to adjust the 2-cycle of the trajectory in the *phi* phase portrait, C_1^* and C_2^* is decided according to the following equation:

$$C_1^* = C_1^d \tag{9.197}$$
$$C_2^* = C_2^d + g_\phi\left(\phi_1[k] - \phi_1[k+1] + (-1)^k \sigma \phi_1^d\right) \tag{9.198}$$

where

$$\sigma = \begin{cases} 1 & \text{: Left direction(or right direction)} \\ -1 & \text{: Right direction(or left direction)} \end{cases} \tag{9.199}$$

C_1^* and C_2^* are the desired PDAC Constants. g_ϕ is the gain of direction adjusting feed-back. $\phi_1[k]$ and $\phi_1[k-1]$ are ϕ_1 value at present and previous step. ϕ_1^d is the desired ϕ_1^d, which is decided from the desired walking direction, e.g. ϕ_1^d=0 if the desired walking direction is straight. Note that ϕ_1 is the yaw angle at $\dot{\theta} = 0$ (see Fig. 9.41).

Fig. 9.59-9.62 shows the the return map of $\delta\phi[k]$ with respect to C_1^* under the condition of g_ϕ=1.0, 2.0, and C_2^*=456.0, 465.0. From these figures, it can be seen that $\delta\phi$ possesses the stable fixed point at 0. Hence, the walking direction can be controlled stably by Eqs. (9.197) and (9.198).

Fig. 9.63-9.66 shows the return map of $\delta\phi[k] := \phi_1[k] - \phi[k-1]$ with respect to the gain, g_ϕ, under the condition of $(C_1^*, C_2^*) = (0.02, 456.0)$, $(C_1^*, C_2^*) = (0.02, 465.0)$, $(C_1^*, C_2^*) = (0.05, 456.0)$, and $(C_1^*, C_2^*) = (0.05, 465.0)$ respectively. As can be seen in these figures, the larger g_ϕ is, the higher the convergence speed of $\delta\phi$ is. However, high g_ϕ causes the large temporal error of C_2 ($C_2^d - C_2$) since C_2^* becomes high temporally. Consequently, as shown in Fig. 9.67, if g_ϕ is set at high gain, the convergence speed of PDAC Constants, i.e. that of the trajectory in the 3-dimensional space composed of $(\theta, \dot{\theta}, \dot{\phi})$, is slow. Therefore, it is necessary to adjust and set g_ϕ adequately depending on the environment and objective.

Fig. 9.68 depicts the phase portrait of θ and ϕ under the condition that C_1^*=0.02, C_2^*=456.0, h=0.428, l_0=0.455, g_ϕ=0.3, and σ=1. As can be seen from this figure, the proposed walk direction control achieves desired 2-cycle trajectory of yaw phase. In practical use, the robot updates ϕ_1^d regularly according to the information of target direction given by a camera or other sensor. As a result, it is conceivable that the robot might get close to the target gradually, and reach there finally.

In the proposed control framework, step-length, walking period, and walking velocity are decided by desired PDAC Constant C_1^d, C_2^d, and h. In addition, h determines the convergence speed of PDAC Constants. Thus in the future, it is necessary to clarify how to decide C_1^d, C_2^d, and h according to the desired walking velocity, such that the energy consumption is minimized and the convergence speed is maximized.

9.4.4 Experiment

9.4.5 Simulation

The validity of proposed control was tested by numerical simulation under the condition that l_0=0.51[m], h=0.488. Fig. 9.69 shows the simulation snapshots of dynamic walking control based on 3-D dynamics with only geometrical constraint and with stabilization based on PDAC Constants. Without stabilization based on PDAC Constants, 3-D walking was converged on not sole trajectory but 2-cycle one, hence

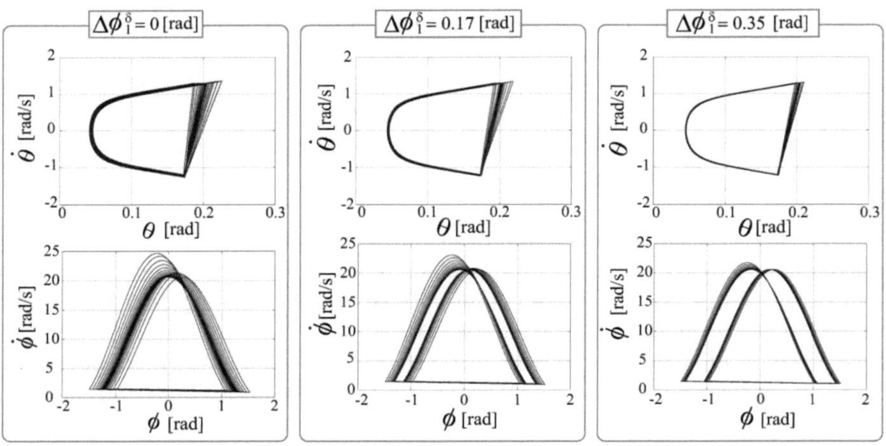

Fig. 9.68 Phase portrait of θ and ϕ (C_1^*=0.02, C_2^*=456.0, h=0.428, l_0=0.455, g_ϕ=0.3, σ=1). (Left) ϕ_1^d=0[rad] (Middle) ϕ_1^d=0.17[rad] (Right) ϕ_1^d=0.35[rad]

the robot could not walk straight and walked leftwards as can be seen in Fig. 9.69. Meanwhile, with stabilization, it was confirmed that the robot can walk straight. After convergence, step-length, walking period, and walking velocity were 0.18[m], 0.65[s], and 0.28[m/s] respectively.

Fig. 9.70 shows COG trajectory and tip trajectory of swing-leg for 1-step in converged trajectory. Joint angles of support- and support-leg are depicted in Fig. 9.71.

9.4.5.1 Experimental Result

We validated the proposed algorithm with the Gorilla Robot III introduced in Chapter 3. The experiment was conducted on the level ground with maximum ± 1.0[cm] irregularity. As a result of the experiment, three-dimensional dynamic walking in 0.14[m] step-length and 0.26[m/s] walking velocity was realized. Although the ground has maximum ± 1.0[cm] irregularity in the experimental environment and the information of the ground shape was not given to the robot, the robot achieved the stable walking without information of the ground. Figure 9.79 shows snapshots of the experiment.

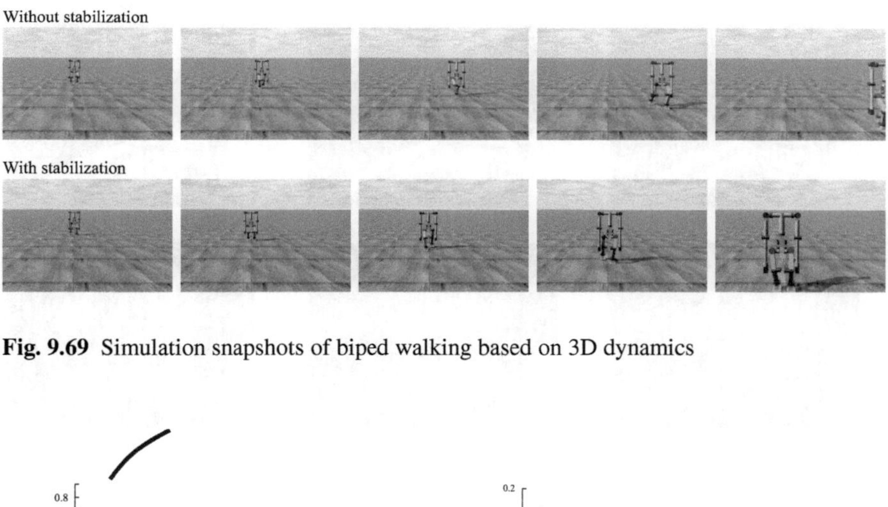

Fig. 9.69 Simulation snapshots of biped walking based on 3D dynamics

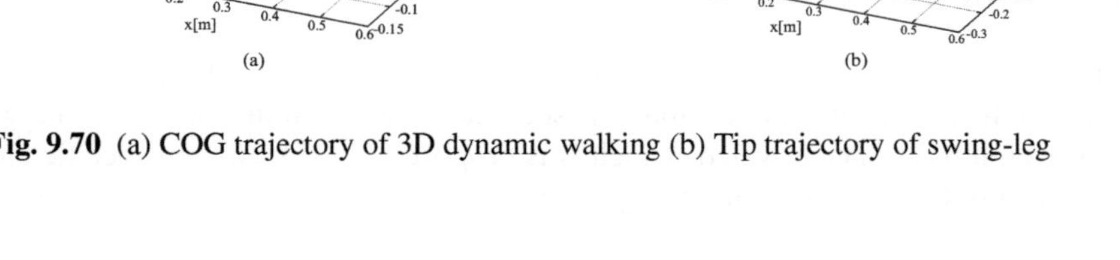

Fig. 9.70 (a) COG trajectory of 3D dynamic walking (b) Tip trajectory of swing-leg

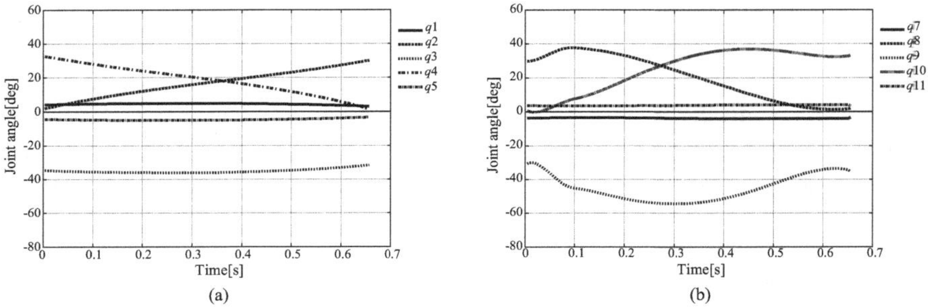

Fig. 9.71 (a) Joint angles of support leg (b) Joint angles of swing leg

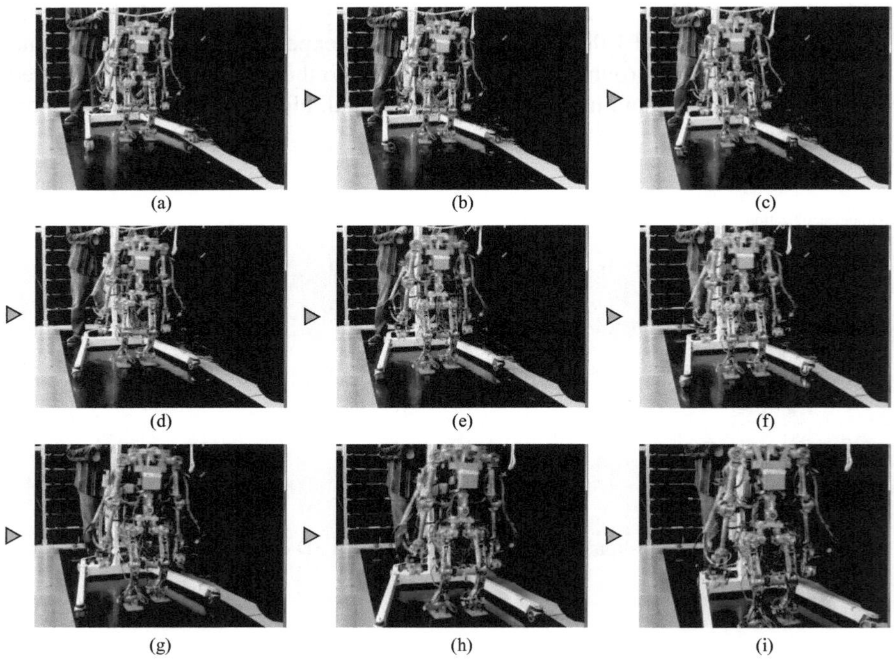

Fig. 9.72 Snapshots of the Bipedal Walking Experiment. Each figure shows the snapshots at (a)1st (b)4th (c)7th (d)10th (e)13th (f)16th (g)19th (h)22th (i)25th step.

9.4.5.2 Energy Efficiency

In this section, in order to compare an energy efficiency between robots of different sizes, the dimensionless specific cost of transport, C_t=(energy used)/(weight×distance traveled) which was proposed in [45], and the specific energetic cost of transport, C_{et}, and the specific mechanical cost of transport, C_{mt} were

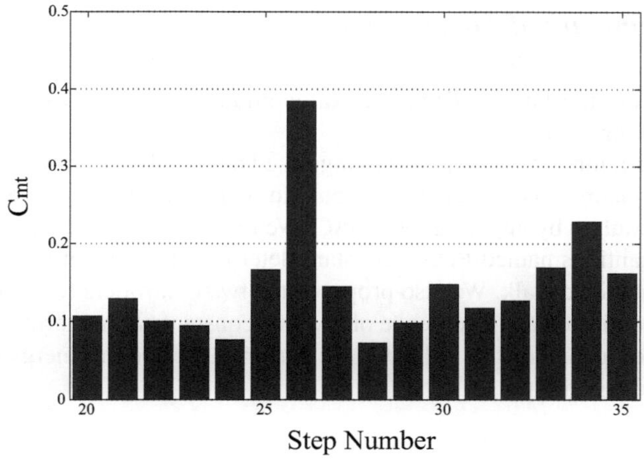

Fig. 9.73 C_{mt} estimated from experimental data.

defined. Whereas C_{et} uses the total energy consumed by the system, C_{mt} considers the positive mechanical work of the actuators. Since a total energy cannot be evaluated in our system, this section uses C_{mt} as the energetic cost of transport. The mechanical work of the actuators in one cycle of the walking E is calculated as follows:

$$E = \int_0^T \sum_{i=1}^N \delta\left(\tau_i \dot{\theta}_i\right) dt, \tag{9.200}$$

$$\delta(x) = \begin{cases} x, \; if \; x > 0 \\ 0, \; if \; x \leq 0 \end{cases}, \tag{9.201}$$

where T is the cycle time of a walk, N is the number of actuators, τ_i and $\dot{\theta}_i$ are the joint torque and the angular velocity of i-th joint number. Then, the mechanical cost of transport C_{mt} in one cycle of a walk is calculated as follows:

$$C_{mt} = \frac{E}{S \times M}, \tag{9.202}$$

where S is the step length of the robot, M is the mass of the robot. After 20 steps, the motion of the robot converges to the stable one. The experimental C_{mt} of each step are shown in Fig. 9.73, and the average of C_{mt} is 0.15. The C_{mt} of Honda humanoid ASIMO [246] which realized a stable three-dimensional dynamic walking applied ZMP-based control is estimated 1.6 in [45], thus our walking algorithm is efficient more than ten times compare to the ZMP-based control.

9.4.6 *Summary of This Section*

In this section, the Passive Dynamic Autonomous Control (PDAC) is applied to bipedal walking controls.

The application is a 3-D biped walking based on 3-D dynamics without dividing. The robot dynamics is modeled as an autonomous system of a three-dimensional inverted pendulum by applying the PDAC. We numerically presented that two conservative quantities named PDAC constant determine the velocity and walking direction of the biped walk. We also proposed the two controllers for two PDAC constants and confirmed a convergence of the two constants in numerical simulations. Finally, experimental results validated the performance and the energy efficiency of the proposed algorithm.

9.5 3-D Biped Walking Adapted to Rough Terrain Environment

9.5.1 *Foot-Contact Model*

In this section, it is assumed that there is an uneven step with the Δh height at kth step. Fig. 9.74 shows the geometric condition at a foot-contact. Yaw angle of a swing-leg is shifted by α from the symmetrical position with a stance-leg at a foot-contact, i.e. it is $\phi_0[k+1] = -\phi_2[k] + \alpha$, where $\phi_0[k+1]$ and $\phi_2[k]$ denote ϕ_0 and ϕ_2 at $k+1$th and kth step respectively. At a foot-contact, l_0, l_2, and θ_2 are set to constant value, that is, COG height of just before a foot-contact, $h = l_2 \cos\theta_2$ is constant. Also, it is assumed that θ_0 is valuable decided by the uneven height Δh. Then, we can get the following equation:

$$h = l_0 \cos(\theta_0[k]) + \Delta h \tag{9.203}$$

$$h = l_0 \cos(\theta_0[k+1]) - \Delta h. \tag{9.204}$$

In this section, perfectly inelastic collision is assumed between the ground and a foot occurs for a moment, same as our previous work [11]. Since angular momentum around a new contact point is conserved, the yaw and roll angular velocity of the pendulum just after $k+1$th foot-contact, $\dot{\phi}_0[k+1]$ and $\dot{\theta}_0[k+1]$ are derived as follows from Eqs. (9.104), (9.107) and the assumption of the perfectly inelastic collision:

$$\dot{\theta}_0[k+1] = \frac{l_2}{l_0}\Big(\dot{\theta}_2[k]\big(\sin\theta_2\sin(\theta_0[k+1]) - \cos(\theta_0[k+1])$$

$$\cos\theta_2\cos\alpha\big) + \dot{\phi}_2[k]\cos(\theta_0[k+1])\sin\theta_2\sin\alpha\Big), \quad (9.205)$$

$$\dot{\phi}_0[k+1] =$$

$$\frac{l_2}{l_0\sin(\theta_0[k+1])}\Big(\dot{\theta}_2[k]\cos\theta_2\sin\alpha + \dot{\phi}_2[k]\sin\theta_2\cos\alpha\Big), \quad (9.206)$$

where $\phi_0[k+1]$ and $\theta_0[k+1]$ are the yaw and roll angle just after $k+1$th foot-contact, $\dot{\phi}_2[k]$ and $\dot{\theta}_2[k]$ are the yaw and roll angler velocity just before $k+1$th foot-contact, ϕ_2 and θ_2 are the yaw and roll angle just before a foot-contact (see Fig. 9.41 and Fig. 9.74).

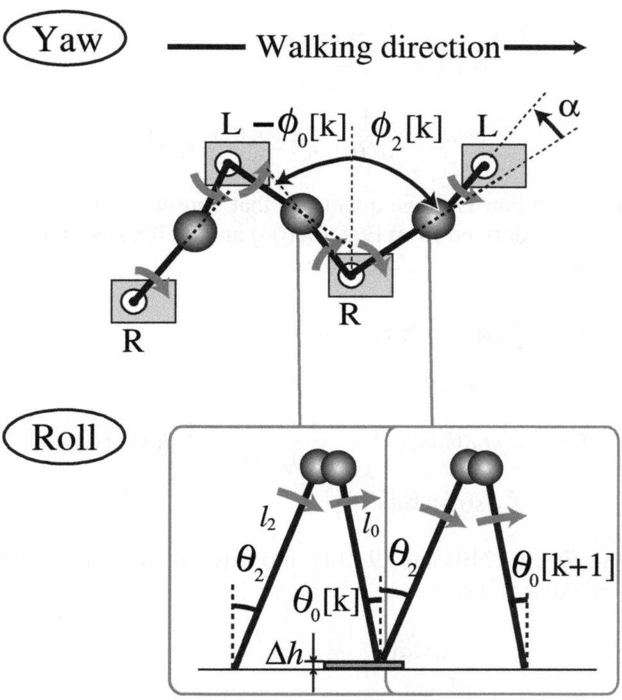

Fig. 9.74 Geometrical constraints at foot-contact.

9.5.2 Stability Analysis

9.5.2.1 Poincaré Map of the Walking Cycle

In this section, the walking cycle is expressed as a nonlinear discrete system with a foot-contact; then the stability analysis around the fixed point is conducted by use of Poincaré map. $\dot{\phi}$ and $\dot{\theta}$ just after kth foot-contact that compose the autonomous system Eqs. (9.104) and (9.107) are selected as discrete state:

$$v_k = \begin{bmatrix} \dot{\phi}_0[k] \\ \dot{\theta}_0[k] \end{bmatrix}. \tag{9.207}$$

From Eqs. (9.103) and (9.106) the angular velocities of $k+1$th step just before a foot-contact $\dot{\phi}_2[k]$, $\dot{\theta}_2[k]$ are derived as follows:

$$\dot{\phi}_2[k] = \frac{\sqrt{2C_1}}{ml_2^2 \sin\theta_2}, \tag{9.208}$$

$$\dot{\theta}_2[k] = \frac{1}{M(\theta_2)}\sqrt{2(D(\theta_2)+C_2)}, \tag{9.209}$$

where C_1 and C_2 are conservative quantities that conserve during a single support phase. C_1 and C_2 are derived from Eqs. (9.103) and (9.106) using the discrete state of kth step as follows:

$$\begin{aligned} C_1 &= \frac{1}{2}\left(\dot{\phi}_0[k]ml_0^2\sin(\theta_0[k])\right)^2 \\ &:= \xi_1(\dot{\phi}_0[k]), \end{aligned} \tag{9.210}$$

$$\begin{aligned} C_2 &= 2(ml_0^2\dot{\theta}_0[k])^2 + \frac{C_1}{\sin^2(\theta_0[k])} + m^2 gl_0\cos(\theta_0[k]) \\ &:= \xi_2(\dot{\phi}_0[k], \dot{\theta}_0[k]). \end{aligned} \tag{9.211}$$

By substituting Eqs. (9.210) and (9.211) into Eqs. (9.208) and (9.209), $\dot{\phi}_2[k]$ and $\dot{\theta}_2[k]$ are expressed as follows:

$$\dot{\phi}_2[k] = \frac{\sqrt{2\xi_1}}{ml_2^2\sin^2\theta_2}, \tag{9.212}$$

$$\dot{\theta}_2[k] = \frac{1}{ml_2^2}\sqrt{2\left(\frac{-\xi_1}{\sin^2\theta_2} - m^2 gl_2\cos\theta_2 + \xi_2\right)}. \tag{9.213}$$

From Eqs. (9.203) and (9.204), $\theta_0[k]$ and $\theta_0[k+1]$ are expressed by use of Δh:

$$\theta_0[k] = \arccos\left(\frac{h-\Delta h}{l_0}\right) \tag{9.214}$$

$$\theta_0[k+1] = \arccos\left(\frac{h+\Delta h}{l_0}\right). \tag{9.215}$$

From Eqs. (9.205), (9.206), (9.212), (9.213), (9.214), and (9.215), the discrete state of $k+1$th step is described as the function of the discrete state of kth step and Δh,

$$\dot{\phi}_0[k+1] = \frac{l_2}{\sqrt{l_0^2-(h+\Delta h)^2}}\left[\frac{h\sin\alpha}{ml_0^2 l_2}\sqrt{W_1\left(\dot{\phi}_0[k],\dot{\theta}_0[k],\Delta h\right)} + W_2\left(\dot{\phi}_0[k],\Delta h\right)\cos\alpha\right] \tag{9.216}$$

$$\begin{aligned}\dot{\theta}_0[k+1] = \frac{l_2}{l_0}&\left[\frac{1}{ml_2^2}\left(\sqrt{\left\{1-\frac{(h+\Delta h)^2}{l_0^2}\right\}\left\{1-\left(\frac{h}{l_0}\right)^2\right\}}\right.\right.\\ &\left.\left.-\frac{h(h+\Delta h)}{l_0}\cos\alpha\right)\sqrt{W_1\left(\dot{\phi}_0[k],\dot{\theta}_0[k],\Delta h\right)} + W_2\left(\dot{\phi}_0[k],\Delta h\right)\frac{h+\Delta h}{l_0}\sin\alpha\right],\end{aligned} \tag{9.217}$$

where

$$\begin{aligned}W_1\left(\dot{\phi}_0[k],\dot{\theta}_0[k],\Delta h\right) = 2&\left\{C_1[k]\left(\frac{l_0^2}{l_0^2-(h-\Delta h)^2} - \frac{l_2^2}{l_2^2-h^2}\right)\right\}\\ &+m^2 g\Delta h + 2\left(ml_0^2\dot{\theta}_0[k]\right)^2,\end{aligned} \tag{9.218}$$

$$W_2\left(\dot{\phi}_0[k],\Delta h\right) = \dot{\phi}_0[k]\frac{(h+\Delta h)^2}{l_2}\sqrt{\frac{1}{l_2^2-h^2}}\ . \tag{9.219}$$

By defining the vector function P from Eqs. (9.216) and (9.217), Poincaré map is described as follows:

$$v_{k+1} = P\left(v_k\right). \tag{9.220}$$

The existence of fixed point of this walking system is not proven, because the system is nonlinear complex system. However, we confirmed that this walking model has stable fixed points by use of return map in our previous work [11]. Especially, if the state of the walking cycle converge to fixed point v^*, the following equation is derived,

$$v^* = P\left(v^*\right). \tag{9.221}$$

Considering Taylor developing around the fixed point,

$$\begin{aligned} v^* + \delta v_{k+1} &= P(v^* + \delta v_k) \\ &\approx P(v^*) + J(v^*)\,\delta v_k, \end{aligned} \tag{9.222}$$

where $J(v^*)$ is Jacobian matrix at the fixed point, δv is the small displacement vector.

By substituting Eq. (9.221) into Eq. (9.222),

$$\delta v_{k+1} = J(v^*)\,\delta v_k. \tag{9.223}$$

Thus, if all of the eigenvalues of the Jacobian matrix $J(v^*)$ are inside the unit circle on the complex plane, the walking cycle is asymptotically stable.

9.5.2.2 Stability Analysis

First of all, we analyze the stability of the walking model explained in the section 9.4.1 on a flat terrain, that is, $\Delta h = 0$. By numerical simulation, the fixed point v^* of this walking model when the parameters are set as $h = 0.541$, $l_0 = 0.56$, $l_2 = 0.561$, $\alpha = 0.035$ is calculated as follows:

$$v^* = \begin{bmatrix} 0.9504 \\ -1.140 \end{bmatrix}. \tag{9.224}$$

By numerical calculation, Jacobian matrix $J(v^*)$ is also calculated as follows:

$$J(v^*) = \begin{bmatrix} 0.8230 & -0.1298 \\ -0.0044 & 0.8155 \end{bmatrix}. \tag{9.225}$$

The eigenvalues λ_1, λ_2 of the Jacobian matrix $J(v^*)$ are obtained numerically as $\lambda_1 = 0.8435, \lambda_2 = 0.7949$. Since all of the eigenvalues are less than 1, the designed walking cycle is asymptotically stable on a flat terrain. Fig. 9.75 shows the simulation result of this condition. From this figure, it is confirmed that $\dot{\theta}$ and $\dot{\phi}$ converge to a certain fixed point.

Next, same stability analysis is conducted on rough terrains. In this analysis, the eigenvalues are calculated when the Δh is changed every 1 mm. In Fig. 9.76, some of the eigenvalues are plotted on the complex plane. From the stability analysis, under the condition that the parameters are set same as for flat terrain, the designed walking model is asymptotically stable when the Δh is -18 mm to 8mm; in the range, the state of the walking cycle converge to fixed point of the flat terrain v^*. More stability analysis confirmed that the asymptotically stable range depends on the setting parameters. The result introduced in this section is one example that can be applied to the Gorilla Robot III explained in Chapter 3. Optimization of the parameters to maximize the stable range is one of the future works.

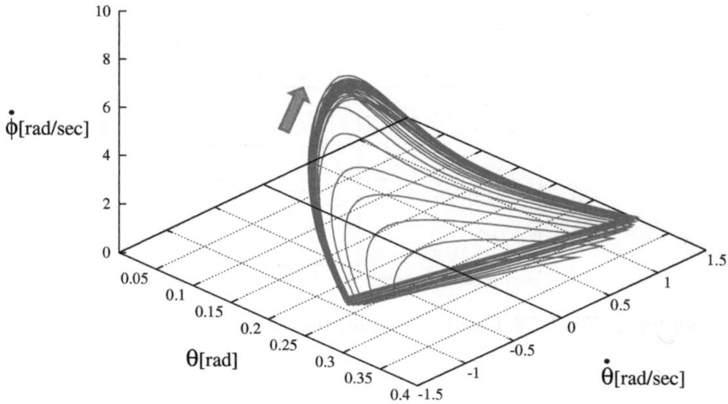

Fig. 9.75 Phase Portrait.

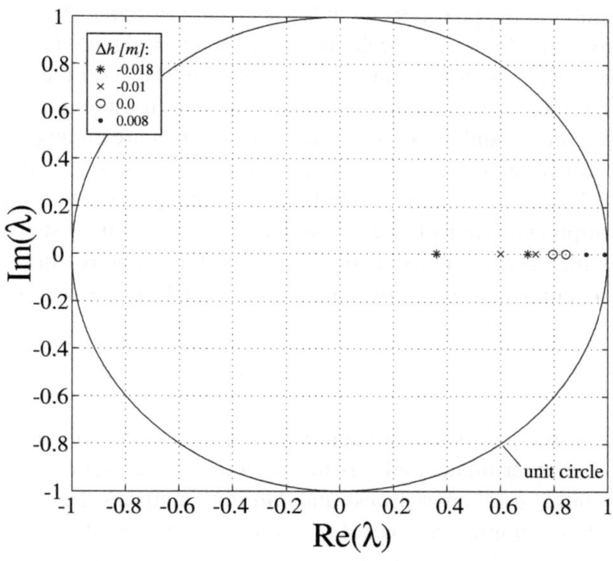

Fig. 9.76 Eigenvalues of the Jacobian.

9.5.3 Experiment

9.5.3.1 Control Architecture

In the experiment, the position and posture of the three-dimensional inverted pendulum modeled in the section 9.4.1 are derived every control step from zero dynamics

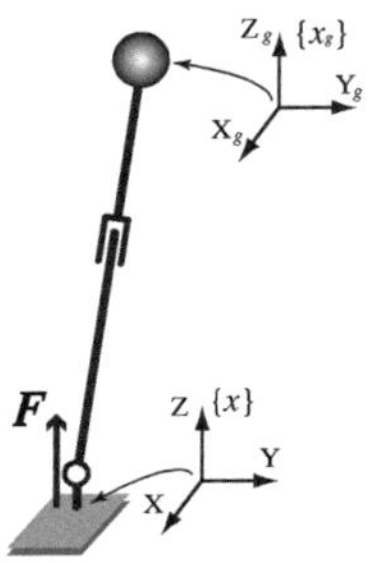

Fig. 9.77 Coordinate System.

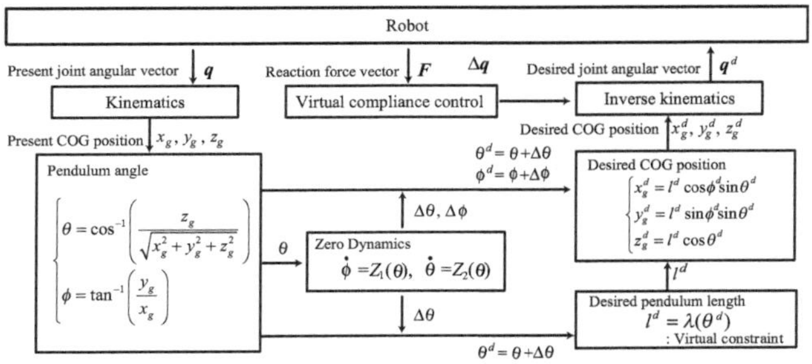

Fig. 9.78 Control Architecture of the Experiment.

Eqs. (9.104), (9.107), and the constraints Eqs. (9.97)-(9.99). By use of inverse kinematics, all the joint angles of the experimental robot are calculated and controlled every control step. In this section, it is assumed point-contact of the robot foot; however most of humanoid type robots include the Gorilla Robot III (It was introduced in Chapter 3.) have feet. In this research, the ankle joints are decided by dynamics Eqs. (9.104), (9.107), and the constraint Eqs. (9.97)-(9.99) base on the assumption that the terrain is flat ($\Delta h = 0$). Therefore, by applying the virtual compliance control to ankle actuators, the pendulum angles of the foot-contact are changed in accordance with the ground irregularity Δh. The virtual compliance control makes ankle joints flexible, and it has effect that the ankle joints become passive like joints; note that the virtual compliance control doesn't stabilize the walking system directly. The virtual compliance control is used for a biped walking control in previous works (e.g. [99]). The condition that foot realizes the virtual compliance is expressed as follows:

$$F = K\Delta x + C\Delta \dot{x}, \tag{9.226}$$

where F is the force and moment vectors acting to foot coordinate, Δx is the displacement of foot coordinate, and $\dot{x}$ is the velocity of the foot coordinate (see Fig. 9.77). $K \in R^6$ and $C \in R^6$ are stiffness and damping matrixes. In this work, the purpose of the virtual compliance control is to modify the pendulum angles; thus only roll and pitch angle of the robot foot are considered. From discrete Eq. (9.226), the compliance position of the foot is derived as follows:

$$\Delta x(t+\Delta t) = \left[K + \frac{C}{\Delta t}\right]^{-1} \left\{F + C\frac{\Delta x(t)}{\Delta t}\right\}. \tag{9.227}$$

where Δt is control cycle, $\Delta x(t+\Delta t)$ is the present error of foot coordinate from ideal position, $\Delta x(t)$ is the error of foot coordinate from ideal position before one control cycle. The compliance position are add to desired trajectory of the COG derived from zero dynamics Eqs. (9.104) and (9.107), by use of inverse kinematics the desired trajectories of all actuator are modified. This control architecture is shown in Fig. 9.78.

9.5.3.2 Experimental Results

We validated the proposed algorithm with the Gorilla Robot III. The experiment was carried out on the basis of the condition that the parameters are equal to those used in the stability analysis. The terrain has 5 mm irregularity on the level ground. As a result of the experiment, three-dimensional dynamic walking was realized. Although the ground has 5 mm irregularity in the experimental environment and the information of the ground shape was not given to the robot, the robot achieved the stable walking without information of the ground. Fig. 9.79 shows snapshots of the experiment and Fig. 9.80 shows the joint angles the experiment. Fig. 9.81 shows the compliance displacement of the experiment. After about four seconds, it is confirmed that change of the compliance displacement is larger when the robot walks on a rough terrain. Since the displacement value is small, the compliance control

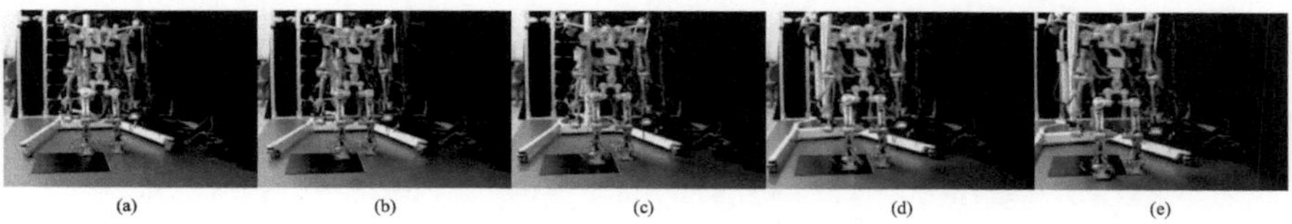

Fig. 9.79 Snapshots of the Bipedal Walking Experiment. Each figure shows the snapshots at (a)7th (b)9th (c)11th (d)13th (e)15th step.

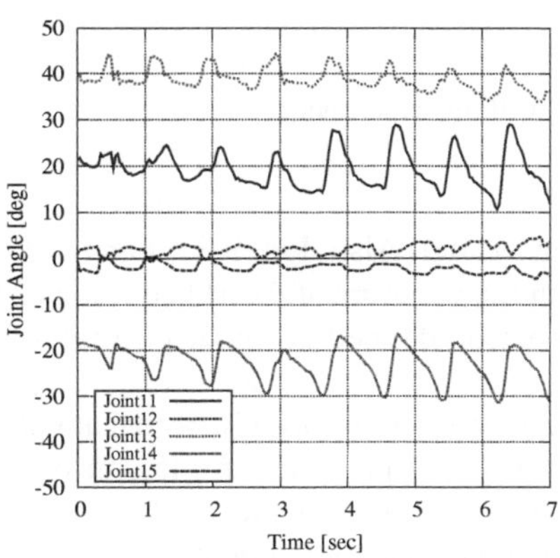

(a) Joint Angles of the Left Leg.

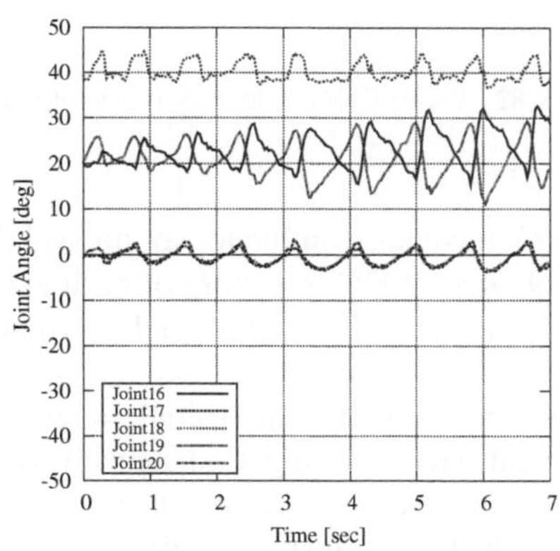

(b) Joint Angles of the Right Leg.

Fig. 9.80 Joint angle of the bipedal walking experiment.

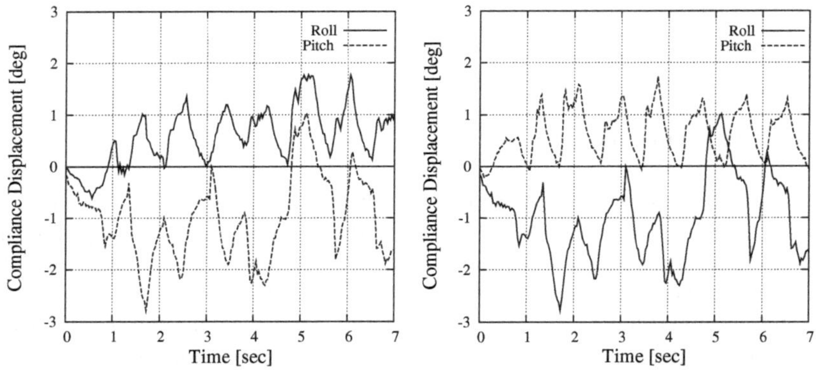

(a) Compliance displacement of the Left Leg.
(b) Compliance displacement of the Right Leg.

Fig. 9.81 Compliance displacement of the bipedal walking experiment.

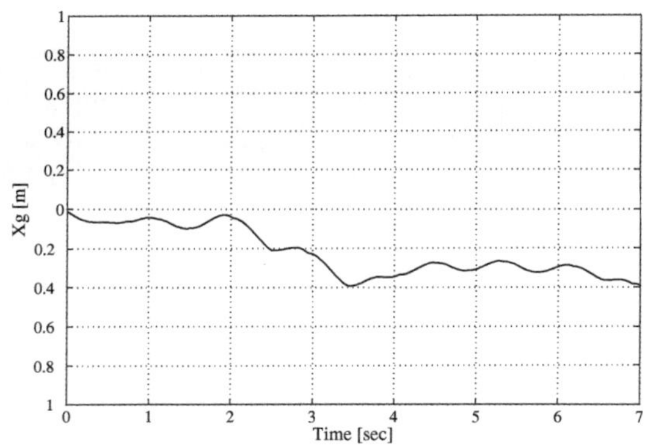

Fig. 9.82 X axis COG trajectory of the bipedal walking experiment.

doesn't have large influence to this walking system. Also, Fig. 9.82, Fig. 9.83, and Fig. 9.84 are COG trajectory of each axis in the experiment. World coordinate system is defined in Fig. 9.38 and original point is defined at initial position of the robot.

We calculate the dimensionless specific cost of transport, C_{mt}=(consumed mechanical energy)/(weight$\times$distance traveled) that is proposed in [45], in order to confirm the energy efficiency that is an advantage of pint-contact method. Table 9.5 shows the C_{mt} of the Gorilla Robot calculated from the experimental result and C_{mt} of the Honda humanoid ASIMO estimated in [45]. The C_{mt} of ASIMO that realized stable three-dimensional biped walking by use of ZMP paradigm is 1.6. On the other hand, the C_{mt} of the Gorilla Robot applied PDAC-based control that we have

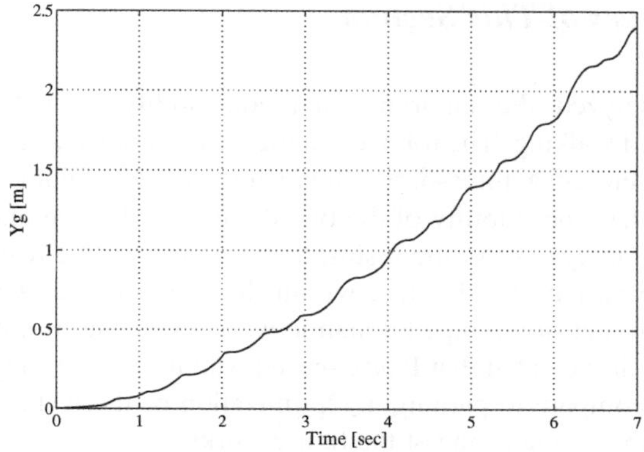

Fig. 9.83 Y axis COG trajectory of the bipedal walking experiment.

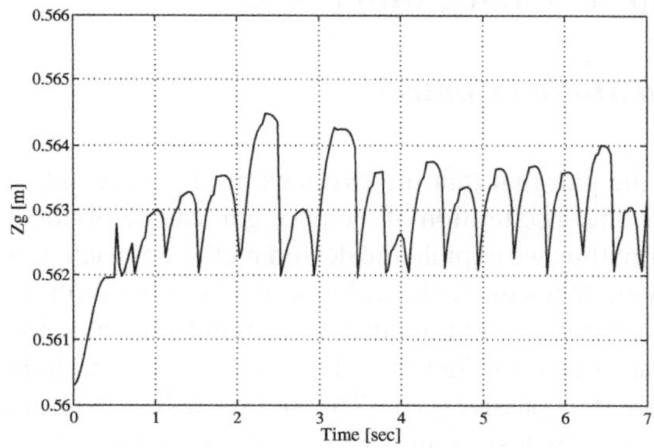

Fig. 9.84 Z axis COG trajectory of the bipedal walking experiment.

Table 9.5 Comparison of the C_{mt}

Method (Hardware)	C_{mt}
PDAC-based method (Gorilla Robot)	0.28
ZMP paradigm (ASIMO)	1.6 (from [45])

proposed is 0.28. The result indicates that PDAC-based walking on the rough terrain is also energy efficient compare to ZMP paradigm.

9.5.4 Summary of This Section

This section analyzed the environmental adaptability of PDAC-based three-dimensional biped walking. The robot dynamics was modeled as a two-dimensional autonomous system of a three-dimensional inverted pendulum by applying the PDAC concept; then the stability of the two-dimensional autonomous system with foot-contact was analyzed. As the result, the system has an asymptotically stable range due to the parameters. Finally, within the stable range, we experimentally realized a three-dimensional biped dynamic walking on the irregular ground. The experimental result indicated that PDAC-based walking on the rough terrain is also energy efficient compare to paradigm. Optimization of the walking parameters to maximize the stable range is one of the future works.

9.6 Quadruped Walking with PDAC

9.6.1 Lateral Motion Control

The pace gait is the gait that that the two legs on the same side move forward together (See Fig. 9.85). The motion of the pace gait can be divided as lateral motion and sagittal motion; thus we explain the design method of each motion controller respectively. When we focus on the lateral motion, the robot motion can be expressed as a motion of an inverted pendulum that the length from ankle joint of the supporting leg changes as shown in Fig.9.91. Here, we consider realizing stable foot pat by setting ankle joint as passive joint. The motion will stop on the condition of no energy input, because the loss of angler momentum happens when the leg switches. In this work, the motion continues by getting mechanical energy from extending motion of the inverted pendulum modeled in above.

Fig.9.87 shows the feet pat in lateral plane. The state (a) and (c) express the state that the inverted pendulum moves up perfectly, that is, the angular velocity is 0. The state (b) and (d) express the moment that the supporting leg changes. In the case that supporting leg is left, when we consider the state ((d)→(a)→(c)) in Fig.9.87, l_0 is the pendulum length just after the supporting leg change to the left leg (d); and also θ_0 is the pendulum angle θ of this moment. In the state (d) to (a), the pendulum length is constant value l_0, and the pendulum angle of the state (a) is θ_1. In the state (a) to (b), the pendulum length is changed. We define the pendulum length in the state (b) as $l_0 + \Delta l$, the pendulum angle as θ_2. In the case that the supporting leg is right, the relationship between pendulum angle and length is same.

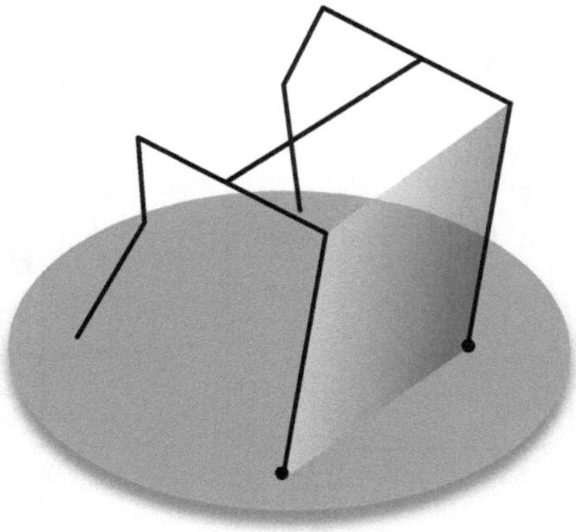

Fig. 9.85 A pace gait

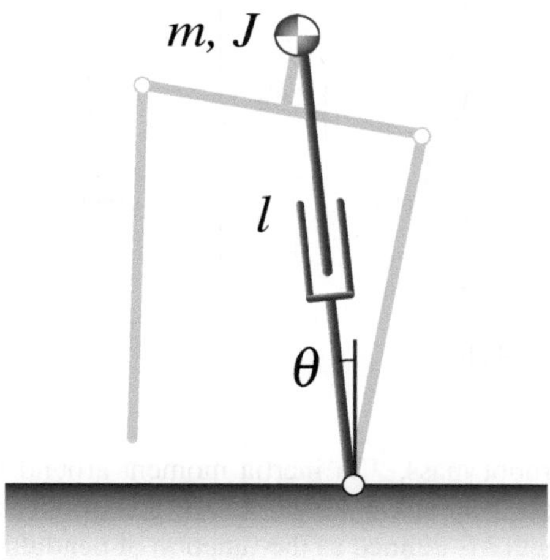

Fig. 9.86 Inverted pendulum model

9.6.2 Design of Pendulum Length

Coefficients of the pendulum length l corresponding to the angle of the passive joint θ is defined so that foot pat shown in Fig.9.87 continue. By assumption that the ankle joint is passive joint and there is no viscously torque, the dynamic equation of the inverted pendulum shown in Fig.9.91 is expressed as follows:

$$\frac{d}{dt}\left((ml^2+J)\dot{\theta}\right) - mgl\sin\theta = 0, \tag{9.228}$$

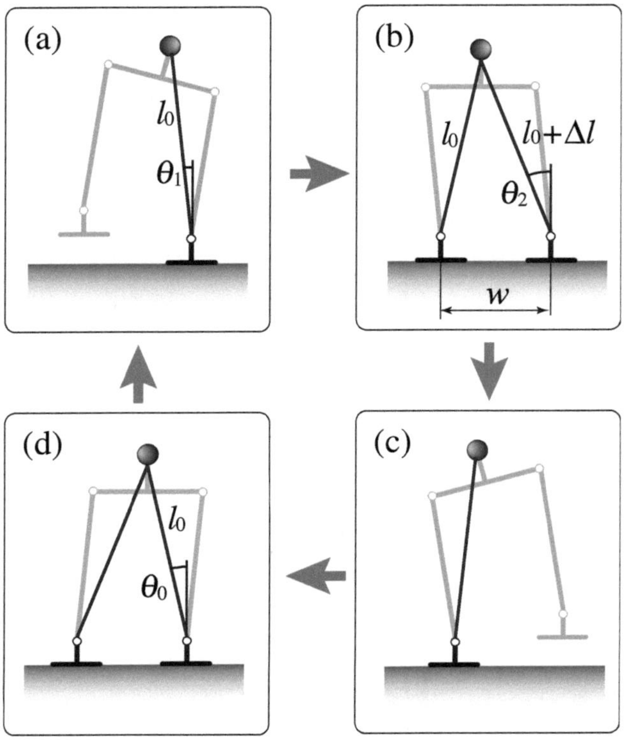

Fig. 9.87 Motion in lateral plane

where, m is whole robot mass, J is inertia moment around the center of gravity (COG), l is the length from ankle joint to COG, g is gravity acceleration. The inverted pendulum length l is defined as the function of pendulum angle $f(\theta)$ and by substituting $l = f(\theta)$ to Eq. (9.228), the dynamics is expressed as follows:

$$\begin{aligned}\dot{\theta} &= F(\theta), \\ &= \frac{1}{mf(\theta)^2 + J}\sqrt{\int \left(2(mf(\theta)^2 + J)mgf(\theta)\sin\theta\right) d\theta}.\end{aligned} \tag{9.229}$$

The function of the pendulum length interlocked by passive joint is defined as follows:

$$l = f(\theta) = a_2\theta^2 + a_1\theta + a_0. \tag{9.230}$$

The coefficients a_0, a_1, a_2 are derived from initial and termination conditions of the pendulum length,

$$f(\theta_1) = l_0, \tag{9.231}$$

$$f(\theta_2) = l_0 + \Delta l, \tag{9.232}$$

$$\left.\frac{df}{d\theta}\right|_{\theta=\theta_2} = 0. \tag{9.233}$$

Substituting Eq. (9.230) into Eq. (9.229), the dynamics is expressed as follows:

$$\dot{\theta} = \frac{1}{m(a_2\theta^2 + a_1\theta + a_0)^2 + J}\sqrt{2mgA(\theta)}, \tag{9.234}$$

where

$$A(\theta) = \frac{A_8}{8}\theta^8 + \frac{A_7}{7}\theta^7 + \frac{A_6}{6}\theta^6 + \frac{A_5}{5}\theta^5 + \frac{A_4}{4}\theta^4 + \frac{A_3}{3}\theta^3 + \frac{A_2}{2}\theta^2 + C,$$

$$\begin{aligned}
A_8 &= m{a_2}^3,\ A_7 = 3m{a_2}^2a_1,\\
A_6 &= m(3a_2a_0 + 3a_2{a_1}^2),\\
A_5 &= m(6a_2a_1a_0 + {a_1}^3) + Ja_3,\\
A_4 &= m(3a_2{a_0}^2 + 3{a_1}^2a_0) + Ja_2,\\
A_3 &= 3ma_1{a_0}^2 + Ja_1,\\
A_2 &= m{a_0}^3 + Ja_0,
\end{aligned}$$

C is integral constant. Considering Taylor developing around $\theta = 0$, the derived dynamics is expressed as two-dimensional polynomial equation as follows:

$$\begin{aligned}
\dot{\theta} &= F(0) + \left.\frac{dF}{d\theta}\right|_{\theta=0}\theta + \frac{1}{2!}\left.\frac{d^2F}{d\theta^2}\right|_{\theta=0}\theta^2\\
&:= F_{app}(\theta),
\end{aligned} \tag{9.235}$$

where

$$\begin{aligned}
F(0) &= \frac{1}{m{a_0}^2 + J}\sqrt{2mgC},\\
\left.\frac{dF}{d\theta}\right|_{\theta=0} &= -\frac{2ma_1a_0}{(m{a_0}^2+J)^2}\sqrt{2mgC},\\
\left.\frac{d^2F}{d\theta^2}\right|_{\theta=0} &= \frac{1}{(m{a_0}^2+J)^4}\Big(((2mgC)^{-\frac{1}{2}}mga_0(m{a_0}^2+J)^2\\
&\quad -2m(a_2a_0+{a_1}^2)\sqrt{2mgC})(m{a_0}^2+J)^2\\
&\quad +8m^2{a_1}^2{a_0}^2(m{a_0}^2+J)\sqrt{2mgC}\Big).
\end{aligned}$$

The exchange of angular velocity at a foot-contact is expressed as follows:

$$\omega_+ = \frac{j + ml_0(l_0 + \Delta l)\cos(\theta_0 + \theta_2)}{j + ml_0{}^2}\omega_-, \tag{9.236}$$

where ω_- is the angular velocity just before a foot-contact, ω_+ is the angular velocity just after a foot-contact. We can get ω_+ from Eq. (9.228) as follows:

$$\omega_+ = \sqrt{\frac{2mgl_0}{J + ml_0{}^2}(\cos\theta_1 - \cos\theta_0)}. \tag{9.237}$$

Foot pat continue based on the motion design using the variation of pendulum length Δl derived from Eqs. (9.244), (9.237), and $\omega_- = F_{app}(\theta_2)$.

9.6.3 Sagittal Motion Control

Lateral motion is designed on the basis on the passive joint. On the other hand, the motion of sagittal plan is designed as time dependently, because there is no rotation between the robot and contact-point in sagittal plane when the robot walk with the pace gait.

The motion of swing leg is modeled as an inverted pendulum as shown in Fig. 9.92. It is assumed that constant torque τ_k is input around axis of rotation. This constant torque compensates the energy loss from the collision of foot-contact. The dynamics of the inverted pendulum is given as follows:

$$ml^2\ddot{\theta} = mgl\sin\theta + \tau_k. \tag{9.238}$$

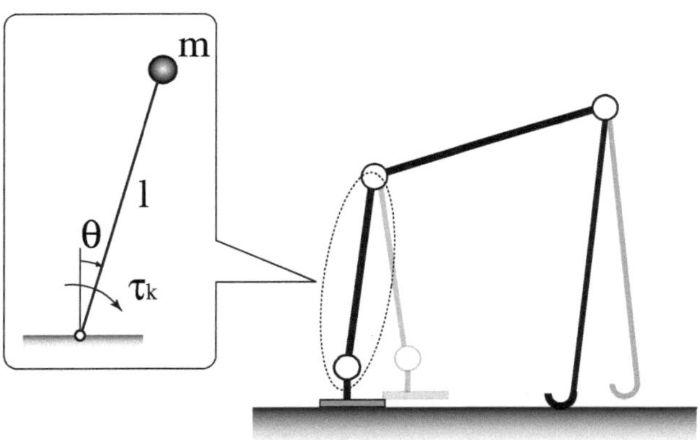

Fig. 9.88 Inverted pendulum model

The constant torque τ_k is derived from walking cycle. From Eq. (9.238), we can get following equation.

$$\theta(t) = Ae^{\omega t} + Be^{-\omega t} - \frac{\tau_k}{mgl}, \tag{9.239}$$

where, A and B are integral constants, $\omega = \sqrt{\frac{g}{l}}$.

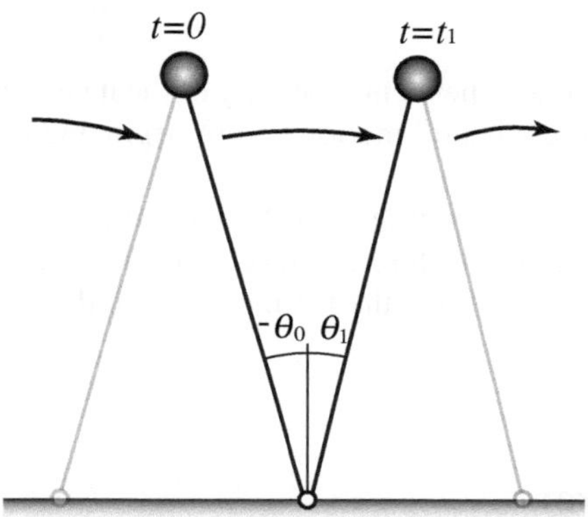

Fig. 9.89 Starting and ending point of the inverted pendulum

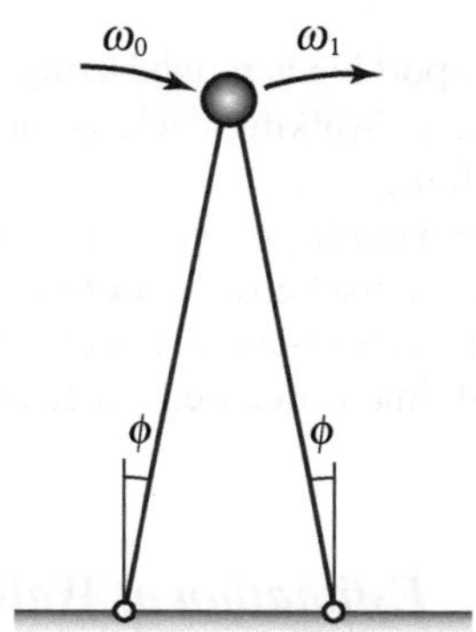

Fig. 9.90 Foot-contact model

We set $\theta(0)$ and $\theta(t_1)$ as θ_0 and θ_1 as shown in Fig. 9.89; then, the integral constants A and B are given as follows:

$$A = \frac{1}{e^{\omega t_1} - e^{-\omega t_1}} \left(\theta_1 - \left(\theta_0 + \frac{\tau_k}{mgl} \right) e^{-\omega t_1} + \frac{\tau_k}{mgl} \right), \tag{9.240}$$

$$B = \frac{1}{e^{\omega t_1} - e^{-\omega t_1}} \left(-\theta_1 + \left(\theta_0 + \frac{\tau_k}{mgl} \right) e^{\omega t_1} - \frac{\tau_k}{mgl} \right). \tag{9.241}$$

By differentiating Eq. (9.239),

$$\dot{\theta}(t) = A\omega e^{\omega t} + B\omega e^{-\omega t} - \frac{\tau_k}{mgl}. \tag{9.242}$$

We set $\dot{\theta}(0)$ and $\dot{\theta}(t_1)$ as ω_0 ω_1; then following equation is derived:

$$\frac{\omega_1}{\omega_0} = \frac{Ae^{\omega t_1} - Be^{-\omega t_1}}{A - B}. \tag{9.243}$$

On the condition of foot-contact shown in Fig. 9.90, the following equation is given:

$$\frac{\omega_0}{\omega_1} = \cos 2\phi. \tag{9.244}$$

τ_k is derived from Eqs. (9.240), (9.241), (9.243), and (9.244):

$$\tau_k = mgl\frac{\left(-2\cos 2\phi + \left(e^{\omega t_1} + e^{-\omega t_1}\right)\right)\theta_0 + \left(-2 + \cos 2\phi\left(e^{\omega t_1} + e^{-\omega t_1}\right)\right)\theta_1}{\left(1 + \cos 2\phi\right)\left(2 - \left(e^{\omega t_1} + e^{-\omega t_1}\right)\right)}. \tag{9.245}$$

The support leg is moved using the derived time dependent trajectory of the inverted pendulum. Walking cycle is measured from the lateral motion before beginning of the walking.

The swing leg moves symmetrically from the support leg. The horizontal velocity of the foot just before a foot-contact is set as zero. The horizontal velocity of the foot just after a foot-contact is also zero. After proper time, the leg move forward. The elapsed time is decided experimentally.

9.6.4 Estimation of Walking Energy

The walking cycle and stride length should be decided according to desired walking velocity, when the robot moves by quadruped walking. In this research, the walking cycle and stride length are decided in order to minimize the energy consumption. Thus, we need energy map according to the walking cycle and stride length. In this subsection, the walking energy of pace gait is estimated. The waking energy is calculated by dividing lateral motion and sagittal motion. The energy of sagittal motion is calculated support leg and swing leg respectively.

9.6.4.1 Lateral Plane

In lateral plane the robot is modeled as inverted pendulum shown in Fig. 9.91. The pendulum length is expressed as the function of joint angle,

$$l = f(\theta). \tag{9.246}$$

One-dimensional autonomous system is derived by use of PDAC,

$$\dot{\theta} = F(\theta) \tag{9.247}$$

By differentiating Eq. 9.246,

$$\dot{l} = \frac{\partial f(\theta)}{\partial \theta}\dot{\theta} = \frac{\partial f(\theta)}{\partial \theta}F(\theta). \tag{9.248}$$

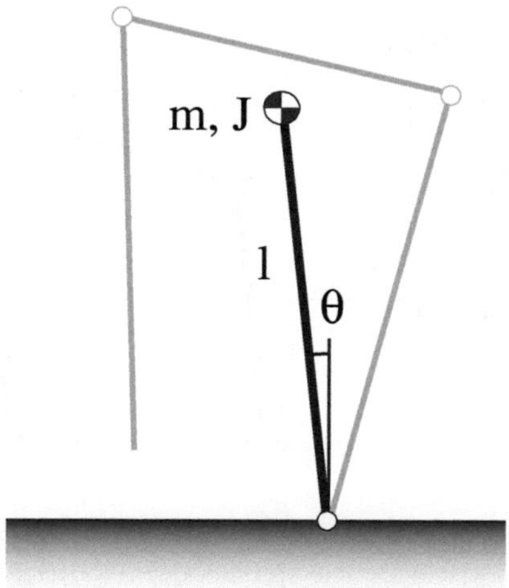

Fig. 9.91 Inverted pendulum model in Lateral plane

Second order differential of Eq. 9.246 is following:

$$\begin{aligned}\ddot{l} &= \frac{d}{dt}\left(\frac{\partial f(\theta)}{\partial \theta}F(\theta)\right)\\ &= \left(\frac{\partial^2 f(\theta)}{\partial \theta^2}F(\theta)+\frac{\partial f(\theta)}{\partial \theta}\frac{\partial F(\theta)}{\partial \theta}\right)F(\theta).\end{aligned} \tag{9.249}$$

Dynamic of the pendulum axis is follows:

$$N = m\ddot{l} + mg\cos\theta, \tag{9.250}$$

where N is the force of the pendulum axis. N is the function of θ from Eq. (9.249). Then, energy is given as follows:

$$\begin{aligned}E_{late} &= \int N\dot{l}dt\\ &= \int N\frac{\partial f(\theta)}{\partial \theta}\dot{\theta}dt\\ &= \int N\frac{\partial f(\theta)}{\partial \theta}d\theta.\end{aligned} \tag{9.251}$$

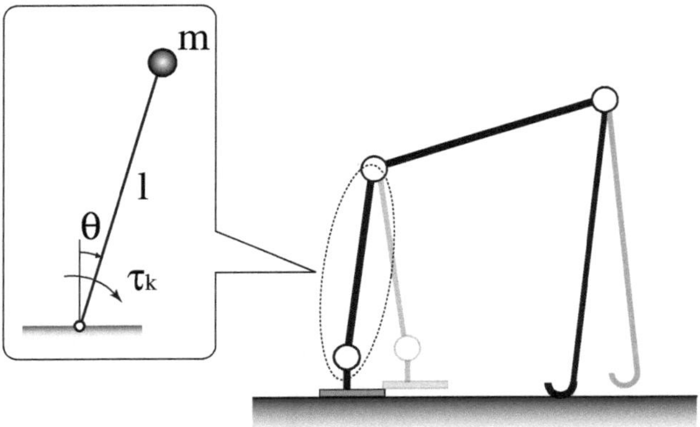

Fig. 9.92 Model of support leg

9.6.4.2 Sagittal Plane

Support leg

The support leg is modeled as an inverted pendulum. The energy is estimated on the condition that joint ankle output the constant torque τ_k. The dynamics is expressed as follows:

$$ml^2\ddot{\theta} = mgl\sin\theta + \tau_k. \tag{9.252}$$

Energy is given as follows:

$$\begin{aligned} E_{sagi}^{sp} &= \int \tau_k \dot{\theta} dt. \\ &= \int \tau_k d\theta \end{aligned} \tag{9.253}$$

Swing leg

The swing leg motion is same as the support leg. The trajectory of swing leg is given from Eq. (9.252),

$$\theta = A\exp^{\sqrt{\frac{g}{l}}t} + B\exp^{-\sqrt{\frac{g}{l}}t} - \frac{\tau_k}{mgl}. \tag{9.254}$$

By differentiating this equation,

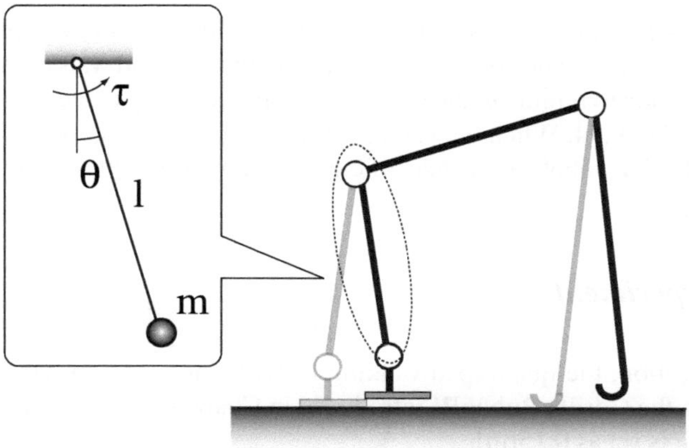

Fig. 9.93 Model of swing leg

$$\dot{\theta} = A\sqrt{\frac{g}{l}}\exp^{\sqrt{\frac{g}{l}}t} - B\sqrt{\frac{g}{l}}\exp^{-\sqrt{\frac{g}{l}}t} \quad (9.255)$$

$$\ddot{\theta} = A\frac{g}{l}\exp^{\sqrt{\frac{g}{l}}t} + B\frac{g}{l}\exp^{-\sqrt{\frac{g}{l}}t} \quad (9.256)$$

Torque is given by the dynamics of the simple pendulum as follows:

$$\tau = ml^2\ddot{\theta} + mgl\sin\theta. \quad (9.257)$$

The energy of the swing leg is given as follows:

$$\begin{aligned} E^{sw}_{sagi} &= \int \tau\dot{\theta}dt \\ &= \int (ml^2\ddot{\theta} + mgl\sin\theta)\dot{\theta}dt \\ &= \int \left(ml^2\left(A\frac{g}{l}\exp^{\sqrt{\frac{g}{l}}t} + B\frac{g}{l}\exp^{-\sqrt{\frac{g}{l}}t}\right) + mgl\sin\theta\right) \\ &\qquad \cdot \left(A\sqrt{\frac{g}{l}}\exp^{\sqrt{\frac{g}{l}}t} - B\sqrt{\frac{g}{l}}\exp^{-\sqrt{\frac{g}{l}}t}\right)dt. \end{aligned} \quad (9.258)$$

9.6.4.3 Energy Map

The energy for waking E is expressed as follows:

$$E = E_{late} + 2E^{sp}_{sagi} + 2E^{sw}_{sagi}. \quad (9.259)$$

Fig. 9.94 shows the 3-D map of the energy according to stride length and walking velocity calculated from above equation. In this map, the continuous line shows the minimizing points according to walking velocity; also dotted line shows the projected line on the plane composed of stride length axis and walking velocity axis. There is the stride to minimize the walking energy according to the walking velocity as shown in Fig. 9.94. When we chose optimum stride length, the waking cycle is also decided. Thus robot can realize efficient walk by use of these parameters.

9.6.5 Experiment

In this subsection, the quadruped walking experiment based on pace gate is conducted using the Gorilla Robot III introduced in Chapter 3. The distance from right leg to left leg is set as 0.15[m].

The robot makes foot pat in the initial position first 10 steps; then walking cycle is estimated. The temporal trajectory of the sagittal motion is made using this estimated walking cycle.

The desired walking velocity is set as 0.10[m/s]. The optimum stride length and walking cycle are given form the energy map. The stride length and walking are 0.30[m] and 3.0[s] respectively.

In order to be same as desired walking cycle, the walking cycle is adjusted by changing the amplitude of the inverted pendulum in the lateral plane. And also, the trajectory of the sagittal plane is designed so that the stride length is same as desired value.

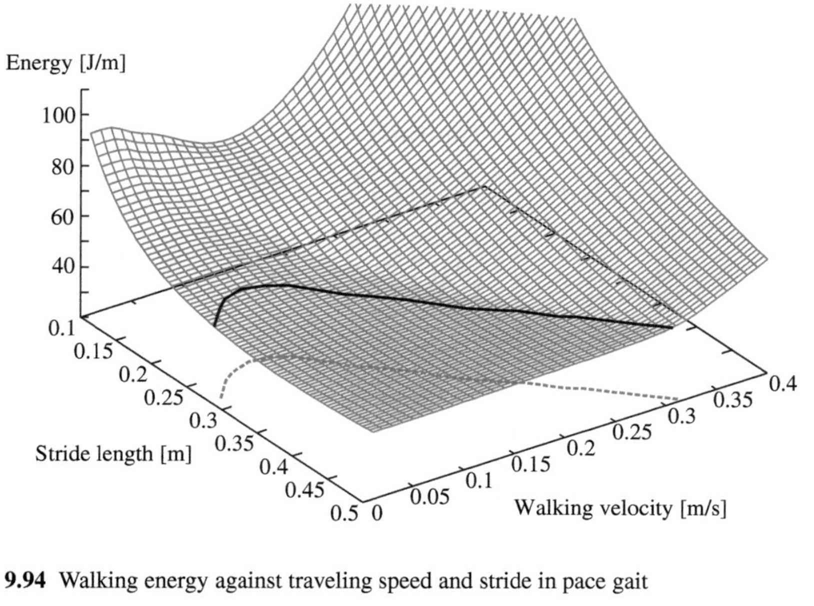

Fig. 9.94 Walking energy against traveling speed and stride in pace gait

Fig. 9.95 shows the joint angles of this experiment. The walking energy calculated from experimental data is 108.7[J/m]. The snapshots of the experiment is shown in Fig. 9.96.

9.6.6 Summary of Quadruped Walking Control

In this section, PDAC is applied to pace gate of a quadruped walking. The motion design is based on the lateral motion and sagittal motion. In the lateral motion, the robot is modeled as inverted pendulum and all joint is interlocked as the angle of passive joint; then the foot pat is designed. The sagittal motion is designed as the temporal trajectory based on the assumption that the motion can be controlled using pitch joint angle, because the closed link is composed in the sagittal plane. The

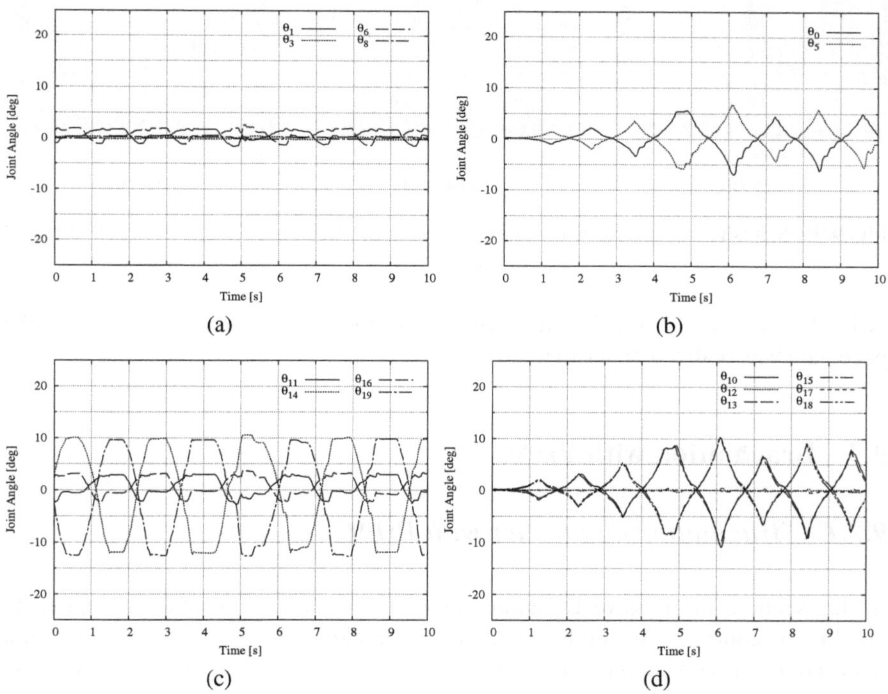

Fig. 9.95 Experimental results of quadruped walking in pace gait (desired walking velocity: 0.10[m/s]). (a) time vs. upper joint angle in the lateral plane, (b) time vs. upper joint angle in the sagittal plane, (c) time vs. lower joint angle in the lateral plane, (d) time vs. lower joint angle in the sagittal plane

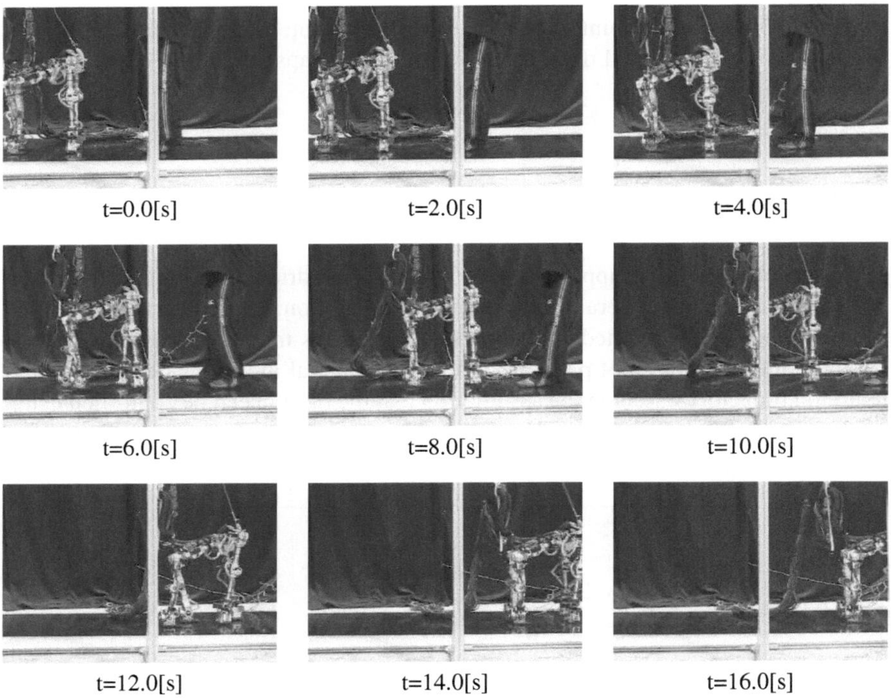

Fig. 9.96 Snapshots of quadruped walking in pace gait (desired walking velocity: 0.10[m/s])

swing leg is designed based on the dynamics of the inverted pendulum. Finally, the pace gate was realized experimentally.

9.7 Brachiation with PDAC

9.7.1 Brachiation Controller with PDAC

In this section, the Passive Dynamic Autonomous Control (PDAC) is applied to brachiation controls. The model of a Brachiation motion is shown in Fig. 9.97. First point-contact is considered. During Fig. 9.97 (a) and (b), robot consists closed link. Thus, cubital joint of forward arm is made passive in order to make robot to opened link. By doing so, contact point among ladder and hand of backward arm. During Fig. 9.97 (b) and (d), point-contact is naturally filled. Secondly, we consider the state in Virtual holonomic Constraint is considered. Virtual holonomic Constraint is applied to robot at someone's discretion. Here the designed method of brachiation controller with PDAC is explained. First, initial condition around contact-point is set in the phase around contact-point (Fig. 9.98-step 1). Secondly desired condition is set in this phase (Fig. 9.98-step 2). At the end, proper Virtual holonomic Constraint is derived in order to connect the initial condition (A) around contact-point to desired condition (d) through (c). Phase plane is drawn. And so, the problem is that proper Virtual holonomic Constraint is estimated by state of robot in condition (a):(θ_1^a, θ_3^a) and desired state of robot in condition (d):($\tilde{\theta_1^d}$). In breif, Eq. 9.260 is derived.

$$\eta = A\left(\theta_1^a, \theta_3^a, \tilde{\theta_1^d}\right) \tag{9.260}$$

For simplify of the controller design, the whole robot is modeled as one-link robot and a three-link robot in Fig. 9.97. The support-arm grasps the ladder and supports the whole body during brachiation. The swing-arm releases the backward ladder in the direction of locomotion and tries to catch the next target ladder. Although the elbow of the swing-arm will bend during the actual locomotion, the effect of dynamics is assumed to be negligible since the bending angle is sufficiently small. Since cubital joint of forward arm is made passive and back-arm does not have mass, the robot modeled one-link model. Each mass of the support-arm and the swing-arm is ignored since it is sufficiently smaller than that of torso. Both legs are actuated to remain stationary with respect to the torso; hence, the legs can be regarded as links. Since cubital joint of forward arm is made passive during condition (a)-(b) and each mass of the support-arm and the swing-arm is ignored, the robot is modeled one-link model. L is length of pendulum. ϕ is angle around wrist. m is center of mass of the robot. In condition (b), rotational canter of forward arm is changed to backward arm. Therefore after rotational canter of forward arm is changed to backward arm, the parameter is added ($\tilde{\cdot}$). Since each mass of the support-arm and the swing-arm is ignored during condition (b)-(c), the robot is modeled one-link model. Next, since the fixed angle between the torso and the horizontal line for gravity is kept during condition (c)-(d), the robot is modeled three-link model. l_1, l_2 and l_3 are length of the support-arm, the swing-arm and the torso respectively, and m is center of mass of the robot. θ_1 is the angle between the support-arm (link l_1) and the downward vertical line, θ_2 and θ_3 are the angles from link l_2 to l_1 and from l_3 to l_1 respectively. Upper subscript shows condition.

9.7.2 *Analysis of the Robot Dynamics*

The proposed brachiation controller has three Virtual holonomic constraints. During the brachiation, the Virtual holonomic constraint is changed by turns. Thus, we analyze by turns.

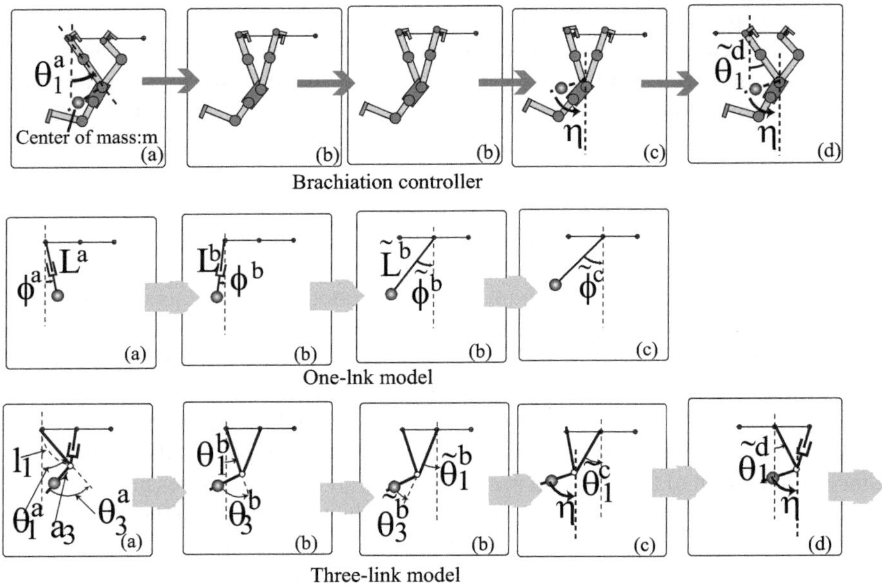

Fig. 9.97 Brachiation model.

9.7.2.1 Pendulum with Variable Lenght

During Fig. 9.97-(a) and (b) in one-link model, joint of forward arm is made passive. In doing so, it is possible to neglect the forward arm, so the robot is made Virtual 2-link: backward-arm and torso. Since torso move a little distance, the robot is modeled pendulum with variable length. Firstly, the first dynamic equation around the contact point O is derived as follows, where the angle around the contact point cannot be actuated; i.e. τ_1=0.

$$\frac{d}{dt}(M_1) - G_1 = 0, \tag{9.261}$$

$$M_1 = mL^2\dot{\phi}, \tag{9.262}$$

$$G_1 = mgL\sin\phi. \tag{9.263}$$

In this, the robot is applied following virtual holonomic constraint (Fig. 9.97 (a)-(b)):

$$L(\phi) = \left(\alpha\phi^2 + \beta\phi + \gamma\right)^{\frac{1}{3}} \tag{9.264}$$

Three coefficients (α, β, γ) is derived from following reservations.

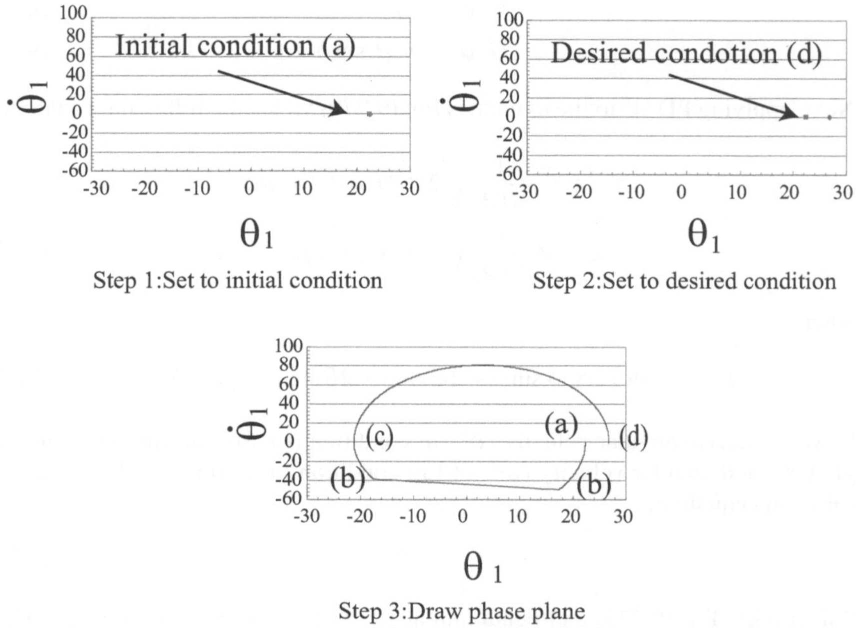

Fig. 9.98 Phase plane.

$$L(\phi^a) = L^a, \dot{L}(\phi^a) = 0 \tag{9.265}$$

$$L(\phi^b) = L^b, \dot{L}(\phi^b) = 0 \tag{9.266}$$

Therefore,

$$\alpha = \frac{(L^a)^3 - (L^b)^3}{(\phi^a - \phi^b)^2} \tag{9.267}$$

$$\beta = -2\frac{\phi^b\left((L^a)^3 - (L^b)^3\right)}{(\phi^a - \phi^b)^2} \tag{9.268}$$

$$\gamma = \frac{(L^a)^3 (\phi^b)^2 + (L^a)^3\left((\phi^b)^2 - 2\phi^a\phi^b\right)}{(\phi^a - \phi^b)^2} \tag{9.269}$$

Substituting Eq. (9.261) into Eq. (9.264) yields the following equation.

$$\frac{d}{dt}\left(M(\phi)\dot{\phi}\right) - G(\phi) = 0, \tag{9.270}$$

where

$$M(\phi) = mL^2 \tag{9.271}$$
$$G(\phi) = mgL\sin\phi \tag{9.272}$$

Next, applying PDAC to the dynamics Eq. (9.270) yields the following equation.

$$\begin{aligned}\dot{\phi} &= \frac{1}{M(\theta)}\sqrt{2\int M(\phi)G(\phi)\,d\phi} \\ &= \frac{1}{M(\phi)}\sqrt{2(D(\phi)+C)},\end{aligned} \tag{9.273}$$

where

$$D(\phi) = mg\,(2\alpha\,(\phi\sin\phi + 2\cos\phi) + 2\beta\sin\phi - \gamma\cos\phi) \tag{9.274}$$

Next, we derive angular velocity ($\dot{\phi^b}$) in condition (b). Substituting pendulum angle (ϕ^a) and angular velocity ($\dot{\phi^a} = 0$) in condition (a) into Eq. (9.273) yields the following equation.

$$C = -D(\phi^a) \tag{9.275}$$

Substituting Eq. (9.275) and pendulum angle ϕ^b in condition (b) into Eq. (9.273) yields the following equation.

$$\dot{\phi^b} = \frac{1}{M(\phi^b)}\sqrt{2(D(\phi^b) - D(\phi^a))} \tag{9.276}$$

Substituting Eqs. (9.267), (9.268), and (9.269) into Eq. (9.276) yields the following equation.

$$\dot{\phi^b} = B\left(L^a, L^b, \phi^a, \phi^b, m, g\right) \tag{9.277}$$

9.7.2.2 Changed Pendulum

In condition (b), rotational canter of forward arm is changed to backward arm (Fig. 9.97-(b)). In this time, angular velocity of pendulum is decrease as following equation.

$$\dot{\tilde{\phi}}^b = \dot{\phi}^b\cos\left(\phi^b - \tilde{\phi}^b\right) \tag{9.278}$$

9.7.2.3 Pendulum

Since the motor of shoulder is fixed during condition (b) and (c), the robot is considered to be a state of virtual simple pendulum (Fig. 9.97). The angle of pendulum in condition (c):$\tilde{\phi}^c$ is calculated.

$$\tilde{\phi}^c = E\left(m, g, \tilde{L}^c, \tilde{\phi}^b, \dot{\tilde{\phi}}^b\right) \tag{9.279}$$

9.7.2.4 Relations of 1Link Model and 3Link

We consider relations of one-link model and three-link during condition (a) and condition (c)(Fig. 9.97:one link model during condition (a) and (b), and three-link model). Detail of equation is left out but following equation is derived.

$$L^a = F\left(\theta_3^a, l_1, a_3\right) \tag{9.280}$$

$$L^b = H\left(\theta_3^b, l_1, a_3\right) \tag{9.281}$$

$$\begin{aligned}\tilde{L}^b &= \tilde{L}^c \\ &= I\left(\tilde{\theta}_3^b, \tilde{l}_1, a_3\right)\end{aligned} \tag{9.282}$$

$$\phi^a = J\left(\theta_1^a, \theta_3^a, l_1, a_3\right) \tag{9.283}$$

$$\phi^b = K\left(\theta_1^b, \theta_3^b, l_1, a_3\right) \tag{9.284}$$

$$\tilde{\phi}^b = N\left(\tilde{\theta}_1{}^b, \tilde{\theta}_3{}^b, \tilde{l}_1, a_3\right) \tag{9.285}$$

$$\tilde{\theta}_1{}^c = O\left(\tilde{\phi}_1{}^c, \tilde{\theta}_3{}^b\right) \tag{9.286}$$

$$\tilde{\theta}_3{}^b = P\left(\tilde{\theta}_3{}^b\right) \tag{9.287}$$

$$\tilde{\theta}_1{}^b = -\theta_1^b \tag{9.288}$$

$$\tilde{l}_1 = l_1 \tag{9.289}$$

Substituting Eq. (9.278) into Eqs. (9.280), (9.281), (9.283) and (9.284) yields the following equation.

$$\dot{\phi}^b = Q\left(m, g, l_1, a_3, \theta_1^a, \theta_1^b, \theta_3^a, \theta_2^b\right) \tag{9.290}$$

Substituting Eqs. (9.279), (9.284), (9.288), (9.289) and (9.290) into Eq. (9.286) yields the following equation.

$$\tilde{\theta}_1^c = R\left(\tilde{l}_1, a_3, m, g, \theta_1^a, \theta_1^b, \theta_3^a, \tilde{\theta}_3^b\right) \tag{9.291}$$

9.7.2.5 Estimated Proper Virtual Holonomic Constraint

First, dynamic equation as three-link model is derived during condition (c)-(d) of three-link model.

$$0 = M_1^{'}\ddot{\theta}_1 + M_2^{'}\ddot{\theta}_2 + h_1 - g_1 \tag{9.292}$$

$$M_1^{'} = m_3\left(l_1^2 + a_3^2 + 2l_1a_3\cos\theta_3\right) \tag{9.293}$$

$$M_2^{'} = m_3\left(a_3^2 + l_1a_2\cos\theta_2\right) \tag{9.294}$$

$$h_1^{'} = l_1a_3\sin\theta_3\left(2\dot{\theta}_1 + \dot{\theta}_3\right) \tag{9.295}$$

$$g_1^{'} = m_3g\left(l_1\sin\theta_1 + a_3\sin\left(\theta_1 + \theta_3\right)\right) \tag{9.296}$$

We introduce the following relation between passive joint and active joint based on the calculation of Eq. (9.297) and requirement for energy efficient brachiation. We leave out detail of reason that is energy efficient brachiation.

$$\eta = \tilde{\theta}_3^{c\to d} + \tilde{\theta}_1^{c\to d} \tag{9.297}$$

Substituting Eq. (9.297) into Eq. (9.292) yields the following equation.

$$\frac{d}{dt}\left(M^{'}\left(\tilde{\theta}_1\right)\dot{\tilde{\theta}}_1\right) - G^{'}\left(\tilde{\theta}_1\right) = 0, \tag{9.298}$$

where

$$M^{'}\left(\tilde{\theta}_1\right) = m\left(\tilde{l}_1^2 + \tilde{l}_1a_3\cos\left(-\tilde{\theta}_1 + \eta\right)\right) \tag{9.299}$$

$$G^{'}\left(\tilde{\theta}_1\right) = mg\left(\tilde{l}_1\sin\left(\tilde{\theta}_1\right) + a_3\sin\left(\eta\right)\right) \tag{9.300}$$

Next, applying PDAC to the dynamics Eq. (9.298) yields the following equation.

$$\begin{aligned}\dot{\tilde{\theta}}_1 &= \frac{1}{M^{'}\left(\tilde{\theta}_1\right)}\sqrt{2\int M^{'}\left(\tilde{\theta}_1\right)G^{'}\left(\tilde{\theta}_1\right)d\tilde{\theta}_1}\\ &= \frac{1}{M^{'}\left(\tilde{\theta}_1\right)}\sqrt{2D^{'}\left(\tilde{\theta}_1\right) + C^{'}},\end{aligned} \tag{9.301}$$

where

$$\begin{aligned}D^{'}\left(\tilde{\theta}_1\right) =& mg\Big(l\tilde{l}_1^3\cos\tilde{\theta}_1 + \tilde{l}_1^3\left(\frac{3}{2}\sin\eta\cdot\tilde{\theta}_1 - \frac{1}{4}\cos\eta\cos2\tilde{\theta}_1 - \frac{1}{4}\sin\eta\sin2\tilde{\theta}_1\right)\\ &+ \tilde{l}_1a_3^2(\sin\eta\cos\eta\sin\tilde{\theta}_1 - \sin^2\eta\cos\tilde{\theta}_1)\Big)\end{aligned} \tag{9.302}$$

Substituting condition (c): Eq. (9.291), and $\tilde{\theta}_1^c$ into Eq. (9.301) yields the following equation.

$$C^{'} = -D^{'}\left(\tilde{\theta}_1^c\right) \tag{9.303}$$

Substituting desired condition (d): $\tilde{\theta}_1^d$, and $\dot{\tilde{\theta}}_1^d = 0$, and Eq. (9.303) into Eq. (9.301) yields the following equation.

$$R\left(\tilde{l}_1, a_3, m, g, \tilde{\theta}_1^c, \tilde{\theta}_1^d, \eta\right) = \frac{1}{M'\left(\tilde{\theta}_1^d\right)} \sqrt{2D'\left(\tilde{\theta}_1^d\right) + C'}$$
$$= 0 \quad (9.304)$$

Substituting Eqs. (9.291) and (9.297) into Eq. (9.304) yields the following equation.

$$S\left(\tilde{l}_1, a_3, m, g, \theta_1^a, \theta_1^b, \theta_3^a, \tilde{\theta}_3^b, \tilde{\theta}_1^d\right) = 0 \quad (9.305)$$

Calculating Eq. (9.305) with Newton-Rapthon method,

$$\tilde{\theta}_3^b = T\left(\tilde{l}_1, a_3, m, g, \theta_1^a, \theta_1^b, \theta_3^a, \tilde{\theta}_1^d\right). \quad (9.306)$$

Substituting Eq. (9.291) into Eq. (9.306) yields the following equation.

$$\tilde{\theta}_1^c = U\left(\tilde{l}_1, a_3, m, g, \theta_1^a, \theta_1^b, \theta_3^a, \tilde{\theta}_1^d\right) \quad (9.307)$$

Substituting Eq. (9.297) into Eqs. (9.306) and (9.307) yields the following equation.

$$\eta = V\left(\tilde{l}_1, a_3, m, g, \theta_1^a, \theta_1^b, \theta_3^a, \tilde{\theta}_1^d\right) \quad (9.308)$$
$$= A\left(\theta_1^a, \theta_3^a, \tilde{\theta}_1^d\right) \quad (9.309)$$

Since $\tilde{l}_1$, a_3, m, g, θ_1^b are decided by robot construction, Eq. (9.309) is derived by Eq. (9.308).

9.7.3 Experiment

Hardware experiment was conducted with the Multi-locomotion robot to verify the proposed energy-saving swing-back motion control. The experimental setup is firstly described, and then the experimental results are discussed.

Table 9.6 Parameter Setting

$D_{bar}(D_{bar}^*)$	θ_1^{ini}	θ_1^{end}	θ_2^{ini}	θ_2^{end}	β
0.4	-17.37	22.37	214.74	156.97	-10.56

9.7.3.1 Continuous Brachiation in Regular Ladder

Experimental Setup

The experiment is conducted using Gorilla Robot III. The force sensor measures the reaction forces from contact bars in order to judge whether the robot successfully grasps the bar or not. The distance between the ladder bars D_{bar} is set constant with 0.4[m] interval, and their height are set in the same height of 2.7[m]. The mass of arms $m_{support}$ and m_{swing} is 7.0 [kg], and that of the torso m_{torso} is 15.0 [kg]. The length of the stretched arm $l_{support}$ and the torso l_{torso} is 0.64 [m] and 0.85 [m] respectively. The other parameters are listed in Table 9.6.

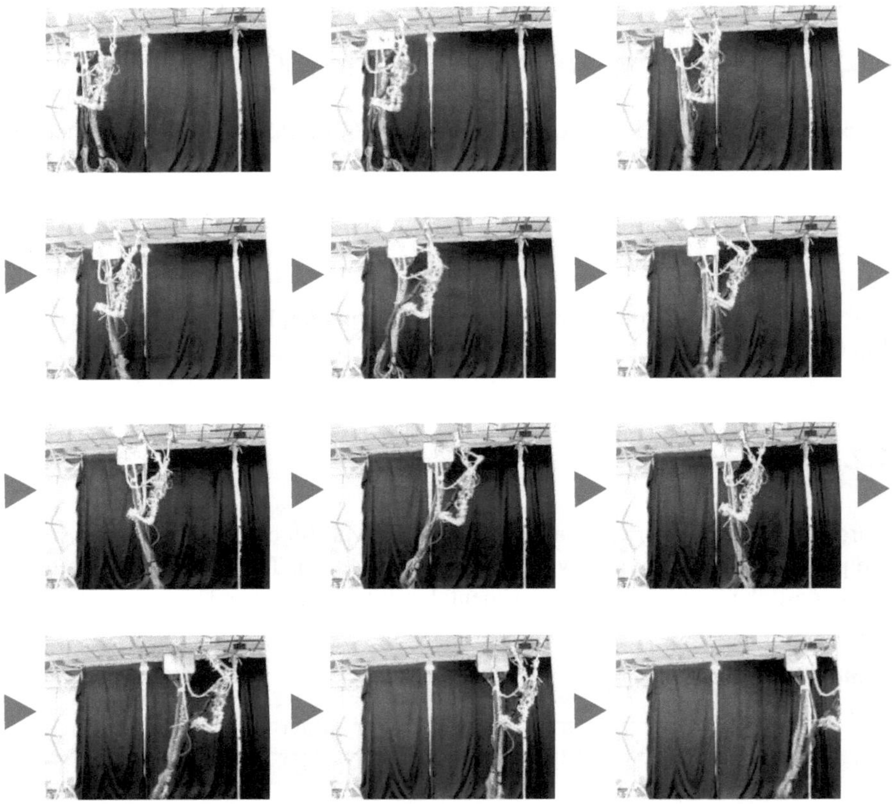

Fig. 9.99 Snapshots of brachiation behavior in regullar ladder.

Table 9.7 Energy Consumption and Moving Speed

	Previous control	w/o Swing back
Energy consumption	422.878[J]	289.188[J]
Movement speed	0.220[m/s]	0.267[m/s]

Experimental Result

The snapshots of brachiation behavior, the angle around wrist of support-arm, and the phase plane of the angle are shown Fig. 9.99, Fig. 9.100, and Fig. 9.101. Stable

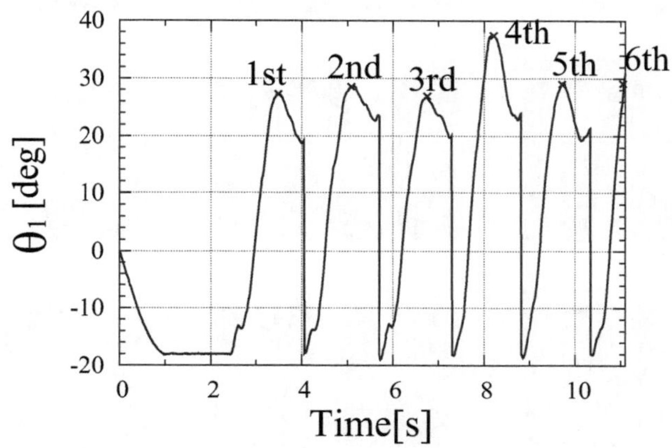

Fig. 9.100 The angle around wrist of support-arm in regular ladder.

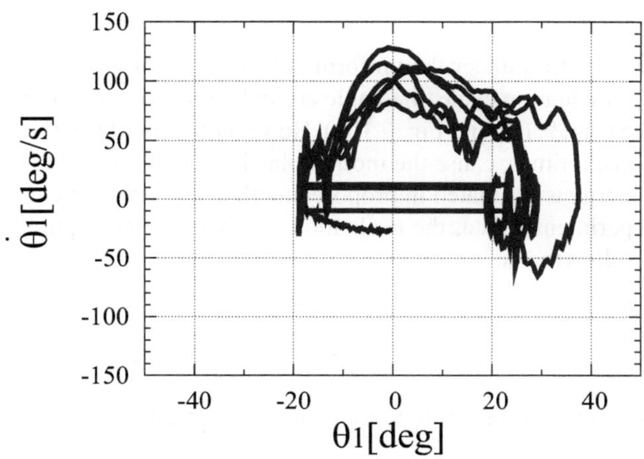

Fig. 9.101 Phase plane of the angle around wrist of support-arm.

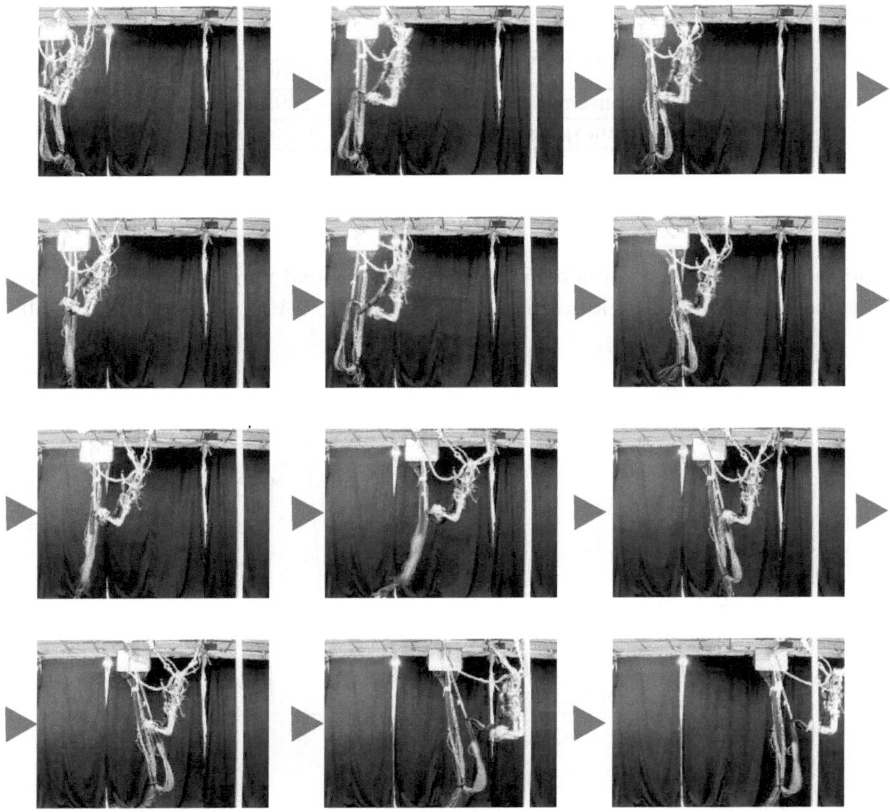

Fig. 9.102 Snapshots of brachiation behavior in irregullar ladder.

continuous brachiation under the uniform ladder setting was successfully done. In four times of brachiation cycle, the angle around wrist is increased due to press the robot by cable (Fig. 9.100 and Fig. 9.101). However the robot is returned normal desired angle in next times, cause the motion plan is set in gripped the ladder in every times. Next energy is estimated in proposed method and proposed method. According to the experimental data, the evaluation on energy consumption per second is given by the following definition;

$$\dot{E} = \sum_{i=0}^{24} |\tau_i \dot{\theta}_1|, \tag{9.310}$$

where i is joint-number. In addition, estimated total energy consumption is given by

$$\dot{E} = \sum_{i=0}^{24} \int |\tau_i \dot{\theta}_1| dt. \tag{9.311}$$

From result, it is confirmed that locomotion speed in the proposed method is improved and the maximum energy consumption per sec is also curbed compared with the previous method.

Table 9.7 shows the estimated energy consumption and the movement speed for single brachiation. It is confirmed that the total motion energy is saved over 30% and the movement speed is improved about 20% in the proposed control.

9.7.3.2 Continuous Brachiation in Irregular Ladder

Experimental Setup

The distance between the ladder bars D_{bar} is set 0.4, 0.4, 0.4,0.5, 0.5, and 0.6[m] interval.

Experimental Result

The snapshots of brachiation behavior and the angle around wrist of support-arm are Fig. 9.102 and Fig. 9.103. The more distance of ladder is, the more angle around the wrist is (Fig. 9.102). The successful braciation is confirmed by experiment.

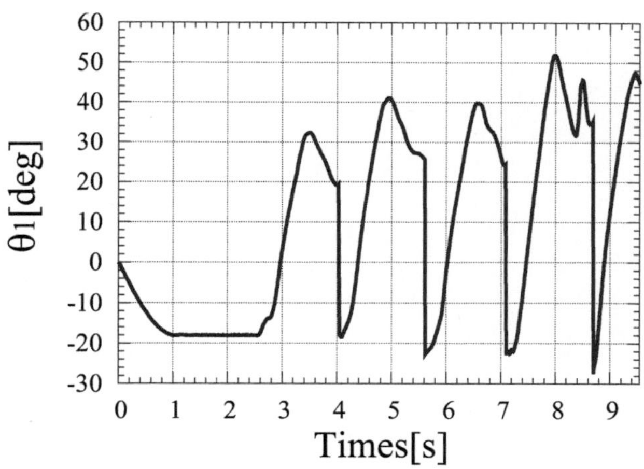

Fig. 9.103 The angle around wrists of support-arm in irregular ladder.

9.7.4 *Summary of Brachiation Control*

In this section, we also applied the PDAC to a brachiation motion and designed the brachiating controller on the regular and irregular ladder. We validated the proposed controller by realizing the continuous regular and irregular ladder brachiation with the Gorilla Robot III. Experimental results show that the PDAC-based method can improve the locomotion action as much as over 30% in every consumption.

9.8 Summary

In this chapter, the PDAC was applied to multiple motions. First application is the 3-D biped walking of 2-D dynamics such as lateral and sagittal motion. Second one is the heel-off biped walking that makes it possible to avoid the problem for the impact shock at a foot-contact. Third one is the 3-D biped walking based on 3-D dynamics without dividing. Fourth one is the quadruped walking, and final one is the brachiation. All motion controllers applied PDAC were validated through numerical simulations and experiments.

Chapter 10
Conclusion

10.1 Summary

A first, this book described robotic locomotion systems such as multi-legged locomotion, brachiation, hopping, and snake locomotion. Next, the novel concept of bio-inspired robotics named Multi-Locomotion Robot was introduced. Multi-locomotion robot is the novel bio-inspired robot which can perform in stand-alone several kinds of locomotion. Finally, control methods and demonstrations of the Multi-Locomotion Robot were described.

In Chapter 1, various robot locomotion were mentioned at first, and then the novel concept of bio-inspired robotics named Multi-Locomotion Robot was introduced. In order to realize a life-supporting robot in the future, it is necessary that robots can ambulate autonomously in the various environments. Multi-Locomotion Robot has the high ability to ambulate by achieving several kinds of locomotion in stand-alone. This diversity of locomotion is inspired by living creatures on earth. We mentioned about diversity of locomotion in animals and described biped walking, quadruped walking, and brachiation as main locomotion mode of Multi-Locomotion Robot.

In Chapter 2, basics with respect to robot locomotion were explained. Especially, the Passive Dynamic Autonomous Control (PDAC) is novel control method that we have proposed [51]. We presented two following basic concepts of PDAC: 1) point-contact 2) Virtual constraint. Point-contact denotes that a robot contacts the ground at a point. Virtual constraint was defined as a set of holonomic constraints on the robot's actuated DoF parameterized by the robot's unactuated DoF [87, 298].

In Chapter 3, three types of Gorilla Robot developed as the Multi-Locomotion Robot were described. Gorilla Robot was designed to be able to perform biped locomotion, quadruped locomotion, and brachiation etc. The hardware configuration of each Gorilla Robot and control system architecture were presented. In the subsequent chapter, Gorilla Robot was employed in some experiments.

In Chapter 4, several brachiation controllers were described. First, we applied the learning algorithm to achieve two types of brachiation: overhand brachiation and side-hand brachiation by Gorilla Robot II; and then we made a controller that can

T. Fukuda et al.: Multi-Locomotion Robotic Systems, STAR 81, pp. 291–295.
springerlink.com © Springer-Verlag Berlin Heidelberg 2012

periodically repeat the obtained side-hand brachiation and Gorilla Robot II was able to achieve a continuous locomotion that moved twice without suspending. Next, we presented a control method to realize smooth brachiation. The action of bending the elbow joint of the swinging arm was proposed in order to solve the gripping problem. The elbow-bending action makes it possible for the robot to decrease the impact forces and use the excess mechanical energy after the end of the locomotion phase. Then, experimental results showed that the robot can successfully achieve smooth, continuous brachiation. Finally, we applied the PDAC to a brachiation and designed the brachiating controller on the irregular ladder. The controller estimates the energy difference caused by symmetry breaking. We validated the proposed controller by realizing the continuous irregular ladder brachiation with the Gorilla Robot III.

In Chapter 5, the motion design of crawl gait was conducted and crawl gait using Gorilla Robot was realized at first. The motion is designed based on COG trajectory maintaining a static stability. The walking energy is analyzed and we gave the optimum walking cycle and stride length according to the desired walking velocity. Then, the optimum crawl gait was realized by use of the energy map. Next, the optimal ratio of limb length of a quadruped robot designed to reduce joint torque on a slope was derived by analysis. Numerical simulations analyzed the torque cost function C_s in each limb length and each slope angle for a robot walking on a slope and the optimal Ratio of Rear leg Length (RRL) was derived. The results indicated that the C_s increases as the slope angle increases if the length of the rear leg is longer than that of the front leg. Alternatively, the C_s decreases as the slope angle increases, if the length of the rear leg is shorter. In other words, the robot that has rear leg lengths that are shorter than its front leg lengths, such as Gorilla Robot III, requires reduced C_s as going forward upslope and going backwards downslope. Additionally, experimental results conducted with an actual robot showed that results of numerical simulation were sufficiently precise to be of future use.

In Chapter 6, we achieved the various types of vertical ladder climbing locomotion by using Multi-Locomotion Robot: static gait, pace gait with continuous velocity and trot gait with acceleration. The stability of vertical ladder climbing of MLR in static gait and pace gait is maintained with the posture control considering the momentum around axis of yawing. Even if the axis has inclination, the reference COG trajectory and acceleration is calculated to determine the motion which can maintain the stability of posture on the ladder. The control flow with the error recognition from output voltage and recovery motion was well operated and the MLR realized the continuous ladder climbing in the trot gait. Additional use of some other sensors would improve the stability of climbing furthermore. Also, we are attempting to realize a transition from the biped walking posture to the ladder climbing mode using an external sensor.

In Chapter 7, a transition motion was designed in which the multi-locomotion robot begins the brachiation motion from the ladder climbing posture. A load-allocation algorithm was proposed to cope with the unbalanced load distribution among the supporting limbs. The robot is successfully prevented from falling down from the ladder by setting a suitable torque at the key joint. The position errors in the initial posture are compensated by the key joint. The difference between the angle of the key joint and the expected angle is used to evaluate the position error. We confirmed that the transition motions with the load-allocation algorithm were robust and stable through the experiments with the multi-locomotion robot. In Chapter 8, following two propositions were made.

The first one was a method to recognize an unknown test course with the laser range finder and to plan for locomotion. The second one was the way of estimating the uncertainty as the risk of falling down. And the robot decides its gait based on the recognition and the risk. By experiments, we verified the proposed methods can be applied to various cases, and showed that stable locomotion with transition between biped and quadruped walk have been achieved.

In Chapter 9, the PDAC explained in Chapter 2 was applied to multiple motions. First application is the lateral motion of 3-D walking. The robot was modeled as an inverted pendulum, length of which is variable, in the lateral plane. Then, the pendulum length was designed to express the function of the pendulum angle (Virtual constraint) and lateral Converged dynamics was found. We proposed the stabilizing method and motion period controller for synchronization with sagittal motion by means of Converged dynamics. Validity of the stabilizing control and motion period control was confirmed by experiment. Second one is the sagittal motion of 3-D walking. We modeled the robot as three-link robot in the sagittal plane and derived Converged dynamics and PDAC Constant by use of this model. Then, discrete dynamics at foot-contact was found based on the assumption of perfectly inelastic collision between sole and the ground. We designed the stabilizing control by use of the constraint of constant step-length and proved its stability by the Liapunov Theory. In addition, the upper layer controller deciding step-length and walking period according to desired walking velocity was built. The experiment of 3-D dynamic walking was conducted and the validity of proposed control architecture was confirmed. Third application is the heel-off biped walking that makes it possible to avoid the problem for the impact shock at a foot-contact. The control method of heel-off walking by means of PDAC was proposed. We built the stabilizing control and demonstrated the convergence of walking dynamics. We compared the proposed heel-off walking with previous method and confirmed that heel-off motion reduced impact shock at foot-contact. Energy efficiency of heel-off walking was examined and the mechanical system to improve the energy efficiency was proposed. Forth application is the 3-D biped walking based on 3-D dynamics without dividing. We described robot dynamics by use of a 3D inverted pendulum model in the polar coordinate system and applied PDAC to the dynamics. Then, 3-D discrete dynamics with foot-contact was derived based on the assumption of perfectly inelastic collision between foot and the ground at foot-contact. We numerically presented that two conservative quantities named PDAC constant determine the velocity and walking direction of the biped walk. We also introduced the two controllers for two PDAC constants and confirmed a convergence of the two constants in numerical simulations. The validity of proposed control architecture was tested by numerical simulations and experiments. Fifth application is the 3-D biped walking adapted to rough terrain. We analyzed the environmental adaptability of PDAC-based three-dimensional biped walking. The robot dynamics was modeled as a 2-D autonomous system of a 3-D inverted pendulum by applying the PDAC concept; then the stability of the two-dimensional autonomous system with foot-contact was analyzed. As the result, the system has an asymptotically stable range due to the parameters. Within the stable range, we experimentally realized a 3-D biped dynamic walking on the irregular ground. The experimental result indicated that PDAC-based walking on the rough terrain is also energy efficient compare to paradigm. Sixth application is the quadruped walking. The motion design was based on the lateral motion and sagittal motion. In the lateral motion, the robot was modeled as inverted pendulum and all joint is interlocked as the angle of passive joint; then the foot pat is designed. The sagittal motion is designed as the temporal trajectory based on the assumption that the motion can be controlled using pitch joint angle, because the closed link was composed in the sagittal plane. The swing leg was designed based on the dynamics of the inverted pendulum. The pace gate was realized experimentally. Final application is the brachiation motion on on the regular and irregular ladder. We validated the proposed controller by achieving the continuous regular and irregular ladder brachiation with the Gorilla Robot III. Experimental results showed that the PDAC-based method can improve the locomotion action as much as over 30% in every consumption.

10.2 Future Perspective

We can point out industrial plants, works on rubble, welfare facilities, and so on as the future stages for the Multi-locomotion robot (Fig. 10.1 and 10.2). Especially, working of the Multi-locomotion robot is expected in the field that human is difficult to come at. At the current moment, we remain some issues such as to build up a systemized control system to achieve several kinds of locomotion as the Multi-locomotion robot, to realize an intelligent system to select suitable locomotion form autonomously according to the situation, and so on. We aim to put the Multi-locomotion robot to practical use and work on remaining issues.

Fig. 10.1 Future perspective

Fig. 10.2 Future perspective

References

1. Agrawal, S.K.: Motion control of a novel planar biped with nearly linear dynamics. IEEE Transactions on Mechatronics 11(2), 162–168 (2006)
2. Albiez, J., Luksch, T., Berns, K., Dillmann, R.: An activation-based behavior control architecture for walking machines. International Journal of Robotics Research 22(3-4), 203–211 (2003)
3. Albus, J.S.: A new approach to manipulator control: The cerebellar model articulation controller (cmac). Transactions of the ASME, Journal of Dynamic Systems, Measurement, and Control 97, 220–227 (1975)
4. Alexander, R.M.: Exploring Biomechanics: Animals in Motion. W. H. Freeman and Co. Publishers (1992)
5. Amit, D.J.: Modeling Brain Function. Cambridge University Press (1989)
6. Amit, D.J., Gutfreund, H., Sompolinsky, H.: Strong infinite number of patterns in a spin glass model of neural networks. Physical Review Letters 55, 1530–1533 (1985)
7. Andrew, A.G., Boahen, K.A., Pouliquen, P.O., Pavasovic, A., Jenkins, R.E., Strohbehn, K.: Current-mode subthreshold mos circuits for analog vlsi neural systems. IEEE Transactions on Neural Networks 2(2), 205–213 (1991)
8. Aoi, S., Tsuchiya, K.: Control of a planar underactuated biped on a complete walking cycle. IEEE Transactions on Robotics and Automation 49(5), 838–843 (2004)
9. Aoi, S., Tsuchiya, K.: Bifurcation and chaos of a simple walking model driven by a rhythmic signal. International Journal of Non-linear Mechanics 41, 438–446 (2006)
10. Aoyama, T., Hasegawa, Y., Fukuda, T., Sekiyama, K.: Control of lateral motion in bipedal walking based on pdac. International Journal of Mechatronics and Manufacturing Systems 3(3/4), 274–290 (2010)
11. Aoyama, T., Hasegawa, Y., Sekiyama, K., Fukuda, T.: Stabilizing and direction control of efficient 3-d biped walking based on pdac. IEEE/ASME Transactions on Mechatronics 14(6), 712–718 (2009)
12. Aoyama, T., Sekiyama, K., Hasegawa, Y., Fukuda, T.: Optimal limb length ratio of quadruped robot minimising joint torque on slopes. Applied Bionics and Biomechanics 6(3-4), 259–268 (2009)
13. Aoyama, T., Sekiyama, K., Hasegawa, Y., Fukuda, T.: Bipedal Walking Control based on the Assumption of the Point-contact: Sagittal Motion Control and Stabilization. In: Climbing and Walking Robots, pp. 185–200. INTECH (2010)
14. Aoyama, T., Sekiyama, K., Hasegawa, Y., Fukuda, T.: Pdac-based 3-d biped walking adapted to rough terrain environment. Journal of Robotics and Mechatronics 24(1), 37–46 (2012)

15. Asano, F., Luo, Z.W., Yamakita, M.: Virtual gravity and coupling control for robotic gait synthesis. IEEE Transactions on Systems, Man, and Cybernetics 31(6), 737–745 (2001)
16. Asano, F., Luo, Z.W., Yamakita, M.: Biped gait generation and control based on a unified property of passive dynamic walking. IEEE Transactions on Robotics and Automation 21(4), 754–762 (2005)
17. Asano, F., Yamakita, M., Furuta, K.: Virtual passive dynamic walking and energy-based control laws. In: The Proceeding of International Conference on Intelligent Robots and Systems, pp. 1149–1154 (2000)
18. Asano, F., Yamakita, M., Kamamichi, N., Luo, Z.W.: A novel gait generation for biped walking robots based on mechanical energy constraint. IEEE Transactions on Robotics and Automation 20(3), 565–573 (2004)
19. Aström, K.J., Borisson, U., Ljung, L., Wittenmark, B.: Theory and application of self-tuning regulators. Automatica 13, 457–476 (1977)
20. Azevedo, C., Poignet, P., Espiau, B.: Artificial locomotion control: from human to robots. Robotics and Autonomous Systems 47, 203–223 (2004)
21. Baae, M., Rutherford, D.A.: Fuzzy relations in a control setting. Kybernetes 7, 185–188 (1978)
22. Baba, N.: A new approach for finding the global minimum of error function of neural networks. Neural Networks 2, 367–373 (1989)
23. Barriére, G., Leblond, H., Provencher, J., Rossignol, S.: Prominent role of the spinal central pattern generator in the recovery of locomotion after partial spinal cord injuries. The Journal of Neuroscience 28(15), 3976–3987 (2008)
24. Barto, A.G., Sutton, R.S., Anderson, C.W.: Neuro-like adaptive elements that can solve difficult learning control problems. IEEE Transactions on Systems, Man, and Cybernetics SMC-13, 834–846 (1983)
25. Benbrahim, H., Franklin, J.A.: Biped dynamic walking using reinforcement learning. Robotics and Autonomous Systems 22, 283–302 (1997)
26. Bertram, J.E.A., Chang, Y.H.: Mechanical energy oscillations of two brachiation gaits: Measurement and simulation. American Journal of Physical Anthropology 115, 319–326 (2001)
27. Bertram, J.E.A., Ruina, A., Cannon, C.E., Chang, Y.H., Coleman, M.J.: A point-mass model of gibbon locomotion. Journal of Experimental Biology 202, 2609–2617 (1999)
28. Bessonnet, G., Chesse, S., Sardain, P.: Optimal gait synthesis of a seven-link planar biped. International Journal of Robotics Research 23(10-11), 1059–1073 (2004)
29. Bessonnet, G., Seguin, P., Sardain, P.: A parametric optimization approach to walking pattern synthesis. International Journal of Robotics Research 24(7), 523–536 (2005)
30. Carpenter, C.R.: Naturalistic Behavior of Nonhuman Primates. The Pennsylvania State University Press (1964)
31. Carpenter, G.A., Grossberg, S. (eds.): Pattern Recognition by Self-Organizing Neural Network. MIT Press (1991)
32. Cham, J.G., Bailey, S.A., Clark, J.E., Full, R.J., Cutkosky, M.R.: Fast and robust hexapedal robots via shape deposition manufacturing. International Journal of Robotics Research 21(10-11), 869–882 (2002)
33. Cham, J.G., Karpick, J.K., Cutkosky, M.R.: Stride period adaptation of a biomimetic running hexapod. International Journal of Robotics Research 23(2), 141–153 (2004)
34. Chang, T., Hurmuzlu, Y.: Sliding control without reaching phase and its application to bipedal locomotion. ASME Journal of Dynamic Systems, Measurement, and Control 115(3), 447–455 (1993)

35. Chang, Y.H., Bertram, J.E.A., Lee, E.V.: External forces and torques generated by the brachiating white-handed gibbon (hylobates lar). American Journal of Physical Anthropology 113, 201–216 (2000)
36. Channon, P.H., Hopkins, S.H., Pham, D.T.: A variational aproach to the optimization of gait for a bipedal robot. Journal of Mechanical Engineering Science(Part C) 210, 177–186 (1996)
37. Chemori, A., Alamir, M.: Multi-step limit cycle generation for rabbit's walking based on a nonlinear low dimensional predictive control scheme. Mechatronics 16, 259–277 (2006)
38. Chemori, A., Loria, A.: Control of a planar underactuated biped on a complete walking cycle. IEEE Transactions on Automatic Control 49(5), 838–843 (2004)
39. Chen, W., Low, K.H., Yeo, S.H.: Adaptive gait planning for multi-legged robots with an adjustment of center-of-gravity. Robotica 17, 391–403 (1999)
40. Cherouvim, N., Papadopoulos, E.: Energy saving passive-dynamic gait for a one-legged hopping robot. Robotica 24, 491–498 (2006)
41. Chevallereau, C.: Time-scaling control for an underactuated biped robot. IEEE Transactions on Robotics Automation 19(2), 362–368 (2003)
42. Chevallereau, C., Formalsky, A., Djoudi, D.: Tracking a joint path for the walk of an underactuated biped. Robotica 22, 15–28 (2004)
43. Chevallereau, C., Westervelt, E.R., Grizzle, J.W.: Asymptotically stable running for a five-link, four-actuator, planar bipedal robot. The International Journal of Robotics Research 24(6), 431–464 (2005)
44. Coleman, M.J., Ruina, A.: An uncontrolled toy that can walk but cannot stand still. Physical Review Letters 80(16), 3658–3661 (1998)
45. Collins, S., Ruina, A., Tedrake, R., Wisse, M.: Efficient bipedal robots based on passive-dynamic walkers. Science 307, 1082–1085 (2005)
46. Collins, S.H., Ruina, A.: A bipedal walking robot with efficient and human-like gait. In: The Proceeding of International Conference on Robotics and Automation, pp. 1983–1988 (2005)
47. Collins, S.H., Wisse, M., Ruina, A.: A three-dimensional passive-dynamic walking robot with two legs and knees. The International Journal of Robotics Research 20(7), 607–615 (2001)
48. Cotsaftis, M., Vibet, C.: Control law decoupling for 2-d biped walking system. IEEE Engineering in Medicine and Biology Magazine 3, 41–45 (1989)
49. Cotsaftis, M., Vibet, C.: Decoupled control for 2d n-link biped walking system. Robotics and Autonomous Systems 5, 97–107 (1989)
50. Dawson, D.M., Qu, Z., Lewis, F.L., Dorsey, J.F.: Robust control for the tracking of robot motion. International Journal of Control 52(3), 581–595 (1990)
51. Doi, M., Hasegawa, Y., Fukuda, T.: Passive dynamic autonomous control of bipedal walking. In: Proceedings of IEEE-RAS International Conference on Humanoid Robots, pp. 811–829 (2004)
52. Doi, M., Hasegawa, Y., Fukuda, T.: Control of smooth bipedwalking by means of heel-off motion. Journal of Robotics and Mechatronics 19(3), 353–360 (2007)
53. Donelan, J.M., Kram, R., Kuo, A.D.: Simultaneous positive and negative external mechanical work in human walking. Journal of Biomechanics 35, 117–124 (2002)
54. Dubowsky, S., DesForges, D.T.: The application of model referenced adaptive control to robotic manipulators. Transaction of ASME, Journal of Dynamic Systems, Measurement and Control 102, 193–200 (1979)
55. Ehara, Y., Yamamoto, S.: Analysis of Gait and Gait Initiation. Ishiyaku Publishers, Inc. (2002) (in Japanese)

56. Eimerl, S., DeVore, I.: The Primates. TIME-LIFE BOOKS (1966)
57. Filipovic, M., Potkonjak, V., Vukobratovic, M.: Humanoid robotic system with and without elasticity elements walking on an immobile/mobile platform. Journal of Intelligent and Robotic Systems 48(2), 157–186 (2007)
58. Fleagle, J.: Dynamics of a brachiating siamang [Hylobates (Symphalangus) syndactylus]. Nature 248, 259–260 (1974)
59. Formal'sky, A., Chevallereau, C., Perrin, B.: On ballistic walking locomotion of a quadruped. International Journal of Robotics Research 19(8), 743–761 (2000)
60. Francois, C., Samson, C.: A new approach to the control of the planar one-legged hopper. International Journal of Robotics Research 17(11), 1150–1166 (1998)
61. Franklin, J.A.: Reinforcement of robot motor skills through reinforcement learning. In: Proceedings of the IEEE Conference on Decision and Control, pp. 1096–1101 (1988)
62. Fukuda, T., Shibata, T., et al.: Neural servo controller; adaptation and learning. In: Proceedings of the IEEE International Workshop on Advanced Motion Control, pp. 107–115 (1990)
63. Fukuda, T., Aoyama, T., Hasegawa, Y., Sekiyama, K., Kojima, S.: PDAC-Based Brachiating Control of the Multi-Locomotion Robot. In: Fodor, J., Kacprzyk, J. (eds.) Aspects of Soft Computing, Intelligent Robotics and Control. SCI, vol. 241, pp. 101–121. Springer, Heidelberg (2009)
64. Fukuda, T., Hosokai, H., Kondo, Y.: Brachiation type of mobile robot. In: Proceedings of the IEEE International Conference on Advanced Robotics, pp. 915–920 (1991)
65. Fukuda, T., Ishigami, H.: Recognition and counting method of mammalian cells on micro-carrier using image processing and neural network. In: Proceeding of the JAACT 1991, p. 84 (1991)
66. Fukuda, T., Kojima, S., Sekiyama, K., Hasegawa, Y.: Design method of brachiation controller based on virtual holonomic constraint. In: Proceedings of the IEEE/RSJ International Conference on Intelligent Robots and Systems, pp. 450–455 (2007)
67. Fukuda, T., Saito, F., Arai, F.: A study on the brachiation type of mobile robot(heuristic creation of driving input and control using cmac). In: Proceedings of the IEEE/RSJ International Workshop on Intelligent Robots and Systems, pp. 478–483 (1991)
68. Fukuda, T., Shibata, T.: Neural network application for robotic motion control. In: Proceedings of the IEEE International Workshop on Intelligent Motion Control (1990)
69. Fukuda, T., Shibata, T.: Research trends in neuromorphic control. Journal of Robotics and Mechatronics 2(4), 4–18 (1991)
70. Fukuda, T., Shibata, T., Tokita, M., Mitsuoka, T.: Neural network application for robotic motion control: Adaptation and learning. In: Proceedings of the International Conference on Neural Networks, pp. 447–451 (1989)
71. Fukuoka, Y., Kimura, H., Cohen, A.H.: Adaptive dynamic walking of a quadruped robot on irregular terrain based on biological concepts. The International Journal of Robotics Research 22(3-4), 187–202 (2003)
72. Fukushima, K., Miyake, S., Ito, T.: Neocognitron: a neural network model for a mechanism of visual pattern recognition. IEEE Transactions on Systems, Man, and Cybernetics SMC-13, 826–834 (1983)
73. Funahashi, K.: On the approximate realization of continuous mapping by neural networks. Neural Networks 2-3, 183–192 (1989)
74. Furusho, J.: A control study of dynamical locomotion robot—a low order model and a hierarchical control strategy. Journal of the Robotics Society of Japan 1(3), 182–190 (1983)
75. Furusho, J., Sano, A.: Sensor-based control of a nine-link biped. The International Journal of Robotics Research 9(2), 83–98 (1990)

76. Garcia, M., Chatterjee, A., Ruina, A.: Efficiency, speed, and scaling of passive dynamic walking. Dynamics and Stability of Systems 15, 75–100 (2000)
77. Garcia, M., Chatterjee, A., Ruina, A., Coleman, M.J.: The simplest walking model: Stability, complexity, and scaling. ASME, Journal of Biomechanical Engineering, 281–288 (1998)
78. Geng, T., Porr, B., Worgotter, F.: Fast biped walking with a sensor-driven neuronal controller and real-time online learning. The International Journal of Robotics Research 25(3), 243–259 (2006)
79. Ghigliazza, R.M., Altendorfer, R., Holmes, P., Koditschek, D.E.: A simply stabilized running model. SIAM Journal of Applied Dynamical Systems 2(2), 187–218 (2003)
80. Goddard, R., Zheng, Y.F., Hemami, H.: Control of the heel-off to toe-off motion of a dynamic biped gait. IEEE Transactions on Systems, Man, and Cybernetics 22(1), 92–102 (1992)
81. Goldberg, D.E.: Genetic Algorithms in Search, Optimization, and Machine Learning. Addison-Wesley (1989)
82. Gorce, P., Guihard, M.: On dynamic control of pneumatic bipeds. Journal of Robotic Systems 15(7), 421–433 (1998)
83. Goswami, A.: Postural stability of bipedal robots and the foot-rotation indicator (fri) point. International Journal of Robotics Research 18(6), 523–533 (1999)
84. Goswami, A., Espiau, B., Keramane, A.: Limit cycles in a passive compass gait biped and passivity-mimicking control laws. Autonomous Robots, 274–286 (1997)
85. Graham, B.T.: The intrinsic factors in the act of progression in the mammal. Proceedings of the Royal Society of London. Series B, Containing Papers of a Biological Character 84, 308–319 (1911)
86. Grishin, A.A., Formal'sky, A.M., Lensky, A.V., Zhitomirsky, S.V.: Dynamic walking of a vehicle with two telescopic legs controlled by two drives. The International Journal of Robotics Research 13(2), 137–147 (1994)
87. Grizzle, J.W., Abba, G., Plestan, F.: Asymptotically stable walking for biped robots: Analysis via systems with impulse effects. IEEE Transactions on Automatic Control 46(1), 51–64 (2001)
88. Grossberg, S.: Adaptive pattern classification and universal reaching: Ii, feedback, expection, olfaction, and illusions. Biological Cybernetics 23, 121–134 (1976)
89. Grossberg, S.: Adaptive pattern classification and universal recording: I. parallel development and coding neural feature detectors. Biological Cybernetics 5, 527–532 (1976)
90. Grossberg, S., Kuperstein, M.: Neural Dynamics of Adaptive Sensory motor control. Pergamon Press (1989)
91. Guez, A., Selinsky, J.: A trainable neuromorphic controller. Journal of Robotic Systems 5(4) (1988)
92. Gupta, M., Qi, J.: On fuzzy neuron models. In: Proceedings of the IJCNN 1991-Seattle International Joint Conference on Neural Networks, vol. 2, pp. 431–436 (1991)
93. Haruna, M., Ogino, M., Hosoda, K., Asada, M.: Yet another humanoid walking—passive dynamic walking with torso under simple control. In: The Proceeding of International Conference on Intelligent Robots and Systems, pp. 259–264 (2001)
94. Hasegawa, Y., Arakawa, T., Fukuda, T.: Trajectory generation for biped locomotion robot. Mechatronics 10(1-2), 67–89 (2000)
95. Hasegawa, Y., Arakawa, T., Fukuda, T.: Trajectory generation for biped locomotion robot. Mechatronics 10, 67–89 (2000)
96. Hasegawa, Y., Fukuda, T., Shimojima, K.: Self-scaling reinforcement learning for fuzzy logic controller-applications to motion control of two-link brachiation robot. IEEE Transactions on Industrial Electronics 46(6), 1123–1131 (1999)

97. Hasegawa, Y., Ito, Y., Fukuda, T.: Behavior coordination and its modification on brachiation-type mobile robot. In: Proceedings of the IEEE International Conference on Robotics and Automation, pp. 3984–3989 (2000)
98. Hasegawa, Y., Ito, Y., Fukuda, T.: A study of the brachiation type of mobile robot (7th report, behavior learning for hierarchical behavior-based controller). Transaction of JSME C-67, 3204–3211 (2001) (in Japanese)
99. Hashimoto, K., Sugahara, Y., Lim, H.O., Takanishi, A.: Biped landing pattern modification method and walking experiments in outdoor environment. Journal of Robotics and Mechatronics 20(5), 775–784 (2008)
100. Hass, J., Herrmann, J.M., Geisel, T.: Optimal mass distribution for passivity-based bipedal robots. International Journal of Robotics Research 25(11), 1087–1098 (2006)
101. Hatton, R.L., Choset, H.: Generating gaits for snake robots: annealed chain fitting and keyframe wave extraction. Autonomous Robots 28, 271–281 (2010)
102. Hauser, J., Chung, C.C.: Converse lyapunov functions for exponentially stable periodic orbits. Systems and Control Letters 23(1), 27–34 (1994)
103. Hebb, D.O.: The first stage of perception: growth of assembly. In: The Organization of Behavior, pp. 66–82. Wily, New York (1949)
104. Hemami, H., Wyman, B.F.: Modeling and control of constrained dynamic systems with application to biped locomotion in the frontal plane. IEEE Transactions on Automatic Control AC-24(4), 526–535 (1979)
105. Herr, H.M., McMahon, T.A.: A galloping horse model. International Journal of Robotics Research 20(1), 26–37 (2001)
106. Herr, H.M., McMahon, T.A.: A trotting horse model. International Journal of Robotics Research 19(6), 566–581 (2001)
107. Hewit, J.R., Burdess, J.S.: Fast dynamic decoupled control for robotics using active force control. Mechanism Machine Theory 16(5), 532–542 (1981)
108. Hinton, G.E., Sejnowski, T.J.: Learning and Relearning in Boltzmann Machine, ch. 7. MIT Press (1986)
109. Hinton, G.E., Sejnowski, T.J., Ackley, D.H.: Boltzmann machines: constraint satisfaction networks that learn. Tech. Rep. CMU-CS-84-119, Department of Computer Science, Carnegie-Mellon University (1984)
110. Hirai, K.: Current and future perspective of honda humanoid robot. In: Proceedings of the IEEE/RSJ International Conference on Intelligent Robots and Systems, pp. 500–508 (1997)
111. Hirai, K., Hirose, M., Haikawa, Y., Takenaka, T.: The development of honda humanoid robot. In: The Proceeding of International Conference on Robotics and Automation, pp. 1321–1326 (1998)
112. Hirose, S., Yamada, H.: Snake-like robots. IEEE Robotics and Automation Magazine 16(1), 88–98 (2009)
113. Hirose, S., Yokoi, K.: The standing posture transformation gait of the quadruped walking vehicle. Journal of the Robotics Society of Japan 4(4), 26–35 (1986)
114. Hirose, S., Yoneda, K.: Dynamic and static fusion control and continuous trajectory generation of quadruped walking vehicle. Journal of the Robotics Society of Japan 14(3), 881–886 (1992)
115. Hitomi, K., Shibata, T., Nakamura, Y., Ishii, S.: Reinforcement learning for quasi-passive dynamic walking of an unstable biped robot. Robotics and Autonomous Systems 54, 982–988 (2006)
116. HONDA (2011), `http://world.honda.com/news/2011/c111108All-new-ASIMO/index.html`

117. Hopfield, J.J.: Neural networks and physical systems with emergent collective computational abilities. Proceedings of the National Academy Sciences 79, 2554–2558 (1982)
118. Hopfield, J.J., Tank, D.W.: Neural computation of decisions in optimization problems. Biological Cybernetics 52, 141–152 (1986)
119. Hoyt, D.F., Taylor, C.R.: Gait and the energetics of locomotion in horses. Nature 292(16), 239–240 (1981)
120. Huang, Q., Nakamura, Y.: Sensory reflex control for humanoid walking. IEEE Transactions on Robotics 21(5), 977–984 (2005)
121. Huang, Q., Yokoi, K., Kajita, S., Kaneko, K., Arai, H., Koyachi, N., Tanie, K.: Planning walking patterns for a biped robot. IEEE Transactions on Robotics and Automation 17(3), 280–289 (2001)
122. Ichihashi, H.: Learning in hierarchical fuzzy models by conjugate gradient method using backpropagation errors. In: Proceeding, vol. 235-240 (1991)
123. Ijspeert, A., Nakanishi, J., Schaal, S.: Learning rhythmic movements by demonstration using nonlinear oscillators. In: The Proceedings of International Conference on Intelligent Robots and Systems, pp. 958–963 (2002)
124. Ishida, T., Kuroki, Y., Yamaguchi, J., Fujita, M., Doi, T.: Motion entertainment by a small humanoid robot based on open-r. In: Proceedings of the IEEE/RSJ International Conference on Intelligent Robots and Systems, pp. 1079–1086 (2001)
125. Vermeulen, J., Lefeber, D., Verrelst, B.: Control of foot placement, forward velocity and body orientation of a one-legged hopping robot. Robotica 21(1), 45–47 (2003)
126. Jalics, L., Hemami, H., Clymer, B.: A control strategy for adaptive bipedal locomotion. In: The Proceeding of International Conference on Robotics and Automation, pp. 563–569 (1996)
127. Jha, R.K., Singh, B., Pratihar, D.K.: On-line stable gait generation of a two-legged robot using a genetic-fuzzy system. Robotics and Autonomous Systems 53, 15–35 (2005)
128. Joacobs, A.R.: Intcreased tates of convergence through learning tate adaptation. Neural Networks 1, 295–307 (1988)
129. Jordan, M.I.: Supervised learning and system with excess degree of freedom. Tech. rep., COINS Technical Report 88-27 (1988)
130. Juang, J.: Fuzzy neural network approaches for robotic gait synthesis. IEEE Transactions on Systems, Man, and Cybernetics 30(4), 594–601 (2000)
131. Kajima, H., Doi, M., Hasegawa, Y., Fukuda, T.: A study on brachiation controller for a multi-locomotion robot—realization of smooth, continuous brachiation. Advanced Robotics 18(10), 1025–1038 (2004)
132. Kajima, H., Hasegawa, Y., Fukuda, T.: Learning algorithm for a brachiating robot. Applied Bionics and Biomechanics 1(1), 57–66 (2003)
133. Kajita, S., Kanehiro, F., Kaneko, K., Fujiwara, K., Harada, K., Yokoi, K., Hirukawa, H.: Biped walking pattern generation by using preview control of zero-moment point. In: The Proceeding of IEEE International Conference on Robotics and Automation, pp. 1620–1626 (2003)
134. Kajita, S., Kanehiro, F., Kaneko, K., Fujiwara, K., Yokoi, K., Hirukawa, H.: A realtime pattern generator for biped walking. In: The Proceeding of IEEE International Conference on Robotics and Automation, pp. 21–37 (2002)
135. Kajita, S., Kanehiro, F., Kaneko, K., Fujiwara, K., Yokoi, K., Hirukawa, H.: Biped walking pattern generation by a simple three-dimensional inverted pendulum model. Advanced Robotics 17(2), 131–147 (2003)
136. Kajita, S., Yamaura, T., Kobayashi, A.: Dynamic walking control of a biped robot along a potential conserving orbit. IEEE Transactions on Robotics Automation 8(4), 431–438 (1992)

137. Kaliyamoorthy, S., Quinn, R.D., Zill, S.N.: Force sensors in hexapod locomotion. International Journal of Robotics Research 24(7), 563–574 (2005)
138. Kaneko, K., Harada, K., Kanehiro, F., Miyamori, G., Akachi, K.: Humanoid Robot HRP-3. In: Proceedings of the IEEE/RSJ International Conference on Intelligent Robots and Systems, pp. 2471–2478 (2008)
139. Kaneko, K., Kanehiro, F., Kajita, S., Hirukawa, H., Kawasaki, T., Hirata, M., Akachi, K., Isozumi, T.: Humanoid Robot HRP-2. In: Proceedings of the IEEE International Conference on Robotics and Automation, pp. 1083–1090 (2004)
140. Kaneko, K., Kanehiro, F., Morisawa, M., Miura, K., Nakaoka, S., Kajita, S.: Cybernetic Human HRP-4C. In: Proceedings of the IEEE/RSJ International Conference on Humanoid Robots, pp. 7–14 (2009)
141. Kato, I., Ohteru, S., Kobayashi, H., Shirai, K., Uchiyama, A.: Information-power machine with senses and limbs. In: First CISM-IFToMM Symposium on Theory and Practice of Robots and Manipulators. Springer (1974)
142. Katoh, R., Mori, M.: Control method of biped locomotion giving asymptotic stability of trajectory. Automatica 20(4), 405–414 (1984)
143. Kawamura, S., Kawamura, T., Fujino, D., Miyazaki, F., Arimoto, S.: Realization of biped locomotion by motion pattern learning. Journal of the Robotics Society of Japan 3(3), 5–15 (1984)
144. Kawato, M., Furukawa, K., Suzuki, R.: A hierarchical neural network model for control and learning of voluntary movement. Biological Cybernetics 57, 169–185 (1987)
145. Keith, A.: On the chimpanzees and their relationship to the gorillas. In: Proceedings of the Zoological Society of London, pp. 296–312 (1899)
146. Kern, M., Woo, P.Y.: Implementation of a hexapod mobile robot with a fuzzy controller. Robotica 23, 681–688 (2005)
147. Kiguchi, K., Kusumoto, Y., Watanabe, K., Izumi, K., Fukuda, T.: Enegy-optimal gait analysis of quadruped robots. Artif. Life Robotics 6, 120–125 (2002)
148. Kimura, H., Kobayashi, S.: Reinforcement learning for continuous action using stochastic gradient ascent. Intelligent Autonomous Systems 5, 288–295 (1998)
149. Kimura, H., Shimoyama, I., Miura, H.: Dynamics in dynamic walk of the quadruped robot. Journal of the Robotics Society of Japan 6(5), 367–379 (1988)
150. Kirkpatrick, S., Gelatt Jr., C.D., Vecchi, M.P.: Optimization by simulated annealing. Science 220, 671–680 (1983)
151. Koditschek, D.E., Buhler, M.: Analysis of a simplified hopping robot. International Journal of Robotics Research 10(6), 587–605 (1991)
152. Kohonen, T.: Self-organized formation of topologically correct feature maps. Biological Cybernetics 43, 59–69 (1982)
153. Kohonen, T.: Self-Organization and Associative Memory. Springer (1984)
154. Koivo, A.J., Guo, T.: Adaptive linear controller for robotic manipulator. IEEE Transactions on Automatic Control AC-28, 162–171 (1983)
155. Kosko, B.: Adaptive bidirectional associative memories. Applied Optics 26, 4947–4960 (1987)
156. Kosko, B.: Fuzzy Associative Memories. In: Fuzzy Expert Systems. Addicon-Wesley (1987)
157. Kun, A.L., Miller, W.T.: Control of variable-speed gaits for a biped robot. IEEE Robotics and Automation Magazine 6(3), 19–29 (1999)
158. Kuo, A.D.: Stabilization of lateral motion in passive dynamic walking. The International Journal of Robotics Research 18(9), 917–930 (1999)
159. Kuo, A.D.: Energetics of actively powered locomotion using the simplest walking model. Journal of Biomechanical Engineering 124, 113–120 (2002)

160. Kurazume, R., Yoneda, K., Hirose, S.: 3d sway compensation trajectory for quadruped walking robot. Journal of the Robotics Society of Japan 19(5), 632–637 (2001)
161. Kwek, L.C., Wong, E.K., Loo, C.K., Rao, M.V.C.: Application of active force control and iterative learning in a 5-link biped robot. Journal of Intelligent and Robotic Systems 37, 143–162 (2003)
162. Larin, V.B.: A note on a walking machine model. International Applied Mechanics 23(4), 484–492 (2003)
163. Li, Q., Takanishi, A., Kato, I.: Learning control of compensative trunk motion for biped walking robot based on zmp stability criterion. Journal of the Robotics Society of Japan 11(4), 557–563 (1993)
164. Lin, B.S., Song, S.: Dynamic modeling, stability, and energy efficiency of a quadrupedal walking machine. Journal of Robotics Systems 18(11), 657–670 (2001)
165. van der Linde, R.Q.: Active leg compliance for passive walking. In: The Proceeding of International Conference on Robotics and Automation, pp. 2339–2344 (1998)
166. van der Linde, R.Q.: Passive bipedal walking with phasic muscle contraction. Biological Cybernetics 81(3), 227–237 (1999)
167. Lippmann, R.P.: An introduction to computing with neural nets. IEEE ASSP Magazine 4, 4–22 (1987)
168. Liu, Z., Li, C.: Fuzzy neural networks quadratic stabilization output feedback control for biped robots via h_∞ approach. IEEE Transactions on Systems, Man, and Cybernetics B33(1), 67–84 (2003)
169. Loffler, K., Gienger, M., Pfeiffer, F.: Sensors and control concept of walking 'johnnie'. International Journal of Robotics Research 22(3-4), 229–239 (2003)
170. Lu, Z., Aoyama, T., Sekiyama, K., Fukuda, T., Hasegawa, Y.: Vertical ladder climbing motion of pace gait with body motion control for a multi-locomotion robot. In: Proceedings of IEEE/SICE International Symposium on System Integration, pp. 48–53 (2009)
171. Lu, Z., Aoyama, T., Sekiyama, K., Hasegawa, Y., Fukuda, T.: Transition motion from ladder climbing to brachiation with optimal load-allocation control. Advanced Robotics 26(8-9) (to appear, 2012)
172. Lum, H.K., Zribi, M., Soh, Y.C.: Planning and control of a biped robot. International Journal of Engineering Science 37, 1319–1349 (1999)
173. Magdalena, L., Fuelin, F.M.: A fuzzy logic controller with learning through the evolution of its knowledge base. International Journal of Approximate Reasoning 16, 335–358 (1997)
174. Mailah, M.: Intelligent active force control of a rigid robot arm using neural network and iterative learning algorithms. Ph.D. thesis, University of Dundee, Dundee (1998)
175. Matsubara, T., Morimoto, J., Nakanishi, J., Sato, M., Doya, K.: Learning cpg-based biped locomotion with a policy gradient method. Robotics and Autonomous Systems 54, 911–920 (2006)
176. McClloch, W.S., Pitts, W.H.: A logical calculus of the ideas immanent in neural nets. Bulletin of Mathematical Biophysics 5, 115–123 (1943)
177. McGeer, T.: Passive dynamic walking. The International Journal of Robotics Research 9(2), 62–82 (1990)
178. McGeer, T.: Passive walking with knees. In: The Proceeding of International Conference on Robotics and Automation, pp. 1640–1645 (1990)
179. McGeer, T.: Dynamics and control of bipedal locomotion. Journal of Theoretical Biology 163, 277–314 (1993)
180. Miller, W.T.: Real-time neural network control of a biped walking robot. IEEE Control Systems Magazine 14(1), 41–48 (1994)

181. Minsky, M.: Perceptrons. MIT Press (1969)
182. Miossee, S., Aoustin, Y.: A simplified stability study for a biped walk with underactuated and overactuated phases. The International Journal of Robotics Research 24(7), 537–551 (2005)
183. Miura, H., Shimoyama, I.: Dynamic walking of a biped. The International Journal of Robotics Research 3(2), 60–74 (1984)
184. Mori, M., Hirose, S.: Development of active cord mechanism acm-r3 with agile 3d mobility. In: Proceedings of IEEE/RSJ International Conference on Intelligent Robots and Systems, pp. 1552–1557 (2001)
185. Mori, M., Hirose, S.: Three-dimensional serpentine motion and lateral rolling by active cord mechanism acm-r3. In: Proceedings of IEEE/RSJ International Conference on Intelligent Robots and Systems, pp. 829–834 (2002)
186. Morimoto, J., Nakanishi, J., Endo, G., Cheng, G.: Acquisition of a biped walking policy using an approximated poincare map. In: The Proceeding of IEEE-RAS/RSJ International Conference on Humanoid Robots, pp. 912–924 (2004)
187. Mu, X., Wu, Q.: Development of a complete dynamic model of a planar five-link biped and sliding mode control of its locomotion during the double support phase. International Journal of Control 77(8), 789–799 (2004)
188. Murray, A.F., Corso, D.D., Trassenko, L.: Pulse-stream vlsi neural networks mixing analog and digital techniques. IEEE Transactions on Neural Networks 2, 193–204 (1991)
189. Muybridge, E.: Animals in Motion. Dover Publications (1957)
190. Nakanishi, J., Fukuda, T.: A leaping maneuver for a brachiating robot. In: The Proceeding of IEEE International Conference on Robotics and Automation, pp. 2822–2827 (2000)
191. Nakanishi, J., Fukuda, T.: A leaping maneuver for a brachiating robot. In: Proceedings of the IEEE International Conference on Robotics and Automation, pp. 2822–2827 (2000)
192. Nakanishi, J., Fukuda, T., Koditschek, D.: Target dynamics controller on a two-link brachiating robot. In: Proceedings of the IEEE International Conference on Robotics and Automation, pp. 787–792 (1998)
193. Nakanishi, J., Fukuda, T., Koditschek, D.: A brachiation robot controller. IEEE Transactions on Robotics and Automation 16(2), 109–123 (2000)
194. Nakanishi, J., Fukuda, T., Koditschek, D.E.: Analytical approach to studies of two-link brachiating robot control. Journal of the Robotics Society of Japan 16(3), 361–368 (1998)
195. Nakanishi, J., Morimoto, J., Endo, G., Cheng, G., Schaal, S., Kawato, M.: A framework for learning biped locomotion with dynamical movement primitives. In: The Proceeding of IEEE-RAS/RSJ International Conference on Humanoid Robots, pp. 925–940 (2004)
196. Nakanishi, J., Morimoto, J., Endo, G., Cheng, G., Schaal, S., Kawato, M.: Learning from demonstration and adaptation of biped locomotion. Robotics and Autonomous Systems 47, 79–91 (2004)
197. Napier, J.: Brachiation and brachiators. In: Napier, J., Barnicot, N.A. (eds.) The Primates: the Proceedings of the Symposium held on 12th-14th April 1962 at the Offices of The Zoological Society of London, pp. 183–195. The Society, London (1963)
198. Napier, J.R., Napier, P.H.: A Handbook of Living Primates. Academic Press (1967)
199. Narazaki, H., Ralescu, A.L.: A sysnthesis method for multi layered neural network using fuzzy sets. In: Proceedings of the IJCNN 1991-Seattle International Joint Conference on Neural Networks, vol. 2, pp. 54–66 (1991)

200. Narendra, K., Parthasarathy, K.: Identification and control of dynamical systems using neural networks. IEEE Transactions on Neural Networks 1, 4–27 (1990)
201. Narendra, K.S.: Adaptive Control using Neural Networks. In: Neural Networks for Control, pp. 115–142. MIT Press (1990)
202. von Neumann, J.: The Computer and Brain. Yale University Press (1958)
203. Nguyen, P.H., Widrow, B.: Neural network for self-tuning control systems. IEEE Control Systems Magazine 10(3), 18–23 (1990)
204. Nielsen, R.H.: Neurocomputing. Addicon-Wesley Publishing Co. (1990)
205. Nishiwaki, K., Kagami, S.: Online walking control system for humanoids with short cycle pattern generation. The International Journal of Robotics Research 28(6), 729–742 (2009)
206. Nishiwaki, K., Kagami, S.: Strategies for adjusting the zmp reference trajectory for maintaining balance in humanoid walking. In: Proceedings of the IEEE International Conference on Robotics and Automation, pp. 4230–4236 (2010)
207. Nishiwaki, K., Kagami, S.: Online design of torso height trajectories for walking patterns that takes future kinematic limits into consideration. In: The Proceedings of IEEE International Conference on Robotics and Automation, pp. 2029–2034 (2011)
208. Nisiwaki, K., Kagami, S., Kuniyoshi, Y., Inaba, M., Inoue, H.: Online generation of humanoid walking motion based on a fast generation method of motion pattern that follows desired zmp. In: Proceedings of IEEE International Conference on Intelligent Robots and Systems, pp. 2684–2689 (2002)
209. Nisiwaki, K., Sugihara, T., Kagami, S., Inaba, M., Inoue, H.: Realtime generation of humanoid walking trajectory by mixture and connection of pre-designed motions-online control by footprint specification. In: Proceedings of IEEE International Conference on Robotics and Automation, pp. 4110–4115 (2001)
210. Norton, R.L.: Design of Machinery: An Introduction to the Synthesis and Analysis of Mechanisms and Machines. McGraw-Hill College (2003)
211. Ogino, M., Hosoda, K., Asada, M.: Acquiring passive dynamic walking based on ballistic walking. In: The Proceeding of the 5th International Conference on Climbing and Walking Robots, pp. 139–146 (2002)
212. Ohno, H., Hirose, S.: Design of slim slime robot and its gait of locomotion. In: Proceeding of the IEEE/RSJ International Conference on Intelligent Robots and Systems, pp. 707–715 (2001)
213. Okada, M., Nakamura, Y.: Development of cybernetic shoulder—a three dof mechanism that imitates human shoulder-motions. Journal of the Robotics Society of Japan 18, 690–698 (2000) (in Japanese)
214. Ono, K., Furuichi, T., Takahashi, R.: Self-excited walking of a biped mechanism with feet. The International Journal of Robotics Research 23(1), 55–68 (2004)
215. Ono, K., Takahashi, R., Shimada, T.: Self-excited walking of a biped mechanism. The International Journal of Robotics Research 20(12), 935–966 (2001)
216. Owen, R.: On the orang, chimpanzee and gorilla. In: Appendix B to On the Classification and Geographical distribution of the Mammalia (Reade Lecture, Cambridge, May 1859), pp. 64–103. Parker, London (1859)
217. Parga, N., Virasoro, M.A.: The ultramatic organization of memories in a neural network. Journal de Physique 47, 1857–1864 (1986)
218. Park, J.H.: Impedance control for biped robot locomotion. IEEE Transactions on Robotics and Automation 17(6), 870–882 (2001)
219. Park, J.H.: Fuzzy-logic zero-moment-point trajectory generation for reduced trunk motions of biped robots. Fuzzy Sets and Systems 134, 189–203 (2003)

220. Park, J.H., Chung, H.: Hybrid control of biped robots to increase stability in locomotion. Journal of Robotic Systems 17(4), 187–197 (2000)
221. Parker, D.: Learning-logic. Tech. Rep. TR-47, Center for Computational Research in Economics and Management Science. MIT (1985)
222. Parsons, P.E., Taylor, C.R.: Energetics of brachiation versus walking: A comparison of a suspended and an inverted pendulum mechanism. Physiological Zoology 50(3), 182–188 (1977)
223. Pennestri, E., Cavacece, M., Vita, L.: On the computation of degrees-of-freedom: A didactic perspective. In: ASME International Design Engineering Technical Conferences and Computers and Information in Engineering Conference, pp. 1733–1741 (2005)
224. Plestan, F., Grizzle, J.W., Westervelt, E.R., Abba, G.: Stable walking of a 7-dof biped robot. IEEE Transactions on Robotics and Automation 19(4), 653–668 (2003)
225. Poggio, T., Girsi, F.: Networks for approximation and learning. Proceedings of the IEEE 78(9), 1481–1497 (1990)
226. Popovic, M., Goswami, A., Herr, H.: Ground reference points in legged locomotion: Definitions, biological trajectories and control implications. International Journal of Robotics Research 24(12), 1013–1032 (2005)
227. Poulakakis, I., Smith, J.A., Buehler, M.: Modeling and experiments of untethered quadrupedal running with a bounding gait: The scout ii robot. International Journal of Robotics Research 24(4), 239–256 (2005)
228. Pratt, J.E.: Exploiting inherent robustness and natural dynamics in the control of bipedal walking robots. Ph.D. thesis, Computer Science Department. MIT (2000)
229. Preuschoft, H., Demes, B.: Biomechanics of brachiation. In: Preuschoft, H., Chivers, D.J., Brockelman, W.Y., Creel, N. (eds.) The Lesser Apes, pp. 96–118. Edinburgh University Press (1984)
230. Psaltis, D., Sideris, A., Yamamura, A.: A multi-layered neural network controller. IEEE Cntrol System Magazine, 17–20 (1988)
231. Quinn, R.D., Nelson, G.M., Bachmann, R.J., Kingsley, D.A., Offi, J.T., Allen, T.J., Ritzmann, R.E.: Parallel complementary strategies for implementing biological principles into mobile robots. International Journal of Robotics Research 22(3), 169–186 (2003)
232. Raibert, M.: Legged Robots That Balance. MIT Press (1986)
233. Raiskila, P., Koivo, H.N.: Properties of a neural network controller. In: Proceedings of International Conference on Automation, Robotics and Computer Vision, pp. 1–5 (1990)
234. McGhee, R.B.: Vehicular Legged Locomotion. Jai Press (1983)
235. Reil, T., Husbands, P.: Evolution of central pattern generators for bipedal walking in a real-time physics environment. IEEE Transaction on Evolutionary Computation 6(2), 159–168 (2002)
236. Reil, T., Massey, C.: Biologically inspired control of physically simulated bipeds. Theory Bioscience 120, 327–339 (2001)
237. Rosenblatt, F.: The perceptron: a probablistic model for information storage and organization in the brain. Psychological Review 65, 368–408 (1958)
238. Rumelhart, D.E., Hainton, G.E., Williams, R.J.: Parallel Distributed Processing. MIT Press (1986)
239. Kagami, S., Nishiwaki, K., Kitagawa, T., Inaba, M., Inoue, H.: A fast generation method of a dynamically stable humanoid robot trajectory with enhanced zmp constraint. In: Proceedings of IEEE International Conference on Humanoid Robotics (2000)
240. Sabourin, C., Bruneau, O.: Robustness of the dynamic walk of a biped robot subjected to disturbing external forces by using cmac neural networks. Robotics and Autonomous Systems 51, 81–99 (2005)

241. Sabourin, C., Bruneau, O., Buche, G.: Control strategy for the robust dynamic walk of a biped robot. The International Journal of Robotics Research 25(9), 843–860 (2006)
242. Saito, F., Fukuda, T.: Brachiation robot. In: The Video Proceeding of IEEE International Conference on Robotics and Automation (1993)
243. Saito, F., Fukuda, T., Arai, F.: Movement control of brachiation robot using cmac between different distance and height. In: IMACS/SICE International Symposium on Robotics, Mechatronics and Manufacturing Systems, pp. 35–40 (1992)
244. Saito, F., Fukuda, T., Arai, F.: Swing and locomotion control for two-link brachiation robot. In: Proceedings of the IEEE International Conference on Robotics and Automation, pp. 719–724 (1993)
245. Saito, F., Fukuda, T., Arai, F.: Swing and locomotion control for a two-link brachiation robot. IEEE Control Systems Magazine 14(1), 5–12 (1994)
246. Sakagami, Y., Watanabe, R., Aoyama, C., Matsunaga, S., Higaki, N., Fujimura, K.: The intelligent asimo: System overview and integration. In: Proceedings of IEEE International Conference on Intelligent Robots and Systems, pp. 2478–2483 (2002)
247. Salatian, A.W., Yi, K.Y., Zheng, Y.F.: Reinforcement learning for a biped robot to climb sloping surfaces. Journal of Robotic Systems 14(4), 283–296 (1998)
248. Sangwan, V., Taneja, A., Mukherjee, S.: Design of a robust self-excited biped walking mechanism. Mechanism and Machine Theory 39, 1385–1397 (2004)
249. Sano, A., Furusho, J.: Realization of dynamic quadruped locomotion in pace gait by controlling walking cycle. In: International Symposium on Experimental Robotics, pp. 491–502 (1991)
250. Sardain, P., Bessonnet, G.: Forces acting on a biped robot. center of pressure—zero moment point. IEEE Transactions on Systems, Man, and Cybernetics A34(5), 630–637 (2004)
251. Sawada, H., Sekiyama, K., Aoyama, T., Hasegawa, Y., Fukuda, T.: Locomotion transition scheme with instability evaluation using bayesian network. In: Proceedings of the IEEE/RSJ International Conference on Intelligent Robots and Systems, pp. 3372–3377 (2010)
252. Sawada, H., Sekiyama, K., Kojo, M., Aoyama, T., Hasegawa, Y., Fukuda, T.: Locomotion Stabilization with Transition between Biped and Quadruped Walk based on Recognition of Slope. In: Proceedings of the IEEE International Symposium on Micro-Nano Mechatronics and Human Science, pp. 424–429 (2008)
253. Schwab, A.L., Wisse, M.: Basin of attraction of the simplest walking model. In: The Proceeding of ASME International Conference on Noise and Vibration (2001)
254. Schwind, W.J., Koditschek, D.E.: Control of forward velocity for a simplified planar hopping robot. In: The Proceeding of International Conference on Robotics and Automation, pp. 691–696 (1995)
255. Seipel, J., Holmes, P.: Three-dimensional translational dynamics and stability of multi-legged runners. International Journal of Robotics Research 25(9), 889–902 (2006)
256. Shih, C.: Ascending and descending stairs for a biped robot. IEEE Transactions on Systems, Man, and Cybernetics 29(3), 255–268 (1999)
257. Shih, C., Gruver, W.A.: Control of a biped robot in the double-support phase. IEEE Transactions on Systems, Man, and Cybernetics 22(4), 729–735 (1992)
258. Shimoyama, I.: Dynamic walk of stilts type biped loco motion. Transactions of JSME C-48(433), 1445–1454 (1982)
259. Shinomoto, S.: A cognitive and associative memory. Biological Cybernetics 57, 197–206 (1987)
260. Simons, E.L.: Primate Evolution: An Introduction to Man's Place in Nature. Macmillan (1972)

261. Simpson, P.: Fuzzy min-max neural networks. i. classification. IEEE Transactions on Neural Networks 3(5), 776–786 (1992)
262. Slotine, J.E., Li, W.: On the adaptive control of robot manipulators. In: Proceedings of the Internatinal Confererenc on Robotics and Automation (1986)
263. Smith, R.: Open dynamics engine (ode) (2008), http://www.ode.org/
264. Song, G., Zefran, M.: Underactuated dynamic three-dimensional biped walking. In: The Proceeding of International Conference on Robotics and Automation, pp. 854–859 (2006)
265. Spong, M.W.: Passivity based control of the compass gait biped. In: The Proceeding of IFAC World Congress, pp. 19–24 (1999)
266. Spong, M.W., Bullo, F.: Controlled symmetries and passive walking. IEEE Transactions on Automatic Control 50(7), 1025–1031 (2005)
267. Stark, L., Okajima, M., Whipple, G.H.: Computer pattern recognition techniques: Electrocardiographic diagnosis. Communications of the ACM 5, 527–532 (1962)
268. Steinbuch, K., Piske, V.A.W.: Learning matrix and their applications. IEEE Transactions on Electronic Computers EC-12, 846–862 (1963)
269. Sugita, S., Ogami, K., Michele, G., Hirose, S., Takita, K.: A study on the mechanism and locomotion strategy for new snake-like robot active cord mechanism –slime model 1 acm-s1. Journal of Robotics and Mechatronics 20(2), 302–310 (2008)
270. Takanishi, A.: Biped walking robot compensating moment by trunk motion. Journal of Robotics and Mechatronics 5(6), 505–510 (1993)
271. Takanishi, A., Naito, G., Ishida, M., Kato, I.: Realization of plane walking by the biped walking robot wl-10r. Robotic and Manipulator Systems, 283–393 (1982)
272. Takegaki, M., Arimoto, S.: A new feedback method for dynamic control of manipulators. Journal of Dynamic Systems, Measurement, and Control 102, 119–155 (1981)
273. Takuma, T., Hosoda, K.: Controlling the walking period of a pneumatic muscle walker. The International Journal of Robotics Research 25(9), 861–866 (2006)
274. Tedrake, R., Zhang, T.W., Fong, M.F., Seung, H.S.: Actuating a simple 3d passive dynamic walker. In: The Proceeding of International Conference on Robotics and Automation, pp. 4656–4661 (2004)
275. Thuilot, B., Goswami, A., Espiau, B.: Bifurcation and chaos in a simple passive bipedal gait. In: The Proceeding of International Conference on Robotics and Automation, pp. 792–798 (1997)
276. Tokioka, M., Mitsuoka, T., Fukuda, T., Kurihara, T.: Force control of robots by neural network models: Control of one-dimensional manipulators. Jounal of the Japan Society of Robotics Engineers 8, 52–59 (1989)
277. Tomiyama, T., Ma, S., Wada, H.: All-directional walking of a quadruped robot based on crawl and rotation gaits. Journal of the Robotics Society of Japan 20(3), 85–92 (2002)
278. TOYOTA (2011), http://www.toyota-global.com/innovation/partnerrobot/
279. Transeth, A.A., Leine, R.I., Glocker, C., Pettersen, K.Y., Liljeback, P.: Snake robot obstacle-aided locomotion: Modeling, simulations, and experiments. IEEE Transactions on Robotics 24(1), 88–104 (2008)
280. Trevor, J.C.: The history of the word "brachiator" and a problem of authorship in primate nomenclature. In: Napier, J., Barnicot, N.A. (eds.) The Primates: the Proceedings of the Symposium held on 12th–14th April 1962 at the Offices of The Zoological Society of London, pp. 197–198. The Society, London (1963)
281. Tsuchiya, K., Tsujita, K.: Locomotion control of a quadruped locomotion robot based on central pattern generator model. Journal of the Robotics Society of Japan 20(3), 243–246 (2002)

282. Tsukagoshi, H., Hirose, S.: The proposal of the intermittent crawl gait and its generation. Journal of the Robotics Society of Japan 17(2), 301–309 (1999)
283. Tsukagoshi, H., Hirose, S., Yoneda, K.: Maneuvering operations of the quadruped walking robot on the slope. In: Proceedings of IEEE International Conference on Intelligent Robotics and Systems, pp. 863–869 (1996)
284. Tzafestas, S.G., Krikochoritis, T.E., Tzafestas, C.S.: Robust sliding-mode control of nine-link biped robot walking. Journal of Intelligent and Robotic Systems 20, 375–402 (1997)
285. Usherwood, J.R., Bertram, J.E.A.: Understanding brachiation: insight from a collisional perspective. The Journal of Experimental Biology 206, 1631–1642 (2003)
286. Vermeulen, J., Verrelst, B., Vanderbought, B., Lefeber, D., Guillaume, P.: Trajectory planning for the walking biped 'lucy'. The International Journal of Robotics Research 25(9), 867–887 (2006)
287. Verrelst, B., Vanderbought, B., Vermeulen, J., Ham, R.V., Naudet, J., Lefeber, D.: Control architecture for the pneumatically actuated dynamic walking biped 'lucy'. Mechatronics 15, 703–729 (2005)
288. Vukobratovic, M., Borovac, B., Potkonjak, V.: Zmp: A review of some basic misunderstandings. International Journal of Humanoid Robotics 3(2), 152–175 (2006)
289. Vukobratovic, M., Brovac, B., Surla, D., Stokic, D.: Biped Locomotion:dynamics, stability, control, and application. Springer (1990)
290. Vukobratovic, M., Timcenko, O.: Experiments with nontraditional hybrid control technique of biped locomotion robots. Journal of Intelligent and Robotic Systems 16, 25–43 (1996)
291. Waibel, A.: Modular construction of time-delay neural networks for speech recognition. Neural Computation 1, 39–46 (1986)
292. Weber, M., Crilly, P.B., Blass, W.E.: Adaptive noise filtering using a errobackpropagation neural network. IEEE Transactions on Instrumentation and Measurement 40(5), 820–825 (1991)
293. Werbos, P.: Beyond regression: New tools for prediction and analysis in the behavior sciences. Ph.D. thesis. Harvard University (1973)
294. Werbos, P.: Consistency of hdp applied to a simple reinforcement learning problem. Neural networks 3, 179–189 (1990)
295. Werbos, P.J.: Backpropagation through time: What it dose and how to do it. Proceedings of the IEEE 78, 1550–1560 (1990)
296. Werbos, P.J.: Neurocontrol and Related techniques. In: Hand Book of Neural Computing Applications, pp. 345–380. Academic Press (1990)
297. Westervelt, E.R., Buche, G., Grizzle, J.: Experimental validation of a framework for the design of controllers that induce stable walking in planar bipeds. The International Journal of Robotics Research 23(6), 559–582 (2004)
298. Westervelt, E.R., Grizzle, J.W., Koditschek, D.E.: Hybrid zero dynamics of planar biped walkers. IEEE Transactions on Automatic Control 48(1), 42–56 (2003)
299. Widrow, B.: The original adaptive neural net broom-balancer. In: Proceedings of the IEEE International Sympsium on Circuit and Systems, pp. 351–357 (1987)
300. Widrow, B., Gupta, N.K., Maitra, S.: Punish/reward: Learning with a criti in adaptive threshold systems. IEEE Transactions on Systems, Man, Cybernetics SMC-3, 455–465 (1973)
301. Widrow, B., Lehr, M.A.: 30 years of adaptive neural networks:perceptron, madaline, and back-propagation. Proceedings of the IEEE 78(9), 1415–1441 (1990)
302. Widrow, B., Hoff Jr., M.E.: Adaptive switching circuits. Tech. Rep. 1553-1. Stanford Electorn Lab. (1960)

303. Widrow, B., Smith, F.W.: Pattern recognizing control systems. In: Proceedings of the Computer and Information Science (COINS) Symposium (1963)
304. Widrow, B., Stearns, S.D.: Adaptive Signal Processing. Prentice Hall (1985)
305. Wieber, P., Chevallereau, C.: Online adaptation of reference trajectories for the control of walking systems. Robotics and Autonomous Systems 54, 559–566 (2006)
306. Wiener: Cybernetics. The Technology Press. John Wiley & Sons, Inc. (1948)
307. Wisse, M., Schwab, A.L., van der Linde, R.Q.: A 3d passive dynamic biped with yaw and roll compensation. Robotica 19, 275–284 (2001)
308. Wisse, M., Schwab, A.L., van der Linde, R.Q., van der Helm, F.C.T.: How to keep from falling forward: Elementary swing leg action for passive dynamic walkers. IEEE Transactions on Robotics 21(3), 393–401 (2005)
309. Yamada, H., Chigisaki, S., Mori, M., Takita, K., Ogami, K., Hirose, S.: Development of amphibious snake-like robot acm-r5. In: Proceedings of 36th International Symposium on Robotics, p. TH3C4 (2005)
310. Yamada, H., Hirose, S.: Development of practical 3-dimensional active cord mechanism acm-r4. Journal of Robotics and Mechatronics 18(3), 305–311 (2006)
311. Yamaguchi, J., Takanishi, A., Kato, I.: Development of a biped walking robot compensating for three-axis moment by trunk motion. Journal of the Robotics Society of Japan 11(4), 581–586 (1993)
312. Yamaguchi, J., Takanishi, A., Kato, I.: Stabilization of biped walking and acquisition of landing surface position information using foot mechanism with shock absorbing material. Journal of the Robotics Society of Japan 14(1), 67–74 (1996)
313. Yang, J.: A control study of a kneeless biped locomotion system. Journal of the Franklin Institute–Engineering and Applied Mathematics 2(331B), 125–143 (1994)
314. Yi, K.Y., Zheng, Y.F.: Biped locomotion by reduced ankle power. Autonomous Robots 4, 307–314 (1997)
315. Yokoi, K., Kanehiro, F., Kaneko, K., Kajita, S., Fujiwara, K., Hirukawa, H.: Experimental study of humanoid robot hrp-1s. The International Journal of Robotics Research 23(4-5), 351–362 (2004)
316. Yoneda, H., Sekiyama, K., Hasegawa, Y., Fukuda, T.: Vertical ladder climbing motion with posture control for multi-locomotion robot. In: Proceedings of the International Conference on Intelligent Robots and Systems, pp. 3579–3584 (2008)
317. Yoneda, K., Iiyama, H., Hirose, S.: Intermittent trot gait of a quadruped walking machine—dynamic stability control of an omnidirectional walk. Journal of the Robotics Society of Japan 14(6), 881–886 (1996)
318. Yoshino, R.: Stabilizing control of high-speed walking robot by walking pattern regulator. Journal of the Robotics Society of Japan 18(8), 1122–1132 (2000)
319. Zheng, Y.F., Shen, J.: Gait synthesis for the sd-2 biped robot to climb sloping surface. IEEE Transaction on Robotics and Automation 6(1), 86–96 (1990)
320. Zhou, C., Meng, Q.: Dynamic balance of a biped robot using fuzzy reinforcement learning agents. Fuzzy Sets and Systems 134, 169–187 (2003)
321. Zhou, C., Ruan, D.: Integration of linguistic and numerical information for biped control. Robotics and Autonomous Systems 28, 53–70 (1999)

MIX
Papier aus verantwortungsvollen Quellen
Paper from responsible sources
FSC® C105338

If you have any concerns about our products,
you can contact us on
ProductSafety@springernature.com

In case Publisher is established outside the EU,
the EU authorized representative is:
Springer Nature Customer Service Center GmbH
Europaplatz 3, 69115 Heidelberg, Germany

Printed by Libri Plureos GmbH
in Hamburg, Germany